PEASANT ORGANISATIONS AND THE DEMOCRATISATION PROCESS IN AFRICA

PEASANT ORGANISATIONS AND THE DEMOCRATISATION PROCESS IN AFRICA

Edited by

Mahmoud Ben Romdhane & Sam Moyo

COUNCIL FOR THE DEVELOPMENT OF SOCIAL SCIENCE RESEARCH IN AFRICA

Avenue Cheikh Anta Diop Angle Canal IV, BP 3304 Dakar, Senegal

ISBN 2-86978-111-3

Cover design by Sulaiman Adebowale

Typesetting by Djibril Fall

Printed in Great Britain by Russell Press Ltd, Russell House, Bulwell Lane, Basford, Nottingham NG6 0BT

Distributed in Africa by CODESRIA

CODESRIA is the Council for the Development of Social Science Research in Africa head-quartered in Senegal. It is an independent organisation whose principal objectives are facilitating research, promoting research-based publishing and creating multiple fora geared towards the exchange of views and information among African scholars. Its correspondence address is:

Po Box: 3304, Dakar, Senegal
E-mail: codesria@sentoo.sn

Distributed elsewhere by
African Books Collective, 27 Park End Street, Oxford, OX1, 1HU UK
Fax: 44-01865-793298 Email: abc@dial.pipex.com
web site: www.africanbookscollective.com

CODESRIA would like to express its gratitude to the Swedish International Development Cooperation Agency (SIDA/SAREC), the International Development Research Centre (IDRC), the Ford Foundation, the MacArthur Foundation, Carnegie Corporation, the Norwegian Ministry of Foreign Affairs, the Danish Agency for International Development (DANIDA), the French Ministry of Cooperation, the Rockefeller Foundation, the United Nations Development Programme, the Netherlands Ministry of Foreign Affairs, the Government of Senegal, FINIDA, NORAD, CIDA, IIEP/ADEA, OECD, IFS, Oxfam America UN/UNICEF, for their support of its research, training and publications programmes.

Content

Part Two

Grassroots Participation and Empowerment

Note on Contributors

Abutudu Musa I. M., obtained a Ph.D from University of Ibadan, Nigeria. He was a member of the School of Social Science, Institute for Advanced Study, Princeton, New Jersey, USA, during the 1997-98 Academic Session. He is currently Senior Lecturer and Acting Head, Department of Political Science, University of Benin, Nigeria.

Asaju Adebakin Simeon, is a Lecturer at the Department of Estate Management, The Federal University of Technology, Akure, Ondo State, Nigeria.

Ayichi Damian, is a Lecturer at the Department of Political Science, University of Nigeria, Nsukka, Nigeria.

Ben Romdhane Mahmoud is a Lecturer in Economics at University of Tunis. He has published or contributed to dozens of books and to about fifty articles on economic policies, social groups and movements, Euro-Mediterranean and the Arab world. His latest works are on globalisation vis-à-vis the new power systems and gender issues.

Eboh Eric, is a Senior Lecturer at the Department of Agricultural Economics, University of Nigeria, Nsukka. He is actively involved in research on socio-economic and institutional aspects of rural development, including gender, rural institutions, natural resource management and environmental sustainability. He has published widely in local and international journals. His recent works include the following books: *Social and Economic Research: Principles and Methods*, *Rural Development in Nigeria: Concepts, Processes and Prospects*, and *Political Economy of National Development: Issues and Perspectives*

El Amin Khalid, is a Lecturer at the Development Studies and Research Centre, Faculty of Economics and Social Studies, University of Khartoum, Khartoum, Sudan.

Ezeani O. Emmanuel, is a Lecturer at the Department of Political Science, University of Nigeria, Nsukka, Nigeria.

Furtado Claudio A., Achada de Santo Antonio, Praia, Cap-Vert.

Kinyanjui Marie N. is a senior research fellow at the Institute for Development Studies, University of Nairobi. Her latest publication is : "Employees in Small Enterprises in Nairobi: Job Search and Career Patterns" in Alila O.P. and Pedersen, P.O., *Negotiating Social Space: East African Micro Enterprises*, Asmara Eritrea, African World Press.

Moyo Sam, (Ph.D) is an Associate Professor of Development Studies and currently an independent Land Policy Analyst. He was formerly director of the Southern African Regional Institute for Policy Studies of the Sapes Trust based in Harare, Zimbabwe. He has published extensively at national, regional and international levels on land policy and land reform issues, agrarian reform, environmental policy, rural development, NGOs, co-operatives and development co-operation.

Komon Jean Paul, is a Lecturer in Economics at University of Yaoundé, Cameroon.

Naccache Gilbert, Rue des Pins 2000 Le Bardo, Tunisie.

Sibanda Arnold E., is a research fellow and currently heads the Department of Agrarian and Labour Studies at the Institute of Development Studies, University of Zimbabwe. He holds an M.Sc Degree in Sociology from the University of London's LSE in development problems. He has published on several topics including industrial relations, structural adjustment, governance, agrarian and rural development issues. He is currently completing his part-time D.Phil at University of Zimbabwe with focus on labour policy in Zimbabwe.

Sy Mouhamadou, is the Director of the Department of Economic Planning and Management of Organisations at the National School of Applied Economics (*Ecole nationale d'économie appliqueé/ENEA*) in Dakar. He has been teaching Management and Analysis of Organisations, Project/ Programme Management and Evaluation since 1993. He also conducts research activities and provides consultancy services. He holds a DESS (Post-Graduate Diploma) in Management (*CESAG*) with specialisation in Project Management and Evaluation, a Certificate in Entrepreneurship (University of Connecticut, USA) and a Secondary School Teacher Training Certificate from Dakar Teacher Training College (*Ecole normale supérieure de Dakar*). His recent research focused on **Perpetuating the Impact of Development Projects** (with a research grant from the World Bank's McNamara Programme), **Gender and Communal Forestry** (a study financed by the FAO Office in Dakar) and **Women's Access to Land Ownership in Senegal** (through a research programme of Emory University School of Law in USA).

Yunusa Mohammed-Bello, Ph.D is the Head of the Department of Urban and Regional Planning, Ahmadu Bello University, Zaria.

Peasant Organisations and Rural Civil Society in Africa: An Introduction

Sam Moyo

Economic Crisis and the Resurgence of Peasant Organisations

Growing academic interest in the study of democratic transitions in Africa has been accompanied by increased research on peasant organisations as a potential agent for promoting democratisation and policy reform. Research on the role of peasant organisations in defending rural livelihoods and in promoting improved social and economic development has also grown. Interest in the diverse range of rural political institutions from both a liberal and social democratic perspective has been stimulated by the limited gains realized from the expansion of multi-party political change in Africa. Political and economic reforms on the continent since the mid-1980s have been accompanied by massive rural social dislocations, further poverty, growing insecurity over land and natural resource property rights, and numerous violent conflicts over the control of resources and the state. Neo-liberal economic policy reforms imposed from above in the context of 'choiceless democracy' (Mkandawire 1998) have added to the erosion of the basic social and political rights of Africans. Peasant organisations are re-emerging on the continent as one of the rural responses to protect people from economic and political crises, and as a potential force in a possible endogenous movement for alternative forms of development.

Peasant organisations, however, have a long history of dispersed struggles against colonial and post-colonial political repression and economic exploitation on the continent. While the achievements of peasant organisations should not be idealised (Yeros 2000), given the empirical record of their utilisation both in progressive and retrogressive struggles and armed rebellions, their importance cannot be underestimated. Even the limited case studies represented in this book remind us of the extensive and leading role peasant organisations have played in various struggles. These include the roles they played in the liberation and resistance struggles of Kenya and Zimbabwe; resistance against commodity market monopolies and price manipulations; resistance against direct exploitation of their labour and the alienation of their resources (by the wealthier and governments), and their resistance against state bullying in the local government arena. Yet, in most recent violent and armed conflicts in Africa the peasantry and their organisations have also been forced to support violent rebellions by armed groups. Having failed to capture the peasants' political support to further current rebellions, unlike the liberation movements which had such support for their loftier nationalist ideals, recent armed rebellions have been violent to peasant communities.

While some of the studies in this book investigate the alternative paths of rural economic growth being promoted by peasant organisations, our concern is more focused on their political responses to the failures of African agriculture and rural development, and the widespread growth of rural poverty among the peasantry today. Moreover, the currently overwhelming focus by donors and governments on rural poverty alleviation aspects of the neo-liberal market-based developmental agenda suggests that the evolution of basic socio-political organisations to defend the mere social reproduction of the peasantry is a critical element of rural development.

Perhaps because of the predominance of urban-based movements for democratisation in Africa over the years, and the real paucity of grand cases of rural revolt and rebellion, besides those liberation struggles led by the middle classes in previous decades, there has been a neglect of the subtle but critical political influences of rural civil society in post-independence democratisation. Indeed, dominant research perspectives, including the urban bias, interest groups and new political economy perspectives, have tended to minimise the role of the peasantry in policy advocacy. The welfare orientation of rural development studies, as well as most donor and NGO interventions in grassroots developments which pretend to be apolitical, underplay the rural constituency in the wider policy advocacy process. Only recently has donor funding increasingly been allocated towards the policy advocacy work of rural peasant organisations.

This CODESRIA book is one more contribution by a network of African scholars to the growing need to understand the nature of rural civil society in Africa and how the evolution of peasant organisations influences the processes of democratisation and economic change. The conceptual framework that undergirds the various contributions and the specific research questions they attempt to address are examined in this introductory chapter.

Concept and Context of Peasant Organisations

Research on the evolution of peasant movements in Africa and indeed the sociology of related social movements is only recently receiving rigorous attention (Mamdani and Wamba dia Wamba 1995). Most of the key concepts underlying this study, such as the definition of peasants, peasantries and peasant organisations, are highly contested let alone standardised (Mafeje 1993). Researchers in this book argue that Africa has peasantries of varied shades, and that these exist in a more nuanced form than some classic Africanist studies (such as Klein 1980; Hyden 1985) have presented them. They debate the meanings of 'peasant organisation' in both ethnographic and political economy frameworks.

The chapters presented in this book question whether there is any democratic content in the peasant organisations' missions. They also debate whether African peasants are in general confronted by the land question in the classical and historical sense of the agrarian question as was found elsewhere. Indeed, the researchers debate whether African peasants are objectively and subjectively conscious of their exploitation and class position, thus elaborating similar debates before them (Williams 1976; Cliffe 1987). They further question whether the peasant organisations they form reflect this consciousness or are merely defensive agencies. In so doing they interrogate the dominant perspective that peasant organisations are based upon survivalist short-term organisational tactics and tend mainly to engage in the 'politics of everyday life' (Scott 1985).

Perhaps the major cleavage in the debate about the nature and significance of peasant organisations and their historic mission is whether their existence is structured around class-based struggles with far-reaching national political implications. The studies here examine whether their purpose is destined to be a localised and dispersed 'everyday form of resistance', using minor 'weapons of the weak' (Scott 1985). Peasant organisations are not expected to be highly or nationally organised as movements, because purportedly they lack the disciplined leadership and the organisational skills necessary to coordinate on a large scale (ibid.). Because of their imputed organisational weakness and their local orientation, peasant organisations are expected not to confront directly the hierarchy and power of the state and the wealthier classes that oppose them. They tend to avoid the ferocious might of their opponents by engaging in the tactics of evasion, foot dragging, petty sabotage and encroaching (ibid.).

But evidence from the literature in general suggests that African peasants have until recently been an important force in many colonial struggles and a few post-colonial rebellions such as in Uganda (Mkandawire 2000). Some researchers argue that when the common actions of many peasant organisations are multiplied they effectively counteract the policies of the ruling classes (Yeros 2000) and can initiate policy reversals. Peasants' resistance to and evasion of natural resource regulation and land tenure policies in Africa, for instance, is a major way in which state policy has been sidelined (Moyo 1995). Indeed, there is an exaggerated sense in which key concepts in the study of peasant organisations have been applied in a mechanistic, oppositional form.

Conceptual vulgarisation in the study of peasant organisations is deployed to criticize the purported over-emphasis of class essentialism and the politics of class in structuralist research. Those research methodologies that focus on exploring struggles to construct social identity in the belief that radical heterogeneity in society disconnects presumably classless communities, which are supposedly isolated from ongoing national struggles for social change, are thus privileged (Veltmeyer 1997). Similarly, the concepts of class and community within peasant societies tend to be presented as rigidly different conceptual categories of opposed schools of thought, in spite of the reality that the concept of 'community' represents a complex of class relations held together by a sense of belonging and shared social identity (ibid.).

In this vein, since the 1980s postmodernism has become the dominant conceptual approach to the study of peasant communities and their social and political dynamics (Veltmeyer 1997). Particular emphasis is placed by this perspective on rejecting the Marxist structural and class-based analyses of the distinctive features of peasant communities'organisational and internal dynamics. The 1980s are said to have spawned a new form of peasant organisation and politics, which highlights the postmodern condition, namely: a radical subjectivity of experience and the self-contribution of a new social subject, a social actor seeking to define and express him/herself on a largely self-constructed stage (ibid., p.140). Thus the multiplicity and diversity of existing peasant organisations is considered to obviate the connectivity between them that class analysis seeks. Nonetheless, class-based social movements are found to be re-emerging, for instance, in the Latin American context, where peasants are leading the present wave of new social movements in struggles which revolve around modernist 'demands for access to land and productive resources, democracy and social justice' (ibid.). The conceptual and empirical work of exploring such class-based struggles and the evolution of the identities of peasant organisations in Africa is still to be done.

Rather than provide standard definitions and a rigid guiding conceptual structure for the research contributions included in this book, the variety of conceptual perspectives and common definitions that have self-evidently emerged among the chapters are left to 'speak for themselves'. Instead this introductory chapter attempts to enrich the conceptual debate by highlighting and synthesising the major conceptual and contextual issues, and the salient empirical trends that seem to characterise peasant organisations in Africa, based upon emerging research results from this book and other relevant ongoing work.

The conceptual framework used in the following to provide a structured rather than an eclectic analysis of the evolution of peasant organisations in Africa centres on their origins and their impacts. The five descriptive-analytic components can improve our understanding of the evolution of peasant organisations and provide building blocks for a common conceptual framework. These are derived from perspectives adopted in various studies (Rahmato 1991; Moyo 2000; Veltmeyer 1997), and are: their social basis and origins; their class and gender identity; their organisational dynamics and democratic content; their social, political and material demands, including the organisational strategies and tactics used to achieve these; and finally the linkages of the peasant organisations to wider society, particularly the state, NGOs and other political actors.

But any attempt to develop an analytic-descriptive framework for the assessment of the evolution of peasant organisations must necessarily be grounded in an understanding of the political and economic context in which the peasant communities function in their different regions of Africa. Here, we suggest that the role of peasants as producers and their relations to markets, the state and others are central features that demonstrate how the local and global contexts confronting peasant families in rural Africa are increasingly converging in both character and physical connection. This context explains why and how it is that relatively similar peasant organisational forms are emerging in the various countries studied. Indeed, we postulate that common forms of political repression and economic exploitation emerging from three decades of economic decline across the continent have provoked similar local organisational responses among peasants. This peasant response has also evoked common rural welfarist support systems, in the name of poverty reduction, rural development and empowerment projects, sponsored by governments, NGOs and donors (Moyo 2000).

The majority of the regions in Africa covered by our case studies of the peasant associations seem to have experienced relatively similar political economic histories, whereby their emerging power structures and relations increasingly represent complex and changing property relations. The bulk of the rural populations are peasant producers of agricultural commodities. To a lesser extent they are engaged in various non-farm and rural-based artisanal activities, which supplement their mainly agricultural incomes, realized from predominantly female family labour. Their agricultural production is enhanced in some cases by irrigation schemes and in most cases through the mobilisation of 'improved seeds' and fertilisers. The case study areas are thoroughly incorporated into the national political economies and into world markets mainly through the export of classic primary commodities whose prices have been declining. These commodities include cotton among the Sudanese and Zimbabwe peasants studied; perennial tree crops such as coffee and palm oil in the Nigerian, Kenya and Cameroon cases; and horticulture in the Senegal and other cases.

An important aspect, which increasingly structures political and economic power relations among the peasants in the regions studied, is the significant trend of land alienation and struggles to restore or gain land rights. Land expropriation on a large scale mainly during the colonial area characterizes the Kenya and Zimbabwe case study areas, while localised and recent land expropriations on a smaller scale underlie the contradictory property relations and struggles in two of the Nigerian case studies and in the Sudan case. Post-colonial, localised land expropriation is emerging almost everywhere. Gender-based struggles for land where men and their male heirs have inequitable control of land is a common problem, which has been the Kenyan and Sudanese experience. Also, disenchantment with local land administration processes managed by state, traditional authorities and local committee structures, with particular reference to unfair land allocation processes, unclear rules and regulations governing land use in projects and inequitable systems of resource use charges, are common in the studied areas.

Struggles for land redistribution, compensation for lost land and for land titles as a means of gaining secure property rights are common in the highly commercially oriented regions studied, given their intensifying class and gender contradictions. Conflicts over inequitable access to water resources and various infrastructures, especially those used for commodity marketing and irrigation, which are all based upon an emerging rural differentiation, are relatively common in all the study areas. Thus peasant associations may be emerging more intensively in those regions where the increased export of commodities results in common struggles against emerging wealthier or kulak classes, further state regulation, increasing resource inequities and the growing marginalisation of some groups, including various women's groups.

Understanding African Peasant Organisations: Trends and Questions

The Social Base of Peasant Organisations

Attempts to understand the typologies and social bases of peasant organisations tend to place greater emphasis upon understanding those peasant organisations that are particularly preoccupied with agricultural economic development projects within a context of market reforms. The study of the politics of rural social mobilisation through a more broadly based conception of the social basis of peasant organisations has tended to be neglected in recent academic works on neo-liberal economic and liberal democratic reform. Most studies of peasant organisations and rural NGOs imbued with a technicist, developmentalist and prescriptive orientation focus on the economic functionalist aspects that characterise peasant organisations and tend to overlook the social class and other sociological categories that define the emergence of contemporary rural civil society and the socio-political roots of their political advocacy.

Since peasant organisations tend to be teleologically but unjustifiably associated with rural institutional changes, which in turn are associated with market-oriented agricultural change and development, it is useful to examine the character of the peasant organisations studied here through an assessment of the social bases from which they emerge. Such social bases can be traced through an assessment of types of organisations they form, their origins, the nature of their memberships and their key functions.

Most of the peasant associations encountered by our researchers were found to be mutual support groups, age grade groups and community project-based groupings, and the occasional protest groups. The case studies show that peasant organisations originate in varied ways. While most grew out of traditional survival institutions of mutual self-help, a few were initiated by charismatic founders while others responded to specific bottlenecks such as marketing, transport or labour problems they confronted in general and/or during the era of structural adjustment. The age grade-based associations were the most varied in social origin and function. They included various labour pooling groups for community infrastructure and personal enterprises involving mainly male associations and a few female associations. 'Manhood' initiation-related activities had spawned some of the age grade associations. Except for a handful of the associations, whose membership and geographic coverage went beyond a single district, the majority of the associations studied are very localised in their membership, functions and aspirations.

The contributions in this book, however, tend to be less inclusive in their definitional framework, because they selected for research only those peasant organisations that are formally constituted in terms of structures, membership and track record. As argued elsewhere, there is need also to examine spontaneous peasant organisations so that the wide dynamics of emerging rural social forces can be captured (see Moyo 1998; Mamdani and Wamba Dia Wamba 1995). An eminent example of this type of informal spontaneous and even illegal form of peasant organisation, which are worthy of study but were not included in the studies here, is the rural land occupation communities, also known as 'squatters'. Such peasant organisations are increasing in a number of southern African countries (see Moyo 1994). Other informal outlawed peasant organisation examples worthy of attention include those community efforts to defend and sustain their rights of access to environmentally sensitive resources controlled by the state, and privately controlled goods such as wildlife, forest and water resources. Indeed, such 'illegal' peasant organisation activities may be found to exist as subterranean offshoots of formally constituted peasant organisations within given peasant communities, or as informal offshoots of political parties.

However, the studies do interrogate the empirically false dichotomy that tends to be used in defining or classifying peasant organisations or social movements, namely: the distinction between autonomous and dependent peasant organisations. As some of the case studies show, the peasant organisations can adopt varying strategies, which temporally shift from dependence upon or resistance to state agencies' impositions, and shifts from strategies of direct confrontation to collaboration with the state. Indeed, peasant organisations have been found to utilise a combination of advocacy and organisational strategies simultaneously, while the state itself, in search of control over peasant organisations, tends to combine strategies such as coercion, persuasion and co-optation through various material incentives (Moyo 1995). This study generally avoids the rigid analytical constructs used to define peasant organisations, such as restricting the scope of mission and strategy in developing the typologies of the peasant organisations or using a limited temporal perspective of their struggles that are invoked to characterise them.

Class and Gender Identities in Peasant Organisations

Any attempt to restrict the definition of peasant organisations to a singular focus such as their economic forms, their geographical focus or the social, ethnic and class basis of their membership, or to exclude those organisations with political ambitions, may miss the rich complexity that surrounds the growth of peasant organisations studied herein. Thus, it is interesting that while most definitions and indeed studies of peasant organisations are focused upon farmers unions or associations moulded within an interest group's conceptual framework (Skalnes 1995; Bratton 1994), which, among other things, correlates the growth of such associations with the rise of the market economy based on Western experiences, there is a tendency to neglect the class or gendered basis of such organisations. Given the distorted growth of African rural markets, in spite of SAP-induced economic liberalisation, and the resilience of undemocratic states on the continent, the growth and multiplicity of peasant organisations in most of the countries reflects the complex socio-political alignment of rural and urban social forces interacting within globalised but marginalised political economies. It is thus critical to understand the precise sociological texture of such organisations and the political processes that underlie the resilience and growth of peasant organisations so as to contextualise their role in the democratisation and economic development process.

A central understanding that informs this study is that peasant organisations are complex and diverse in their surficial character. Thus the emerging multiplicity of peasant organisations at the national, sub-national and even local levels makes them amenable to a variety of classifications or typologies in terms of their organisational features, their programmatic focus, their value systems and their approaches to mobilising human and material resources for development. Such classifications as have been proposed (for example, Rahmato 1991; Moyo 1994) depend upon the heuristic orientation of the researcher concerned. However, a central weakness of such classifications, which these studies address, is that they lack an understanding of the complex set of relationships that prevail among the multiplicity of peasant organisations.

A proper understanding of the nature of relationships between peasant organisations and other actors such as NGOs, state agencies and donors, raises key questions for debate. Does the diversity and multiplicity of peasant organisations merely reflect their geographic and political dispersion and the classic atomisation expected of the peasantry (Scott 1985; Yeros 2000)? Or is there an empirical basis to argue that there is an emergence of a deepening ideological and substantive differentiation of rural civil society? If so, is there a class-based cleavage emerging over mainstream rural development strategy? These questions should provoke further interrogation: Is there evidence of the emergence of a consciously designed and meaningful liberal democratic political tendency in rural Africa?

The argument that peasant organisations are increasingly being homogenised and indeed are perceived to be adapting towards modernised management systems and values is also examined by the studies here. Anthropologistic perspectives, which suggest that peasant organisations in Africa are culturally peculiar, rather than universally defined, are queried. For instance, are peasant organisations surviving because of special cultural bonds of trust, which bind groups of peasant households? Thus do social-psychological and fundamental cultural variables shape the organisation and interests of peasant organisations in Africa? In this vein some of the studies suggest that rural class formation and other functional interests increasingly bind peasant organisations, including those which frequently organise themselves spontaneously as small temporary groupings of peasant households responding to specific crises in their social reproduction system. Hence the role of ethnicity and culture as determinants of the survival and growth of peasant organisations, in particular the significance of primordial ideologies and values in cementing peasant organisation structures and membership, are examined in the context of evolving material and class-based struggles in rural Africa.

Some of the chapters pursue political-economic approaches to the study of peasant households, focusing on the nature and effects of peasant differentiation and the resultant exploitative relations that occur within and among peasant households. The varied political strategies used by differentiated households in their search for state patronage and protection in the agricultural system (Moyo 1995) are also interrogated in respect of their implications for the evolution of peasant organisations. In our analytical framework, therefore, the social base of peasant organisations, and their class and gender based identities, are critical for understanding the political economic basis of peasant household structuration at the micro and macro levels. Such an approach elucidates how institutional mechanisms such as peasant organisations and their socio-political relationships and struggles with state, NGO and donor agencies reproduce or mediate rural differentiation.

The evidence from the studies shows that while there are many different types of localised peasant organisations encountered, there is really no basis to expect the 'merie Africa' scenario, whereby a classless peasantry generates new organisations that are more inclusive than not. Thus the expectation that the multiplicity of peasant organisations reflects the postmodern search for multiple identities among a peasantry that faces radical differentiation (Veltmeyer 1997) is not valid. The peasant organisations examined in this book show how definite social cleavages and identities are being developed and reinforced around class and gender-based exclusiveness. Through interlocking memberships of various peasant organisations led by a few prominent members of the peasant societies studied, class forces remain critical in structuring peasant organisations. In many cases, the leaders of peasant associations who tended to be wealthier males, also tended to be found as leaders in the clan or other forms of traditional institutions.

In a number of peasant associations the poorer peasants, landless labourers and women were excluded from the development project based organisations, and therefore from the material benefits that could be derived from them, because they did not in the first place possess pieces of land, water rights or cash to finance their participation in the projects. Those who do not produce commodity surpluses tended not to be members of market associations. Interlocking memberships and leadership of peasant associations thus tended to concentrate access to resources among a few, and thus to undergird the class and social differentiation of the peasantry rather than merely accentuate the different cultural and social identities of the peasantry.

Wealthier peasants were found to gain disproportionately more from peasant associations than the poorer in terms of access to credit, water rights, land, infrastructure for irrigation, and equipment such as tractors derived from state, NGO and donor support to the associations' projects. The upper peasantry tended to be entrusted by others to lead the lobbying of and interaction with governments and donors, thus reinforcing their longer-term capture of power through their control over the associations and concentration of the material or resource benefits of peasant association projects in them. Their education and urban exposure were commonly used to justify their leadership of associations, although in a number of cases they tended to be appointed to such leaderships by traditional authorities who had influence over the local clan based peasant structures.

A central issue in this empirical context is whether the historical mission and capacity of peasant organisations to promote political and economic changes is being driven by their class vision rather than by local cultural interests. The question is whether peasant organisations can pursue a national project for social justice or even a nationalist project to develop capitalist farming. Some of the studies argue that the peasant organisations are progressive and not individualistic in character, because they can undertake collective action. The studies show that they are a modernising force, as is reflected in the adaptation of peasant organisation strategies to the demands and procedures of the state and of private sector economic and political systems. But the studies suggest that most of the peasant organisations are *ad hoc*, conjunctural and short-term responses to the current economic and political crisis. They argue that they are not a conscious strategic movement attempting to capture 'centre stage' in the state and market system and that they are mostly defensive and survivalist.

The ethnographic basis of peasant leadership and its patriarchal grounding reproduce male dominance of peasant associations' leadership in general and their encroachment even into the leadership ranks of mainly women's peasant organisations. The case studies show emerging gender contradictions where women tend to have the 'secretary post' because it is presumed to be a 'female job'. However, in one women's peasant association a man held this post because of his literacy skills and because of the obvious leadership importance of this 'key job'. Yet this contradicts the fact that many of the women's peasant organisations had been established to resist male or gender-based exploitation and other disadvantages. Indeed, the objectives of the women's peasant organisations studied are intended to enhance female autonomy and to protect the fruits of their labour in financially gainful economic projects. Women capitalist or wealthier women peasants also seem to emerge out of the peasant associations, through their improved private access to land, land titles and other rights and production infrastructures. Interestingly, women in the peasant organisations studied were said to collaborate better between themselves than men's peasant organisations, and in some cases they succeeded better than men's peasant organisations in lobbying the state for various agricultural support inputs.

Women-based or women-dominated peasant organisations, however, tend to be mostly developed around the margins of the mainstream rural economic and political system, focusing on marginal income generation projects mostly instigated by the state, NGOs and donors. Few of the women's peasants organisations were found to have an articulated autonomous and empowering gender-sensitive or feminist ideology. Their economic strategy tends not to be politically articulated. Thus the women peasant organisations studied reflected a 'women in development projects' approach, rather than a truly gendered, let alone feminist, political approach, reflecting their domination by patriarchal systems of the state and society in general.

Organisational Dynamics and Internal Democracy

The above weaknesses of peasant organisations studied can be explained through an analysis of their real sources of power and control. Most male- or female-dominated peasant organisations are essentially dominated by the middle and ruling classes on the continent, most of which are in alliance with external powers through donor development assistance. Thus the leadership, origins, design of programmes and funding of peasant organisations are controlled by political or economic elites in government or the private sector, and by middle-class elites located in NGO and donor organisations. Furthermore, membership within peasant organisations tends to be dominated by the middle peasants, kulaks or better-off small-scale commercial farmers (Bratton 1995), who comprise less than 30 per cent of rural populations. The poor peasantry are not only excluded in peasant organisations but it is this exclusion that enables the state to co-opt peasant organisations to support currently inefficient and inequitable agricultural and rural development strategies

There is widespread optimism in the literature that peasant organisations and pluralism based on such grassroots empowerment can promote democracy in Africa (Rahmato 1991). The proliferation of peasant organisations is indeed impressive: for instance, there are about 35,000 of them in Nigeria, 60,000 in South Africa and over 3,000 in Zimbabwe. However, peasant organisations seem to have failed to bring about democratisation or an alternate development strategy that would benefit the over 70 per cent of Africa's population located in rural areas. The expectation that peasant organisations are becoming a potential nucleus for democratisation, through popular pressure for state accountability, fairness, participatory development, maintenance of basic human and social rights, freedoms of expression and political choice has so far not been well substantiated. Moreover, there are growing doubts among many academics and leaders of NGOs about the efficacy of so-called participatory development approaches, since it is now recognised that participation tends to be controlled outside the peasant organisations, and that it tends to be superficial. In addition, such approaches are considered organisationally less viable because of their weak material and economic basis.

In this respect a major contradiction raised in the literature is that one of the key constraints to the prospect of peasant organisations playing a significant role in promoting democratisation in society or in promoting positive economic policy shifts is their own undemocratic or even anti-democratic internal organisational features and procedures. Yet the literature on this reflects simplistic, functionalist approaches to the identification of internal democracy. The latter tends to be deduced from surficial and formalist organisational forms of democracy, such as the electoral systems used by the peasant organisations, their leadership styles and financial accounting systems, rather than the content of the democracy or process of democratisation. There has been a neglect in the literature of the class and ethnic influences on the structuring of internal peasant organisations and their democratic practices. The literature demonstrates a partial treatment of the role that external control by the state, NGOs and donor elements plays on peasant organisations' strategies, decision-making and membership, and how these can promote corrupt and undemocratic practices among peasant organisations.

The majority of the peasant organisations examined here operate through elected representative executive committee structures. The structures were similar in form across the continent's case studies. A few of the peasant organisations' committees however had their committee leaders being either lineage heads of the clan or men who had assumed such leadership through ascription or through being appointed to such posts by higher level traditional authorities. Even state officials in charge of development projects from which the peasant associations were derived tended to influence the selection of leaders. The type of peasant organisation encountered would influence the nature of its representational politics. The most commonly elected leaders were those who were represented in age-based associations, while clan based mutual support and survival associations tended to have leaders appointed by traditional leaders. The leadership committees often had long-standing leaders, although electoral extensions of incumbency, removal from office and occasional impeachment had been experienced.

Thus, on the criteria of basic liberal democratic representative politics, the peasant associations studied tended to be progressive and to encourage democratic practice locally. The tendency for the peasant associations to demand certain basic social and material rights, fair practices in the market place, social justice and equitable access to resources, as discussed below, also suggests that their existence encourages social democratic values and struggle. However, the tendency for the peasant association leaders to hold interlocking positions in the peasant communities concerned and for resources to be concentrated among a few of them and for the mutual support-type organisations to have appointed leaders, reflected the seeds of anti-democratic practices in the peasant associations. These anti-democratic tendencies therefore suggest that a limited influence on the wider democratisation process can be expected of peasant organisations as a movement so far.

The above trends thus suggest the need for more detailed analyses of the politics of peasant organisations and their fundamental material interests beyond questions of their management capacities or internal democracy and their developmental role measured in terms of project performance and outputs. Indeed, the question of the potentials and constraints of peasant organisations' influence on democratisation or major social change requires an analytic framework that captures both internal and external influences on their organisation and strategies, as well as an understanding of the location of peasant organisation leaderships in the wider arena of competitive bidding for and alliances directed at gaining wider political and economic power.

The Demands, Strategies and Tactics of the Peasant Organisations

The demands of peasant organisations and their strategies to achieve their goals are also a useful conceptual tag around which to improve understanding of rural social mobilisation in the agenda for political and economic change. Many studies of peasant organisations have, however, tended to impose a unilinear conception of peasant organisations, which views them as modernising institutions or adapting into developed interest groups that mainly lobby for specific technical changes in the agricultural support system using particular strategies and negotiation tactics familiar to pluralist democracy and advocacy practices in European and American situations (see various studies such as Farrington and Martin 1988; Bratton 1985; Arnaiz 1994; Arnaiz et al. 1995). In this conception the tendency for peasant organisations to have a diverse range of social and economic demands for services and a multiplicity of activities is considered by some to make them ineffective in policy advocacy (Skalnes 1995). Such analytic approaches tend to be prescriptive regarding the management orientation roles of peasant organisations, given that they tend to be directed at advising donors and NGOs on how to support peasant organisations towards the larger agendas of agricultural extension and technology networks (Farrington and Martin 1988).

But the most critical problem with these prescriptive approaches, which are ensconced in the rational choice school of thought, is their tendency to view the demand for specific development services, technologies and markets in a narrow economistic sense rather than in a more complex analytic framework that appreciates the political demands of peasant organisations and state responses to them. Thus rational peasant organisations are expected to demand discrete development inputs at different points in time rather than call for social transformation, or structural changes such as land redistribution, and other reforms. An interesting thrust of the growing literature on the evolution of African peasant organisations tends to explain the emerging economic crisis, contingent on structural adjustment reforms, especially in agriculture, as encouraging a social movement that seeks to empower grassroots organisation in their search for radical agricultural policy changes (Veltmeyer 1997; Moyo 1995; Yeros 2000). Our research concern here therefore is to understand how peasant organisations deal with the political dynamics of socio-economic change in peasant households, in particular their a response to agricultural economic decline under the rapidly liberalising African economic policy framework.

An understanding of peasant household political economies that does not reduce them to agricultural micro-economic enterprise units, which seek to maximise specific monetary goals through the mobilisation of the correct economistic balance of inputs, is necessary to understand how the socio-political relationships that emerge among peasant households and within peasant organisations shape their demands and strategies. Moreover, there is need to capture the political significance and longer-term demands that emerge from the tendency of peasant organisations to focus their activities on seeking the supply or delivery of particular farm inputs in order to realize economies of scale in the agricultural distribution system. Peasant struggles to attain reduced input prices and optimal output prices, and the tendency of peasant organisations to contest existing marketing channels or financial intermediaries, must be understood in the context of the search for wider policy reform or political change.

The contributions in this book indicate that the demands of peasant organisations and the method used to achieve some gains were varied. Their demands ranged from radical demands for social and political transformation to the more short-term piecemeal demands for discrete benefits. Virtually none of the associations studied were underground or armed groups, nor were they concerned to overthrow the central or local state. In only one case study did a fledgling secessionist 'discourse' seem apparent, and that peasant association in fact was lobbying for the creation of a local self-government structure within the existing state.

Most of the demands of the peasant associations were direct struggles for the redistribution of concrete resources. Resource struggles focused on regaining lost land or water, or to be compensated for the same, or to expand the land or water rights of marginalised groups such as women, were central in most of the peasant associations. In a few cases the struggle to attain better prices for agricultural commodities sold in international markets via marketing agencies or cooperative and state functionaries were the key demand. There were also general struggles along class, gender and ethnic lines against the unfair distribution of access to resources and projects initiated by the state, NGOs and donors.

Interestingly, land struggles took a varied but central place in the demands and travails of the various peasant organisations. Land expropriation was being resisted by peasant organisations in almost every case study, as the state, traditional leaders and private local elites had promoted land alienation at the expense of local community members. Migrant farmers' demands for land and attempts to exclude them or initiate rules which curtail their rights, were common demands among some of the peasant organisations. Struggles for or against land property rights being individuated through title deeds, struggles against the introduction of land rentals or levies, and struggles against exclusion of peasants through the leasing of state forest reserves were common. This demonstrated the intensity of local demands for land reform. New rather than traditional land allocation committees were emerging within the peasant associations to defend against land problems arising from the practices of the state, traditional authorities and an emerging local agrarian bourgeoisie. State-directed land management structures were being challenged because they marginalised local peasants through increasingly discordant land administration systems introduced on community and project lands, and because they now involved local and 'foreign' migrants.

A common approach used in most of the case studies to organise the peasant organisations' demands was the tactic of pooling resources such as labour, finance or savings, public infrastructure and marketing facilities. These resource pools in turn tended to be used to demand subsidies from the state and NGOs or to extract community labour for given services. A philosophy and practice of egalitarianism or self exploitation of labour deployed to fill in service gaps left by the state was also used as a strategy to demand additional state resources and to defend against corrupt officials responsible for local service provision.

A wide range of advocacy or organisational tactics besides the *labour-pooling resource-demand tool* were deployed. Some of the peasant associations wrote letters, petitions and visited government and other officials to lobby for specific resources such as fertilisers. Occasionally parallel marketing channels of commodities whose prices were being contested were established. In one case the boycott and a traditional community 'curse' were used to protest against given practices by traditional leaders or lineage elders.

As a result of these various struggles a range of benefits were attained. Material benefits including farming inputs, water, land and public infrastructures tended to be the most commonly sought for and realized benefits. Improved land and labour productivity and production techniques emerged in some of the case study areas. Some levels of improved social justice, gender equality and equitable access to resources were being achieved in a few of the case studies. Members of peasant associations gained from frequently provided training in terms of specific technical skills, organisational skills and ideological and social cohesion. The peasant organisations also provided various kinds of information to their members. In some cases the peasant associations were a stepping stone into both local party and state positions or even wider political roles for some members.

Thus the peasant organisations studied were indeed filling gaps in the provision of infrastructures and services created by the growing weaknesses of the African state, and the weaknesses of existing markets despite the promises made under structural adjustment. The major areas of rural decline affected by this included the supply of agricultural inputs, finance, information, extension and marketing channels, and the provision of social security in relation to food relief and basic social services. While a few of the peasant organisations enhanced the agricultural productivity and incomes of the peasant households members, many of them were defending against exploitative or extractive practices of governments and private middlemen. The majority of peasant organisations sought to deploy mutual self-reliance strategies and various coping tactics under the threat of environmental pressures, an extractive state and distortive market tendencies, especially the diminishing returns to commodity production for the export market.

Very few of the peasant organisations studied fought directly for democratisation, in the sense of seeking political control over the state machinery. But they did demand limited political and social rights and attained mostly marginal levels of participation in state development programmes into which they tended to be co-opted. The wider implications of the peasant organisations' demands could further be traced from the nature of their formal linkages to the larger society.

Relations of Peasant Organisations to the State and Wider Society

The relationships of peasant associations to each other, between them and the state, and between them and other organisations of civil society such as NGOs, are critical in explaining the role and significance of such associations in the wider society. A conceptual framework, which views peasant organisations as an integral part of rural civil society, of necessity has to understand how peasant organisations are connected to the state. If civil society is not defined as comprising a merely anti-state phenomenon, opposed to or independent of the state (Sachikonye 1995), then it is easier to readily appreciate the meaning of the existing intricate, varied and contradictory linkages between peasant organisations and the state. In one sense peasant organisations increasingly play a critical role in the gradual evolution of rural local government, especially with regard to their increasing role of filling the service provision gaps left by the state.

Indeed, these peasant organisations may be conceived as a new mechanism of promoting the so-called 'indirect rule system', that traditional authorities still effect *vis-à-vis* central governments. This is the case in our case studies where the peasant organisations seem to increasingly substitute the local state and traditional authority in organising community infrastructure building and maintenance, in exacting local taxes in the form of labour contributions and cash charges for the delivery of services provided such as water, land administration and so forth, and in the co-management of state-initiated development projects. Indeed, this tendency begs further questions in the debate on the nature and weaknesses of rural civil society as provoked recently by various scholars (Mamdani 1996).

Thus the evidence that the state responds positively to peasant organisations' lobbying for given resources is also a close reflection of how state officials in turn utilise the leverage they gain from such support to influence the development and organisation of peasant organisations. The studies show in this vein how the state is deeply involved in influencing the leaderships of peasant associations, and how state officials attempt to promote particular development practices and trajectories through the development projects involving peasant associations. The choice of crops grown, taxes charged and specific forms of resource allocations undertaken by some of the peasant associations are clearly in consonance with the wishes of the state. And, as we argued earlier, these state–peasant organisation interactions tend to be in the interest of reproducing and reinforcing particular rural class relations and power structures. Hence state marketing agencies or development authorities that have been foreclosed under structural adjustment regimes are being replaced by state-encouraged or state-facilitated peasant organisations' marketing projects. Some of the peasant organisations have thus become central to the structural adjustment-based strategy of decentralisation and rural market reform agenda. In this way peasant organisations are integral to the intensified incorporation of the peasantry in Africa, just as much as they defend against repression and exploitation related to such incorporation.

The relationship between the state and peasant organisations is fraught with contradictions. Resistance by peasant organisations of direct and indirect state repression, and exploitation through the expropriation of their resources and through extractive regulations, are common tendencies in the state–peasant organisation relationship. Their resistance to regulation and manipulation through traditional leadership and local authority structures is the central *raison d'etre* of many of the peasant organisations encountered. Cooperation, contestation, co-optation and compulsion are interchangeably applicable concepts, which govern the relationship of the peasant associations to the state. The dependence of the former on resources supplied by the latter, including donors and NGOs sanctioned by the state, is critical in explaining this contradictory and evolving relationship.

The emergence of alliances between peasant organisations seems also to relate to the hegemonic tendencies of the state in its relationships with NGOs and donor agencies, as these promote similar peasant organisation development activities across the country case studies. Some of the studies also suggest that the peasant organisations are increasingly created by external agencies such as the state, and that their organisational features reflect the requirements of state funding bureaucracies. The common role of peasant organisations in promoting 'poverty alleviation' suggests that they may be welfare agents intended to contain the growing rural discontentment with the African state and the marginalising tendencies of the global market system.

The expectation that peasant organisations create political space for democratisation (Rahmato 1991) is swamped by the contradictory relationships between the state and peasant organisations. Some of the case studies suggest that the tendency to create instrumental structures by government counteracts alternative strong centres of rural power. Thus peasant organisations could either close democratic space by countervailing chiefs and other traditional leadership that the state is unable to control, or by counteracting influential NGOs and opposition party branches. However, a more flexible understanding of the origins and changing motives of peasant organisations, and how their relations to the state tend to shift in relation to changing economic and political circumstances, is critical in this debate (Moyo *et al.*, 2000). The changing of ruling parties, the effects of changing economic policies and the shifting political and economic fortunes of elite classes are also instrumental in understanding the nature and development of peasant organisation and state relationships. The effects of economic and political liberalisation on the missions of peasant organisations may be more profound than is recognised.

There are indeed inherent difficulties that peasant organisations face in their effort to capture political space so as to further economic policy reform and democratisation. Because of the absence of competitive politics, peasant organisations have to compete for influence with ruling parties and governments that have absolute control in rural areas (Bratton 1995). However, since peasant organisations continue to focus on mobilising the rural political constituency, through delivery of development goods and services, which the state can no longer provide, their importance in generating democratic space in the wider society need not be underestimated. The mere possession by peasant organisation leadership of instruments for mass mobilisation, such as vehicles, communication systems and some skilled operatives, even when these are not used for overt party political purposes or to contradict governments, have become a threat to the monopoly over power held by rural state bureaucrats and regional party leaderships (Moyo 1992). Governments, however, tend to capitalise politically on the delivery by NGOs or peasant organisations of goods and services to their memberships on grounds that such delivery is a result of government policy or support (ibid.). There remains contradictory evidence regarding the relationship between peasant organisations and governments in terms of the extent to which they collaborate in mobilising external donor resources for developing participatory development approaches or for reinforcing repressive and exploitative tendencies.

Moreover, the case studies show how peasant organisation leaderships tend to use peasant organisations for their upward political entry into various levels of government or party leaderships. Examples of the strategically planned entry into government of peasant organisation leaders at national, provincial and district levels are evident in the case of ORAP in Zimbabwe (Moyo 1992), and the absorption of peasant organisation leaderships into the new South African government. The expectation, however, that most organisations of civil society including peasant organisations do not aspire to destroy or capture the state (see Nyangoro 1994; Mamdani and Wamba Dia Wamba 1995) is corroborated by the studies. The expectation that peasant organisations play a politically significant mediation role between the state and peasants in promoting democratisation, as implied by Rahmato (ibid.), does seem to be substantiated by the missions and activities of the peasant organisations studied here.

Study Approach and Layout

The chapters contributed in this volume are the result of research undertaken by network members of a CODESRIA Multinational Working Group (MWG) established in 1995. This project builds upon and is part of the accumulated conceptual and empirical research work on social movements which CODESRIA began earlier (see Mamdani and Wamba Dia Wamba 1995). It also deepens CODESRIA's broader research agenda on the study of democratisation processes in Africa (see Mkandawire and Olukoshi 1995).

This project owes a tremendous debt to Dessaglen Rahmato. He wrote the founding concept paper—known in CODESRIA as the 'Green Book'—entitled *Peasant Organisations in Africa: Constraints and Potentials* (Rahmato 1991). All the research proposals were developed in response to the wide-ranging issues raised by Rahmato. The research issues were also elaborated from the lively debate that ensued as a result of Archie Mafeje's provocative response to Rahmato (Mafeje 1993). In 1995 various discussion papers were presented by the editors and resource persons at a methodology workshop in Dakar, which reviewed the research proposals on peasant organisations. Those inputs and the debates at the final network conference held in late 1997 in Harare shaped the chapters presented here.

One of the major expectations of the methodology workshop on peasant organisations was the need for the studies to focus their research contribution on improving our understanding of emerging social movements and civil society in rural Africa. The main objective of most of the studies is thus to broaden the empirical and comparative basis upon which theory on social movements is developed. The network also agreed to focus this book thematically on primarily examining the role played by peasant organisations in promoting the democratisation process. Such a focus was not, however, intended to exclude the treatment of other related social and economic issues and processes that peasant associations address. Thus while nine of the studies focus directly on issues of democratisation, the rest highlight the political-economic processes underlying the developmental activities of the peasant associations studied.

The structure of the chapters in this book is in three parts, which reflect their heuristic grouping into three sub-thematic areas representing the predominant roles of the peasant organisations. Part one presents studies of those peasant associations that reflect a primary focus on promoting democratisation, especially at the local government level. Thereafter part two presents studies that predominantly examine how peasant organisations empower their members to enhance grassroots participation in social and development activities. Finally, the third part of the book presents studies of peasant organisations that originate in response to specific economic development project initiatives, especially during the structural adjustment reforms since the 1980s.

The twelve studies of peasant organisations presented here are drawn from eight countries, covering the five geographic regions of Africa. Five of the studies are from Nigeria, while one study each is from Cameroon, Cape Verde, Kenya, Senegal, Sudan, Tunisia and Zimbabwe. The studies covered four different types of peasant organisations selected by the researchers. These include: mutual support and welfare-oriented peasant organisations; age grade-based peasant organisations; community development or project based associations; and peasant organisations established to protest against specific concerns. It is hoped that this wide range of case studies in terms of the types of peasant associations, countries and contexts covered provides an adequate basis for a comparative empirical assessment of their experiences.

Given that this book is an exploratory foray into the under-researched and inadequately conceptualised problem of the nature and influences of peasant organisations in rural civil society, the study approach adopted was deliberately based upon diverse methods. The researchers had much leeway to choose suitable case study areas and study approaches. But all the studies do use empirical fieldwork-based data and information gathered from various sources. They deploy a variety of data collection methods and rely on different forms of evidence to substantiate their arguments. Most of the studies administered questionnaire surveys covering peasant family members belonging to the peasant organisations studied and other non-members in the areas served by the organisations. Some of the studies relied upon informal interviews undertaken with sub-committee members of the peasant organisations, with selected focus groups of members and with key informants. In most instances in-depth case studies were undertaken of selected branches of the peasant organisations and/or the villages served by them. Informal interviews and observations were commonly used in such case studies.

The majority of the case study material was derived from formally registered peasant organisations and focused on the evidence provided by leaders and members of the same. This limited the scope of perspectives gained, especially since those belonging to the more 'underground' types of peasant organisations tended to be excluded. Many of the studies also relied on written records, which tended to be more detailed for those peasant organisations involved in relatively large but local development projects. Government records were also an important source of information, as were the records of banks, marketing depots, factories and cooperative unions.

The historical breadth of the study was limited by the nature of peasant organisations selected for study. The majority of the peasant organisations studied and the information utilised are based upon organisations that were formed between the late 1970s and 1997. Only a couple of the studies were grounded in more remote historical sources and in events dating back to the 1950s. Thus the case studies presented here mostly reflect the more recent post-colonial generation of peasant organisations.

References

Arnaiz, M.E. (1994) *Farmer Organisations in the Technology Change Process: An Annotated Bibliography*, ODI Network, Paper 53, London.

Arnaiz, E., Merrill-Sands, D. and Mukwende, B. (1995) *The Zimbabwe Farmers' Union: Its Current Role in Technology Development and Transfer*, ODI Agricultural Research and Extension Paper, London.

Bernstein, H. (1977) 'Notes on capital and peasantry', *Review of African Political Economy*, 10.

Bernstein, H. (1979) 'African peasantries: a theoretical framework', *Journal of Peasant Studies*, 8, 4, 430-1.

Bernstein, H. (1990) 'Taking the peasant', in H. Bernstein *et al.* (eds.), *The Food Question: Profits Versus People*, London, Earthscan.

Boone, C. (1994) 'State and ruling classes in post-colonial Africa: the enduring contradictions of power', in J. S. Miggdal, A. Kholi and V. Shue (eds.), *State and Social Forces: Domination and Transformation in the Third World*, Cambridge, Cambridge University Press.

Boone, C. (1995a) 'Rural interests and the making of modern African states', *African Economic History*, 23.

Boone, C. (1995b) 'The social origins of Ivorian exceptionalism: rural society and state formation', *Comparative Politics* 27, 4.

Brass, T. (1991) 'Moral economists, subalterns, new social movements and the (re) emergence of a (post) modernised (middle) peasant', *Journal of Peasant Studies*, 18, 2.

Bratton, M. (1985) 'Farmers organisations and food production in Zimbabwe', *World Development*, 14, 3.

Bratton, M. (1994) 'Micro-democracy? The merger of farmer unions in Zimbabwe', *African Studies Review*, 37, 1.

Cliffe, L. (1977) 'Rural class formation in East Africa', *Journal of Peasant Studies*, 4, 3.

Cliffe, L. (1987) 'The debate on African peasantries', *Development and Change*, 18, 2.

Farrington, J. and Martin, A. (1988) *Farmer Participation in Agricultural Research: A Review of Concepts and Recent Practices*, ODI, Paper 9, London.

Hyden, G. (1986) 'Discussion: the anomaly of African peasantry', *Development and Change*, 17, 4.

Klein, M. A. (1980) *Peasants in Africa*, Beverly Hills, Sage.

Mafeje, A. (1993) 'Peasant organisations in Africa: a potential dialogue between economists and sociologists — some theoretical and methodological observations', *CODESRIA Bulletin*, 1, 14-17.

Mafeje, A. (n.d.) Land and Agrarian Questions in Africa (unpublished mimeo).

Mamdani, M. (1996) *Citizen and Subject: Contemporary Africa and the Legacy of Late Colonialism*, Princeton, NJ, Princeton University Press.

Mamdani, M. and Wamba Dia Wamba, E. (eds.) (1995) *African Studies in Social Movements and Democracy*, Dakar, CODESRIA.

Mkandawire, T. and Soludo, C. (1999) *Our Continent, Our Future: African Perspectives on Structural Adjustment*, Dakar, CODESRIA/IDRC/AWP.

Mkandawire, T. (2000) "Conflict and Peasants" (unpublished).

Mkandawire, T. and Olukoshi, A. (1995) *Between Liberalisation and Oppression: The Politics of Structural Adjustment in Africa*, Dakar, CODESRIA.

Moyo, S. (1991) *NGO Advocacy in Zimbabwe*, ZERO Working Paper, Harare.

Moyo, S. (1992) *An Impact Assessment of ORAP*, mimeo, Bulawayo, ORAP.

Moyo, S. (1995) *The Land Question in Zimbabwe*, Harare, SAPES Books.

Moyo, S. Makumbe, J. and Raftopolous, B. (2000) *NGOs, The State and Politics in Zimbabwe*, SAPES Books, Harare

Nyangoro, J. E. (1994) 'Reform politics and democratisation policies in Africa', *African Studies Review*, XXXVII, 1.

Rahmato, D. (1991) *Peasant Organisations in Africa: Constraints and Potentials,* CODESRIA Working Paper 1/91.

Sachikonye, L.M. (1995) *Democracy, Civil Society and the State: Social Movements in Southern Africa*, mimeo, Harare, SAPES Trust.

Saul, J. and Woods, R. (1987) 'African peasants', in: T. Shanin (ed.), *Peasants and Peasant Society,* New York, Oxford University Press.

Scott, J. (1985) *Weapons of the Weak: Everyday Forms of Peasant Resistance*, New Haven, CT, Yale University Press.

Skalnes, T. (1995) *The Politics of Economic Reform in Zimbabwe: Continuity and Change in Development*, London, Macmillan Press.

Veltmeyer, H. (1997) 'New social movements in Latin America: the dynamics of class and identity', *The Journal of Peasant Studies*, 25, 1.

Williams, G. (1976) 'Taking the part of peasants: rural development in Nigeria and Tanzania', in P. C.W. Gutkind and I. Wallerstein (eds.), *The Political Economy of Contemporary Africa*, London, Sage.

Yeros, P. (2000) *Peasant Struggles for Land and Security in Zimbabwe; A Global Moral Economy at the Close of the Twentieth Century*, working paper.

Zghal, A. (1995) 'The "bread riot" and the crisis of the one-party system in Tunisia', in M. Mamdani and E. Wamba Dia Wamba (eds.), *African Studies in Social Movements and Democracy*, Dakar, CODESRIA.

Part One

Peasants Associations
and the Democratisation Process

1

Peasant Associations in the Lower Anambra Basin Area of Eastern Nigeria: Grassroots Democratisation and Empowerment

Eric C. Eboh

Introduction

Many African countries, including Nigeria, are facing multiple socio-political and economic crises of tremendous proportions. Economic Structural Adjustment Programmes (SAPs) have eroded living standards while pressures for democratisation, responsible governance and accountable leadership reflect a move towards a new socio-political order across the continent. The peasantry bears the major brunt of the crises, given its economic marginalisation, political exclusion and social alienation (Klein 1980; Rahmato 1991; Ibeanu 1993).

This continuing crisis of the state and economy calls into question the 'modernisation' strategies applied by most African countries since independence. Since the mid-1970s, there has been a fundamental shift away from the dominant modernisation paradigm towards a systematic search for development alternatives (Oakley 1991; Rahmato 1991). The 'new' development thinking focuses on autonomous development strategies that are 'people-centred' rather than 'people by-passing' or 'people-marginalising'. Although there is no single definition or model of this new development, many scholars agree that the basis of development must entail the social, political and economic participation of the people involved (Oakley & Marsden 1984).

The people-centred development perspective suggests that because peasant organisations are deeply rooted in rural society they can expand their influence on the evolution of a new African socio-political and economic order. This argument and its underlying logic assume that peasant organisations have the capacity to become effective building blocks or breeding centres for grassroots democratisation and empowerment. Several factors and conditions could undermine this proposition. Social and economic differentiation and inequalities among the peasantry, which permeate peasant organisations, could place democratic decision-making and peasant participation at risk. The relations of peasant organisations with the state and other outside agencies could weaken them and their ability to defend peasant interests and voice popular initiatives. Where external influences are critical, the transparency and accountability of peasant organisation leadership to their members could become questionable.

In south-eastern Nigeria peasants have for long pursued their social and economic interests and everyday lives using organisational strategies based on co-operation, solidarity and mutual support (Okorie & Obeta 1986). Ideally, peasant organisations provide a forum where peasants discuss common problems and pool efforts and resources in an active process of taking initiatives and action to promote peasant welfare and/or defend themselves against exploitation and alienation (OAU/IILS 1983; Oakley & Marsden 1984). However, various types of peasant organisations based on different origins and goals have focused on promoting their memberships' predominantly agrarian interests. In south-eastern Nigeria such organisations include farmers' mutual support groups, welfare associations, farmers' organisations and neighbourhood or kinship unions involving peasants (Okorie & Obeta 1986; Eboh 1993). They organise practical and flexible activities which facilitate peasants' access to services, assets and resources.

Nigerian socio-economic crises characterised by structural adjustment-related recession pose enormous challenges for peasant organisations. These organisations must operate in a milieu of complex and changing social, political and economic forces. On the one hand, such changing conditions may impinge on the capacity and effectiveness of peasant organisations, while on the other the same conditions could strengthen their organisational linkages and group development. Yet there is also the possibility that the crisis could lead to 'imitation' or 'counterfeit' peasant organisations, which pretend to be serving peasant interests but actually do not do so, and in some cases do the opposite (OAU/ILO 1983).

Given these harsh realities, the enormous expectation that peasant organisations could influence change in the Nigerian socio-political and economic order must be interrogated against the real potentials and limitations of such organisations. This research examines the nature and degree of relationships that have evolved within peasant organisations themselves, and between peasant organisations and the state. The research assesses the potentials and constraints of peasant organisations in enhancing grassroots participation, empowerment and democratisation. The specific objective is to examine how the internal dynamics of peasant organisations are being shaped by social and economic differentiation within the peasant society and by the socio-political forces within the local milieu.

The research questions explored include: under what conditions did the organisations originate, and what primary tasks and issues do they pursue? How do local social, political and economic forces shape their internal governance and cohesion? What are the gender issues in peasant organisations, and how does the local milieu impact on them? The study also investigates how relationships between peasant organisations impact on their organisational features and strategies, and on their ability to voice, project and secure broader peasant interests and demands. This in turn raises research questions regarding the linkages between peasant organisations and the state. These include: what are the forms and degrees of articulation between organisations and the local authority and structures of state power? What issues and concerns are taken up under collaborative arrangements involving peasant organisations? What instruments of collaboration are used and how efficacious are they? What is the impact of their collaboration on internal governance and decision-making?

In addition, the study examines how the relationships between peasant organisations and the state are affected by changing socio-political and economic conditions, and how emerging patterns of relationships affect the strategies, autonomy and credibility of peasant organisations. The relevant research questions here include: how do peasant organisations define or position themselves in the wider socio-political and economic context? What issues do they take up with state authorities, and to what degree have concessions been obtained and peasant demands secured? Are peasant organisations exploited by state organs, and which peasant group membership or leadership act as channels for the state structures to assist or exploit peasant organisations?

Peasants and Peasant Organisations

Debates on whether African smallholder agriculturalists or rural producers fall into the social and historical category called 'peasants' and about how connected they are with 'peasantness' and associated rural power relations remain a central discourse among African social scientists. There is a large body of theoretical literature on African peasantries and the nature of 'peasant societies' in Africa in the context of different agrarian settings (see Mafeje 1993). While one school defines peasants as rural producers securing their livelihood through the use of family labour on family land, another conceptualises peasants in terms of their subordination to a central authority and their dependence on state markets that subject them to unequal exchange (see Klein 1980). Both theorisations, especially that which alludes to asymmetrical relationships to state power, do not conceive of the peasantry as a homogeneous mass or agglomerate. Scholars tend to agree that the peasantry is characterised by internal differentiation along lines such as unequal distribution of power and access to productive resources, and that this breeds exploitative relationships between different segments of the peasantry (Bernstein 1979; Saul & Woods 1979). Even when social relations within the peasantry are predominantly of a non-exploitative nature, peasant exchanges or co-operative practices often entail a transfer of some surplus from the poor and middle peasantry to the rich peasantry, usually in disguised forms (Mamdani 1986; Amin 1991).

A related aspect of this debate concerns the extent to which capitalist penetration of rural Africa and the relationship between the state and peasantry leaves the African peasantry 'uncaptured', lacking initiative and innovation and unable to develop because of its 'economy of affection' (Hyden 1986). Some accounts suggest that African peasants are indeed captured by the state and that class, not communal (or affection) relations are most important in their lives, and that they are productive and innovative (Cliffe 1987).

There is a growing literature and empirical work on the Nigerian peasantry. Peasant studies in Nigeria received a remarkable impetus during the oil boom and illusions about a 'green revolution' (Mafeje 1993). Relatively speaking, earlier studies tended to focus on within-country descriptions and comparisons of the peasantry, particularly its social, economic and political characteristics (Beer & Williams 1975; Beer 1976). State–society relations as applicable to the peasantry, especially in the context of the external relations of peasants to the Nigerian state, have also received theoretical and empirical attention (Beer 1980; Ake 1985; Williams 1991).

Other key issues in the ongoing debate include alienation and displacement of peasant cultivators arising from government expropriations of peasant farmlands as well as massive land acquisitions by capitalist farmers (Omeje 1995); the deteriorating condition of the Nigerian peasantry (Ibeanu 1993); increased peasant vulnerability in the face of recession and structural adjustment (Swindell & Iliya 1992), and the human rights significance of these occurrences (Omeje 1995). Equally important are theoretical and empirical reflections on peasant resistance, violence and 'anti-power' social movements, which simply seek to defend and maintain their autonomy against domination and violation (Amadiume 1995; Omeje 1995).

Recent debates on peasant organisations centre on an appropriate definitional paradigm for the integral features of peasant organisations in different agrarian settings (Rahmato 1991; Mafeje 1993). Peasant organisations have been categorised from a perspective which differentiates primary from secondary organisations or one which distinguishes the informal from the formal. The diverse forms of peasant organisation found on the ground include mutual support associations, age grades, saving and credit associations, farmers' associations, welfare associations, burial societies and co-operatives. This variety defies simple categorisation. In fact, relations in peasant societies, which tend to centre on lineage organisation and labour co-operation groups whose origins are often precapitalist, have changed over time in their social content and form (Amin 1991).

The propensity of peasants to organise has elsewhere been related to the extent of state and market penetration (Bratton 1986) and also to peasants' perception of their social environment (Rahmato 1991). The centrality of social networks, typified generally by peasant organisations in the adaptive processes of African rural peoples, is well acknowledged in the literature (Richards 1986). Contemporary discussions about the nature of peasant organisations are now preoccupied with their potential to expand processes of popular participation, grassroots empowerment and democratisation. Disillusioned by the failure of 'modernisation' strategies and the crisis of development, hopes are focused on a development paradigm that is 'people-based' and which integrates popular participation, grassroots empowerment and democratisation as both means and end (OAU/ILO 1983; Oakley & Marsden 1984; UNECA 1990).

Within this framework, peasant organisations are assigned 'ambitious' roles, such as being intermediaries of social change and rural development through such processes as planning and implementing sustainable development programmes (Esman & Uphoff 1984; Cernea 1987); articulating and projecting people's power (Rahmato 1991); demanding the devolution of power (Aloki 1989); defending peasants' political space (Holmquist 1980); and raising popular consciousness, grassroots involvement, participation and empowerment (Chambers 1983). However, it is debatable whether these expectations are realistic.

The capacity of peasant organisations to promote socio-political and economic change could be undermined by both internal and external influences. Internal factors capable of hindering their prospects include social and economic differentiation, which can impede internal democracy and distort internal organisational features and procedures. Whether peasant organisations can transcend their local interests and defensive survival strategies and advocate broader democratisation, state accountability, participatory development and political space remains unanswered (Moyo 1995).

External influences that limit this prospect include the tendency of the state and donors to seek control of peasant organisations and strategies by manipulating or exploiting peasant differentiation, and their use of instruments of coercion, persuasion and co-optation (Ibid.). Additionally, local authority and power structures restrain the prospects of peasant organisations by frustrating local mechanisms for group decision-making and organisational effectiveness.

Given these contradictions in the potential of peasant organisations to influence wider socio-political and economic change, it is important to understand more clearly their origins, changing motives, composition and strategies in relation to changing economic and socio-political circumstances.

Study Area and Research Approach

The Lower Anambra Basin area consists of 15 peasant communities proportionately shared between the present Anambra and Enugu states in south-eastern Nigeria. The area is endowed with vast stretches of fertile land that is relatively accessible by tarred road from the major urban areas of Nsukka, Enugu, Awka and Onitsha. Its agrarian structure is predominantly smallholder farming, with a large presence of private large-scale farm investments and state-owned agricultural projects like the Ada Rice Production Company. Land tenure in the area is a mixture of communal and individual rights, with a growing trend towards individualised tenure.

The specific location of this study was the Lower Anambra Irrigation Project (LAIP), which is the centre of agricultural activity. The enclave consists of the communities of Omor, Umumbo, Umorum, Ifite-Ogwari, Igbankwu, Omasi, Umueje and Anaku served by LAIP in irrigation water distribution, land clearing (ploughing and harrowing) and produce-processing services.[1] The large-scale farm investments are mainly targeted at rice and maize. The peasants' staple crops include maize, cassava, garri and rice.

There is a traditional gender division in access to land ownership and farm labour. The socio-political setting is such that every community is regarded in the local parlance as 'autonomous', meaning it has a distinct traditional chief called an igwe. The latter is independent of other communities and is the custodian of the traditional culture, beliefs and authority. Every community consists of villages that are structured on descent and kinship.

Peasant organisations in a variety of forms pre-date state-owned and private capitalist farming activities in the area, while others (such as land committees) are a consequence of the increased threats of alienation and displacement posed by government expropriations and massive private acquisitions of peasants' farmlands, especially since the mid-1980s. Additional pressures on the peasantry and peasant organisations have come from the government's divestiture or commercialisation policies regarding projects (like the LAIP) which originally 'subsidised' peasants' access to irrigation water, tractor services and produce processing. How the peasant organisations in the area face the challenges of land alienation, displacement, impoverishment and reduced access to farm services will be discussed below. The LAIP area offers a peculiar but changing agrarian and economic setting to question the potentials, strategies, dynamics and performance of peasant organisations and their relationship with the state.

This research is based on case studies of peasant organisations selected at random in three communities in the LAIP area. Thirteen representative types of peasant organisation were chosen, including peasant mutual support groups, farmers' groups, village land committees and producers' associations. A village appraisal designed to sample the diversity of peasant organisations in the various communities led to the identification of the specific peasant organisations selected.

Information was gathered from multiple sources, including leaders and members of selected peasant organisations and individuals associated with power and authority structures within the local communities. Interviews were held with key informants (that is, old and experienced peasants) with a good knowledge of the past and present conditions of the peasantry and peasant organisations. Informal (unstructured) interviews on a face-to-face basis were conducted with members of the peasant organisations selected to represent different strata, gender and any other form of categorisation used in the locality. Information gathered on the internal dynamics of peasant organisations was broadly outlined as follows: origin; changing motives; membership composition (gender, class, village or community bases); organisational features, strategies and procedures; influences of peasant differentiation (socio-economic inequalities) on internal decision-making (democratic, bureaucratic or dictatorial); nature and extent of rural economic goods and services mobilised; distribution of benefits; the types of issues taken up and the socio-political and economic circumstances shaping them; the way issues were taken up (confrontation, collaboration, resistance or advocacy); and the impact of the above on the capacity and effectiveness of peasant organisations in promoting peasant participation and interests.

On peasant organisations' articulation with similar organisations and local authority structures, the information collected included: the forms, types and levels of relationships with other peasant associations; the relationships of peasant organisations with the community power and authority structures, and the impact of these on their internal governance; the type of issues and concerns taken up by peasant organisations in conjunction with each other, and the strategies (reactive and/or proactive) employed; and how peasant differentiation (class, gender, unequal access, etc) within the local community was shaping these relationships, competition for scarce resources and/or political relevance (space) between different peasant organisations, and the impact of such processes on the broader interests in the local community.

Concerning the relationships between peasant organisations and the state, information gathered focused on: location of peasant organisation leadership in the wider arena of competitive bidding for and alliances directed at gaining political and economic power; struggles to defend and maintain autonomy; resistance to incorporation into the state system, state domination and violation of self-rule (anti-power attribute); strategies used in seeking state patronage and protection in the agricultural system; socio-economic and political concessions and benefits gained from interactions with the state; and the effects of external penetration by the state on peasant organisations' strategies and decision-making.

The Evolution of Peasant Associations in the Lower Anambra Basin Area

An examination of the organisational bases, internal differentiation and operational dynamics of peasant organisations is necessary for an understanding of their potentials for articulating and defending peasant interests. The peasant associations were widely divergent in their period of existence, with a range of 3 to 17 years and a modal age of 7 years. While some of the associations were formed on the basis of the local people's (peasants') initiative, others were apparently founded upon instructions or directives emanating from certain government authorities. A few emerged out of the efforts of non-governmental organisations (NGOs) or agencies. It was observed that whereas some of the associations were formed in response to local people's felt need to rally peasant efforts for promoting mutual co-operation, social security and collective welfare, others originated mainly from a desire to improve local people's chances of receiving government-sourced agricultural credit, production inputs and other farm-related assistance

One association was formed under the auspices of the LAIP, which mandated irrigation water users to be members so that they could be reached as a group on issues concerning water distribution, maintenance and protection of water facilities and related matters. Some all-women associations were also found to have been formed in response to special state-related programmes, such as the Better Life Programme for Rural Women. The associations pursued a variety of activities revolving around the promotion of members' interests and welfare. These involved fostering mutual co-operation between women farmers for self-help purposes and to promote the crop production, processing and marketing activities of members. This process also created a forum for members to obtain assistance from government and NGOs for agricultural credit, production inputs, and so on.

The organisational structures of the peasant associations were based on executive committees that derived legitimacy from the general membership. Open elections to the executive committee were conducted on the expiration of a predetermined tenure, usually three years. About 15 per cent of the associations reported having two-year and four-year periods of leadership tenure. One association was found to be run by executive officers appointed or hand-picked by the LAIP. This was not surprising considering that the association was formed under LAIP auspices. Explanations of this imposition from above suggested that the LAIP intervened to 'save' the association from alleged collapse following 'perennial and incessant leadership conflict and executive feuds'. The authority had to re-establish sanity and order as the necessary means of reviving the association.

The normal size of the executive committee among the associations was found to range from six to eight members. In one of the associations there was a mandatory inclusion of traditional chiefs (igwes) as permanent members of the elective committee, in compliance with the directive of the project authority which initiated the association. One interviewee said the reason why igwes were directors of the association was that the Japanese experts who worked with the LAIP said it must be so.

Most of the associations reported holding monthly general meetings where policy decisions were taken predominantly by 'consensus'. Elsewhere, some associations insisted on a simple majority vote to carry through a decision, especially in circumstances where consensus became too difficult to secure. But this was noted only in a few cases, since the 'spirit of unity and brotherhood' pervaded the deliberations at meetings. Even though most associations reported reaching decisions by consensus, the decision-making process itself implicitly contained the simple majority vote principle.

Membership fees were fixed, but subject to review as deemed necessary by the association. About 30 per cent of the associations currently fixed their membership fee at N10, while 23 per cent required a fee of N30 from new members. Remarkably, one all-women association required new members to pay a subscription fee of N5,000 to the share capital of the group's business venture. There is no doubt that membership fees had been reviewed upwards when compared to the amount paid at inception of the associations. While 31 per cent of the associations had maintained the same membership fee since inception, about 23 per cent had pushed up their fee by up to 100 per cent and 15 per cent by 200 per cent. There was one association which had implemented a 500 per cent increase in membership fees between its inception and the research date.

Membership size is an important factor, with potential effects on the bond of mutual trust and cord of solidarity that glues the membership together. Too large a membership could stretch the effective range of enhancing the esprit de corps among members. The trend in membership size (whether decreasing, increasing or the same) over the years could also reveal certain underlying influences on the associations.

Mixed trends were found. One all-men association experienced a decline of membership by about 40 per cent. With one exception, all women-only associations witnessed a decrease in membership of between 16 and 60 per cent. On the other hand, one all-men association had experienced an increase in membership of 200 per cent, and one gender-neutral association an increase of 142 per cent. It was suggested that the negative trends in the size of membership were largely a consequence of the 'frustration' among members who perennially suffered unfulfilled expectations, especially concerning access to production loans, fertiliser and related assistance through their associations.

The associations exhibited varying forms and degrees of internal social and economic differentiation, as reflected by certain individual members, who although not executive officers wielded considerable power and influence on the decisions and activities of the associations. These powerful and influential individuals, according to the interviewees, were usually owners of relatively larger farms, and much wealthier and more educated than the mass of members. They had privileged access to local and/or other government authorities and project agencies as well as to the traditional chieftaincy institutions. According to one interviewee, 'for example in my association, there is a powerful and influential individual who owns up to 20 plots instead of three plots normally allowed to every farmer-participant under the irrigation project'. Other interviewees identified such individuals to be powerful and influential because of their 'high social ranking', 'money power' and 'political connections'. The respondents expressed a variety of assessments on the desirability, usefulness and impact of this set of individuals.

Whereas some respondents said such individuals played a constructive and facilitative role in the affairs of the association, others expressed negative views about their impact. But respondents who reported positive impressions hinged their views on the fact that such powerful and influential members often brought the interviewees short-run pay-offs. Their intervention, for instance, facilitated the members' access to government assistance and supplies. This tended to obscure the fact that their interventions could have potentially dangerous implications for the stability and democratic culture in the associations. Some interviewees reported instances in which the democratic processes and peasant advocacy role of the peasant associations had faced subtle manipulation and severe strains from wealthy elements.

The interviewees' muted criticisms of (or rather explicit indifference to) the role of these powerful and influential individuals is symptomatic of the apparently helpless acquiescence of the mass of members in response to increased socio-economic distress. Interviewees said they often disapproved but agreed to the activities of the powerful and influential individuals without complaining or arguing (that is, agitating) because of obvious social and economic disadvantages. Some respondents explained that while they might not be unaware of the adverse ramifications of the interventions by certain powerful and influential individuals, they chose to ignore any negative consequences on stability and democratic culture in the associations since 'we do not have the means or capabilities to reverse the situation'. According to them, 'the fact that the mass of the members do not openly challenge their activities and manipulations should not be construed for consent' — 'we are just acquiescent', they said.

The Articulation and Protection of Peasant Interests

This study assessed the empirical validity of the associations' claim that they originated and functioned on the traditional co-operative principle of igwebuike, which means 'collective (or pooled) action produces greater and stronger results and impact than individual or separate actions'. The peasant associations employed five approaches in articulating, projecting and protecting the peasant interests at the centre of this empirical analysis. These approaches lay in providing access to farm credit or loans, fertiliser, improved planting materials, irrigation water and tractor use. In addition, certain special gender-specific interests were promoted.

It was observed that the peasant associations regularly requested farm production inputs, agricultural credit and other assistance from relevant authorities such as local, state and federal government agencies, NGOs, specialised development programmes and projects, and donor funder schemes. Formal applications containing specific details of requests were submitted to the relevant authorities through official communication channels. Follow-up visits were then embarked upon by executive officers and/or selected persons (especially members who were powerful and influential) under delegated authority from the association with the aim of pressing the agencies for positive attention and favourable responses to the requests.

While a number of peasant associations reported zero pay-off in realising their legitimate demands from relevant authorities, others were found to have experienced varied degrees of success. One extreme case was found whereby the peasant association had apparently 'given up' or ceased to make demands upon the authorities. This probably reflected a peak of frustrations arising from perennial failure to secure farm inputs and related assistance. A set of questions regarding how effective peasant associations had been in obtaining members was posed. Respondents pointed out that requests for inputs, especially fertiliser, were made regularly to relevant authorities, but only a fraction of the demands were usually obtained. Requests made for rice milling machines and fertiliser did not succeed most of the time, or lower numbers of machines were obtained.

Focusing on fertiliser-related peasant agitation as the specific object of analysis, it was observed that different associations experienced varied rates of success in obtaining their fertiliser demands from the authorities (Table 1). Of the 13 peasant associations, only one did not request fertiliser that year. This rather 'odd' association has apparently lost faith in the whole exercise of making efforts to obtain fertiliser for its members and so accordingly 'did not make an attempt to get fertiliser because of predictable negative outcomes'. Nonetheless, the fact that only one out of all the associations failed to request and project peasant demands for fertiliser indicates that the majority of the associations were conscious of their responsibility to their members, and in fact did make efforts to actualise those interests or meet the needs of members. It is instructive to note that the association which did not apply for fertiliser had an all-men membership.

Table 1 Peasant Associations' Agitation for Fertiliser and Recorded Success Rates

Identification of association		Requested fertiliser in the year	Amount requested (50kg bags)	Amount obtained (50kg bags)	Per cent of request obtained	Any agitative efforts	Efforts yielded result?
A	all-men	yes	-	0	0	yes	no
B	all-men	yes	50	5	10	yes	no
C	all-men	no	n.a.	n.a.	n.a	n.a	n.a.
D	men & women	yes	-	0	0	yes	no
E	all-women	yes	50	5	10	yes	marginal result
F	all-women	yes	100	5	5	yes	no
G	all-men	yes	-	0	0	no	n.a.
H	all-women	yes	-	0	0	yes	no
I	all-women	yes	50	5	10	no	n.a.
J	all-women	yes	100	15	15	yes	no
K	all-women	yes	200	60	30	yes	no
L	all-women	yes	50	20	40	yes	yes
M	all-women	yes	100	20	20	yes	no

(-) not given
(n.a.) not applicable

Source: Field Survey, 1996-97

Out of the 12 associations that articulated and presented members' requests for fertiliser, four or 33 per cent recorded zero success, that is, they received no amount of fertiliser. Out of these four associations recording no success, two were all-men associations, one was gender-neutral and the other was an all-women's association. So, of all the eight all-women associations in the sample, only one recorded zero success. Computed percentages of success show that three associations obtained about 10 per cent of their total fertiliser requests, one association 15 per cent and another 20 per cent, while two others secured 30 and 40 per cent each. All the associations in the relatively higher success bracket (greater than 20 per cent) were all-women groups.

Of all the peasant associations that made fertiliser requests, only two (or 17 per cent) did not agitate or press the authorities beyond initial articulation and presentation of such requests to the relevant agencies. Furthermore, of all the 10 associations that made agitative and pressurising efforts, eight (or 80 per cent) reported zero pay-off, one (or 10 per cent) reported greater than zero (but very little) pay-off, while another one (or 10 per cent) reported relatively impressive results from its agitative efforts. The fact that up to 80 per cent of the peasant associations did not reap any positive gain (that is, obtain additional fertiliser supplies) as a result of pressurising the authorities can be attributed to one or a combination of the following reasons: the muted nature of the agitations; the use of inappropriate and ineffectual techniques and approaches; and limitations resulting from the rudimentary organisational development and small-sized nature of the associations. Some associations lodged protests through letters that were not followed up with delegated visits to the concerned authorities, apparently because of the lack of funds to pay for transport and related travel costs. Other associations claimed to have made initial delegated visits to the relevant agencies but were unable to sustain the visits through shortage of funds. The associations blamed their plight on the excessive bureaucratic red tape associated with the process of obtaining fertiliser from government authorities, in addition to the rent-seeking practices of fertiliser officials.

The only association which reported relatively positive results from making agitative efforts beyond mere application for fertiliser was an all-women one. Not surprisingly, therefore, this association recorded the highest (that is, 40 per cent) overall success rate in obtaining its fertiliser requests from the relevant agencies. In contrast is the finding that the association that made no effort at agitating for its fertiliser requests did not obtain fertiliser at all. The other association which did not invest efforts in pressurising for fertiliser requests was either content with the mere 10 per cent success already achieved following initial lodgement of requests or lacked faith in the effectiveness of continued agitation for fertiliser. The latter attitude appeared to be predominant.

The all-women association that experienced an outstanding degree of success in obtaining its requests was further interrogated to elicit whether and how formative attributes and extraneous promptings impacted on its performance relative to other associations. Specific factors found to be linked with its success rate included the fact that this peasant organisation had received encouragement from officials of the Women's Commission in terms of the association's formation and in the monitoring of its activities. Also the fact that the association was commodity-focused (that is, formed for the promotion of rice production, processing and marketing among women rice farmers) and that the association was privileged to have been 'specifically advised' by officials of Women's Commission in applying for this fertiliser made the difference.

The underlying factor contributing to the observed remarkable success rate of this association was that it had approached a variety of agencies for specific inputs. This approach was explicitly summed up by the association thus: 'specific requests were made for fertiliser from the Ministry of Agriculture, rice milling machines from the Women's Commission, tractor services from Anambra State Tractor Hiring Company Awkuzu and loans from the Development Education Centre, Enugu; and we obtained satisfactory amounts of these items'.

But an overall assessment of the approach and instruments employed by peasant associations to articulate, project and pursue peasant interests would indicate that even though the associations invested efforts to actualise specific demands, there was a feeling of declining faith and increasing uncertainty regarding the fruitfulness or even desirability of continued agitation for fertiliser.

Besides fertiliser politics, there were other agrarian and membership-related issues that peasant associations promoted in articulating, projecting and securing peasant interests. In at least two instances peasant associations achieved modest successes in obtaining these broader demands. One case was the peasant association under the LAIP which had articulated, mobilised and pushed for a united peasant response to the sharp increases in irrigation water and tractor service charges under the project. The procedure, technique and approaches adopted were systematic. On realising that the charges had been increased by the project authority, the association mobilised and sensitised the members, culminating in the lodging of a formal protest with the project manager, through letters followed by face-to-face contacts. After a series of meetings, contacts and continued agitation for a rethink on the increments, the project authority reviewed the charges downwards, cutting the increment in the charge for irrigation water by 33 per cent.

Elsewhere, one all-women association was observed to be actively involved in articulating and defending or protecting women's interests regarding a traditional custom, which it considered to be inimical and humiliating. According to this custom, when two men are engaged in a conflict over land, and this leads to the death of either one of them that had yet to bear children, a woman is appointed to 'replace' the deceased. This means she has to bear children in the name of the deceased. Fully aware of the debasing and unpleasant consequences of this aspect of a traditional custom, the association in concert with sister associations developed a strategy for intervening in such potentially dangerous disputes to ensure that the dispute did not degenerate into the loss of life of any of the parties. Following their intervention, the incidence of replacing the deceased childless man with a child-bearing woman has been considerably reduced, to the extent that this particular customary practice is virtually redundant.

The strategy of collaboration between sister associations in articulating and projecting peasant demands and interests was common to most of them, especially the all-women ones. Only three of the peasant associations (all of them all-men groups) did not have any form of collaborative arrangement with another association. Of the nine associations that reported a form of collaboration with another association, only one was an all-men group. So, more elaborate and intense collaboration was observed among the all-women than the all-men associations. One area of operation where the collaborative action was most evident was in the joint hosting of visiting loan inspectors, project officials and state officers. This strategy contributed to diminished running costs for individual associations. Other reported cases of collaborative arrangement among the associations included occasional joint meetings of the executive officers of the different associations, joint contributions to community projects and joint transport arrangements for conveying members to and from organised trips or conveying supplies. A number of the all-women associations attributed their collaborative practices to strong promptings by officials of the Women's Commission. Collaborative practices between the associations were observed to have paid off considering that numerous across-the-board favours and concessions had been won on several occasions by all-women associations.

Gender-wise comparison of peasant associations reveals that generally all-women associations exhibited relatively more agitative action, greater effectiveness in agitating for and meeting members' needs and interests, and more subtle collaborative practices with sister associations and in obtaining results from such practices than all-men groups. This finding appears incongruent with the observation that there was a decrease in membership (between inception and the research period) in most (63 per cent) of all-women associations. It may be argued that the greater effectiveness of all-women associations should have led to more people joining them and that few people would disengage from the associations. But the swing of the argument depends upon which of the situations came first — the decrease in membership or the better results and effectiveness experienced.

Our findings suggest that the decrease in membership pre-dated the recorded effectiveness of the associations. At inception, the associations had relatively large memberships because of the initial enthusiasm to use group advantage in meeting individual needs for farm inputs and external assistance. But as time passed and the desired benefits were either not forthcoming or came very slowly and in meagre amounts, members enthusiasm sagged with the result that some started to withdraw or disengage from the association. Ironically, the resultant smaller sizes became an advantage since the associations now had smaller numbers of members to cater for whose interests and needs were more convergent and hence easier to meet.

Peasant Association, Democratic Culture and Empowerment

One important ramification of the external relations of peasant associations that has bearing on their internal processes and disposition in the socio-political milieu is their registration status. All the peasant associations with the exception of two (a Water Users Association and a Young Farmers Association) had formal registration with both the local and state government authorities. Whereas the Water Users Association did not register either with the local or state government, the Young Farmers Association registered only with the local government. It was found that the Water Users Association did not consider registration necessary since the association was essentially derived from the LAIP, which itself was a federal government-sponsored programme. On the other hand, the Young Farmers Association claimed that its non-registered status (in respect of state government) was a result of its inability to raise funds for paying the statutory fees. Consequent upon its non-registration with the state government authorities, the association limited its fertiliser requests to the local government level.

In one association, specifically the Water Users Association in the LAIP, the present executive officers were appointed by project officials in contradistinction to the democratic and autonomous imperatives that ideally should guide such a peasant platform. Not surprisingly, therefore, project officials were in attendance at monthly meetings, to provide 'appropriate advice'. There is no doubt that such interference impacts adversely on the democratic process, autonomy and independence of the association. In fact, part of the reason for the poor membership commitment and membership indifference to the association's activities was the widely prevalent perception by members that the association was undemocratic and highly prone to interference from the state authorities. Not even the abolition of the N20 membership fee was able to significantly rekindle interest among members. Rather than be the defenders and forerunners of peasant interests (in this case water usage issues), the executive officers, having been appointed by the project, were considered to merely serve as agents of the state, with a mandate to ensure that water users were mobilised to maintain and protect water canals and other irrigation facilities.

In another association external state authority structures and the local traditional chieftaincy influenced heavily the functioning and disposition of the association.[2] In addition to exploiting their privileged positions to obtain larger areas of farmland from the project authorities, the traditional chiefs seemed to have shown little or no enthusiasm in following up most of the association's grievances and complaints to the project authorities and state institutions. One interviewee described the traditional chiefs as being 'often engaged in actions and practices that are inimical to the general interests of members'. Other members perceived the executive officers to be a 'rubber stamp' for the actions and decisions of these 'powerful' chiefs.

However, other associations reported no form of overt or covert interference by traditional chieftaincy authorities. Recognising that the traditional institutions constituted the repository of the culture and beliefs of the people, the associations often invited traditional chiefs to 'grace' local receptions for visiting government and/or non-governmental officials. Others said they merely informed the traditional chiefs about such events without inviting them. An all-women's association which engaged prominently in the defence and protection of women's interests with respect to the traditional custom of 'replacing dead men with living women' was found to be in active consultation with the traditional chieftaincy authorities.

Some interviewees observed that powerful members of peasant associations who were not executive officers often used their influence and connections to the state to get larger allocations of items being shared among members, in apparent disregard for pre-set sharing arrangements. Under such circumstances, the ability and effectiveness of peasant associations in defending the collective interests become uncertain. Other members reported instances of disregard for democratic decision-making in areas such as the appointment and selection of individuals to attend seminars or training workshops to which the association had been invited. Many interviewees expressed dissatisfaction with the practice of selecting only 'powerful' and 'highly educated' members to benefit from these training seminars and workshops irrespective of the contrary views of members.

Another sore point in the relations between the membership and executive officers was the handling and use of materials, facilities and items belonging to the association. In one association, for instance, it was alleged that an executive officer had diverted motorcycles meant for the smooth running of the association into his predominantly private use. According to one respondent, 'some officials are more concerned with their personal welfare. They tend to monopolise the benefits, for example, going to workshops.' Another member of an association claimed that 'sharing of fertiliser is often done in the manner that if the fertiliser quantity is large it is shared equally, but if small, the executive officers take it'. Similarly, a respondent from another association asserted that 'if the fertiliser is much, it is shared equally, otherwise the executive officers take a lion's share'.

Since members of the various peasant associations are the primary stakeholders of the associations, how they perceive the relevance and responsiveness of their associations is fundamental. An inquiry was undertaken to explore questions regarding the existence of opportunities to disagree over the decisions and actions of the executive officers and the availability of channels for lodging protests and making petitions and demands. Of the 13 associations studied, a cross-section of members in six (46 per cent) expressed the feeling that their expectations since joining the associations were unrealised, particularly in relation to the associations' inability to meet the farm input, production credit and fertiliser needs of the members. Additionally, members of the different associations reported some dissatisfaction over a number of decisions they considered inadequate, inappropriate and objectionable. In apparent allusion to the inadequate relevance of an association, a member emphasised: 'I expected the association should organise members into an *isusu* for credit purposes, and my complaints regarding this were not sustained by the majority opinion.' A member of another association stated: 'my wish is for our contributions to be shared periodically according to a pre-agreed sequence, rather than for money to be kept in the bank. I complained that I do not like the present arrangement, but the majority opinion did not support my suggestion'.

Conclusion

As a precondition for serving as building blocks of democratic culture and channels of peasant empowerment, peasant associations must themselves possess democratic credentials, grassroots legitimacy and responsible attitudes to peasant needs, interest and welfare. Further interrogation of the peasant associations' members in regard to these attributes is necessary for assessing whether there is sufficient ground or basis for assigning them the enormous responsibility of articulating and projecting democratic values.

From the empirical evidence generated in this study, there are indications that peasant associations exhibit certain characteristics, attitudes and dispositions which make them amenable to democratic culture and which provide congenial opportunities for them to promote grassroots empowerment. These attitudes and practices include openness and transparency of deliberations and decision-making at general meetings, and the conduct of open, free and fair democratic elections into executive offices under a defined tenure. There also exist opportunities for members to lodge protests and make petitions to the state and traditional authorities on issues of concern.

The continued articulation and presentation of demands or requests for fertiliser, credit and other needed forms of assistance from government and non-governmental organisations is another milestone in the effectiveness of peasant organisations. The agitation through follow-up visits, written reminders and various forms of pressurising the state mean that the peasants are able to obtain some positive benefits from collective efforts. The fact that a cross-section of their memberships perceive them to be relevant and useful but severely constrained in effectively securing benefits keeps up the momentum for maintenance of peasant organisations. The existence of subtle forms of joint or collaborative arrangements and practices to project and pursue agendas of common peasant interests, and the apparent solidarity and relative stability of the associations in the face of changing socio-cultural and politico-economic environments also reflect their durability.

On the other hand, a range of practices, attitudes and attributes of peasant associations that was observed have the potential to undermine the democratic credentials of the associations and their role as the genuine voice of peasants. For instance, peasant associations which originated out of 'instructions' from state officials were used as channels for dispensing government patronage and assistance, and as a result members of the peasant association had primarily come together to benefit from such governmental favour. Interference by external actors in the running and operations of peasant organisations, such as the disbandment of elected executive committees and their replacement with appointed persons, was also quite common.

The use by some executive officers of peasant associations of official posts to advance personal rather than group interests and taking undue advantage of opportunities created through their links with external power and authority structures tends to undermine the democratic process. The rent-seeking practices of these powerful members were largely unchallenged by the mass of the members, who often expected short-term pay-offs from influential members of the association. Such executives' weak and sporadic agitation and follow-up of members' interests and demands from relevant authorities undermines peasant organisations. This weakness, made worse by the pervasive low resource base, poor organisational systems, small-sized character and limited political muscle of the associations, contributed to their diminished executive capacity and effectiveness.

In contributing to the debate on the democratic credentials and potentials of peasant associations, our empirical findings suggest that in order to promote the capacity of these associations as reliable and legitimate advocates of peasants interests two considerations should be explored. One is to strengthen and build on those attitudes, practices and attributes of peasant associations that are germane to democratic culture. The other is to diminish and minimise peasant associations' potential to function in an undemocratic style through those attitudes, practices and characteristics that limit the effectiveness of the associations in articulating, projecting and protecting/ defending peasant interests.

Notes

1 The staple food crops included maize, cassava and garri, with rice as the main crop.

2 The association, also subsumed under the LAIP, included traditional chiefs in its executive, according to project instructions.

References

Ake, C. (1985) 'The Nigerian state: Antinomies of a periphery formation', in C. Ake (ed.) *Political Economy of Nigeria*, London, Longman.

Aloki, A. O. (1989) 'Rural development in sub-Saharan Africa: A different view of political and administrative decentralisation', *International Review of Administrative Sciences* 55, 3: 401-32.

Amadiume, I. (1985) 'Gender, political systems and social movements: A West African experience', in M. Mamdani. and E. Wamba dia Wamba (eds.) *African Studies in Social Movements and Democracy, Dakar*, CODESRIA.

Amin, N. (1991) 'Peasant differentiation and food security in Zimbabwe, The Project on African Agriculture', Social Science Research Council, Working Paper no. 1, New York.

Beer, C. (1976) *The Politics of Peasant Groups in Western Nigeria*, Ibadan, Ibadan University Press.

Beer, C. (1980) 'The farmer and the state in Western Nigeria: The role of the farmers' organisation and cooperatives', in P. Collins (ed.) *Administration for Development in Nigeria*, Lagos, African Education Press.

Beer, C. and Williams, G. (1975) 'The politics of Ibadan peasantry', *The African Review* 5, 3: 235-56.

Bernstein, H. (1979) 'African peasantries: A theoretical framework', *Journal of Peasant Studies* 8: 4: 430-1.

Bratton, M. (1986) 'Farmers' organizations and food production in Zimbabwe', *World Development* 14, 3: 367-84.

Cernea, M. (1987) 'Farmer organizations and institution building for sustainable development', *Regional Development Dialogue* 8, 2: 1-19.

Chambers, R. (1983) Rural Development: Putting the Last First, London, Longman.

Cliffe, L. (1987) 'The debate on African peasantries', *Development and Change* 18, 2.

Eboh, E.C. (1993) 'Analysis of indigenous systems of group saving and capital formation among Nigerian rural households', research report submitted to African Development Foundation, Washington, D.C.

Esman, M. and Uphoff, N. (1984) *Local Organizations: Intermediaries in Rural Development*, Ithaca, Cornell University Press.

Holmquist, F. (1980) 'Peasant political space in independent Africa', *Canadian Journal of African Studies* 14, 1: 156-67.

Hyden, G. (1986) 'Discussion: The anomaly of African peasantry', *Development and Change* 17, 4.

Ibeanu, O. (1993) 'The deteriorating condition of the Nigerian peasantry', in O. Nnoli (ed.) *Dead-End to Nigerian Development*, Dakar, CODESRIA.

Klein, M A (1980) (ed.) *Peasants in Africa*, Beverly Hills, Sage.

Mafeje, A. (1993) 'Peasant organisations in Africa: A potential dialogue between economists and sociologists — some theoretical and methodological observations', *CODESRIA Bulletin* 1: 14-17.

Mamdani, M. (1986) 'Extreme but not Exceptional: Towards an Analysis of the Agrarian Question in Uganda', DPP Working Paper Series no. 5, Faculty of Technology, The Open University, Milton Keynes, UK.

Moyo, S. (1995) 'Peasant organisations and research issues on democratisation in Africa', paper presented to CODESRIA Multinational Working Group Methodology Workshop on Peasant Organisations in Africa, 21-30 November, Dakar.

Oakley, P. (1991) *Projects with People: The Practice of Participation in Rural Development*, Geneva, International Labour Organisation.

Oakley, P. and Marsden, D. (1984) *Approaches to Participation in Rural Development*, Geneva, International Labour Office.

OAU/IILS (1983) *Democratisation of National Development in Africa*, proceedings of a symposium in Dakar, 15-18 March 1982, Geneva, International Institute for Labour Studies.

Okorie, A. and Obeta, M. E. (1986) 'Nigerian traditional mutual aid societies and the contemporary with modern cooperatives', *Journal of Rural Cooperation* XIV, 2: 107-17.

Omeje, K. C. (1995) 'Alienation and displacement of peasant cultivators: An example of human rights violation', in E. C. Eboh, C. U. Okoye and D. Ayichi (eds.) Rural *Development in Nigeria: Concepts, Processes and Prospects*, Enugu, Auto Century.

Rahmato, D. (1991) 'Peasant Organisations in Africa: Constraints and Potentials', Dakar, CODESRIA Working Paper 1/1991.

Richards, P. (1986) *Coping with Hunger: Hazard and Experiment in an African Rice Farming System*, London, Allen & Unwin.

Saul, J. S. and Woods, R. (1979) 'African peasantries', in T. Shanin (ed.) *Peasants and Peasant Societies*, Harmondsworth, Penguin Books.

Swindell, K. and Iliya, M. A. (1992) 'Accumulation, Consolidation and Survival: Non-farm Incomes and Agrarian Change in Northwest Nigeria', Project on African Agriculture, Social Science Research Council, Working Paper no. 6, New York.

UNECA (1990) African Charter for Popular Participation in Development and Transformation (Arusha 1990), E/ECA/CM16/11, Addis Ababa.

Williams, D. C. (1991) 'State–society relations in perspective: The politics of land reform in Nigeria', unpublished Ph.D. dissertation, University of Florida.

Williams, G. (1975) 'Political consciousness among the Ibadan poor', in E. Dekadt and G. Williams (eds.) *Sociology and Development*, London, Tavistock.

2

The Internal Dynamics of Peasant-based Co-operatives and Democratisation in Nigeria

Adebakin Simeon Asaju

Introduction

Peasant organisations are voluntary associations set up by citizens to promote their common welfare. Women's associations, development and cultural associations, co-operatives, community welfare unions, social clubs and professional bodies are part and parcel of this movement. These organisations are popular because members freely support their formation and activities that cut across narrow interests, and they are run on the basis of democratic principles. This study focuses on the internal dynamics and potentials of peasant-based co-operatives in search of self-reliance and democracy.

The peasantry is regarded as the neglected rural majority who have suffered most the consequences of Africa's economic and political crisis. The historical roots of the peasant's predicament can be traced to colonialism (Lappe, Collins & Fowler 1977). The colonial state, whose legitimacy was guaranteed by force, did not conceal its extractive role in rural areas. Its marketing boards siphoned surpluses from peasants and their organisations through excessive taxation, elimination of competition, under-payment for cash crops and other monopolistic strategies. While investing massive sums of money for the development of railways and communication it did nothing to improve rural welfare.

The suppression and repression of the peasantry did not, however, end with the attainment of political independence. The postcolonial state inherited the agrarian base of its predecessor and pursued an industrialisation strategy based on the extraction of surpluses from peasant agriculture. In Nigeria, the prevailing feature of the economic and rural development policies was the promotion of monopolies by the economically and politically dominant class over the peasantry. Rural development policies were imposed from above with little input from the grassroots who constituted the target population. Various studies, especially those by Wallace (1980), Forrest (1981), Nkom (1982), Beckman (1985) and Dunmoye (1987), show that despite flowery pronouncements and huge expenditures by government, rural development initiatives, such as the Agricultural Development Projects and the River Basin and Rural Development Authorities, impoverished the rural populace by displacing communities, increasing the numbers of landless and enhancing absentee landlordism.

The new development paradigm today is a bottom-up approach devoid of paternalism and state coercion. It promotes egalitarian and democratic processes. The adoption of the participatory development approach is regarded by many as one of the key solutions to the continent's mounting social, economic and political problems and also as an end in itself (ECA, African Charter 1994). In the words of Lappe, Collins and Fowler (1977), rural people must not only control the land they work but effectively participate in wielding national, political and economic power. They must exercise a strong voice through their co-operatives and political organisations in the control of credit and marketing institutions. As part of this taking charge, they will need to acquire the knowledge and skills to manage their own affairs. It is generally believed that no meaningful self-reliance can be realised without self-sufficiency in food production. To achieve this, the single most important challenge facing Nigeria's agricultural sector revolves around selecting an appropriate strategy for mobilising the over 50 million peasants in the country who have in the past produced the food and export crops that sustained the population and financed national development.

Nigeria is endowed with fertile land resources a huge and productive labour force, a well-researched package of technological innovations for promoting crop production, financial resources through the Nigerian Agricultural and Co-operative Bank (NACB) and commercial banks, and the local production of a variety of appropriate fertilisers. The missing link appears to be effective policies and the weak mobilisation and organisation of small-scale farmers into viable co-operatives and farmers' associations (Gana 1988).

Co-operatives in Nigeria are bedevilled by serious financial, logistic and managerial problems that have reduced their capacity in sustaining meaningful agricultural production. As observed by Osuntogun (1983), the main problems which impede co-operatives' efficiency and effectiveness can be categorised as managerial, structural and financial. More importantly, government intervention and control has been a severe obstacle to the development of a democratically controlled farmers' co-operative movement. The movement cannot accommodate the peasants' needs and aspirations because state control has destroyed self-reliance.

Objectives and Research

The objective of this study was to examine the role of peasant-based co-operatives in the development of a participatory democratic culture and to identify the major problems and constraints confronting efforts of the co-operative movement to democratise. The study explored the internal organisational structure and management procedures of co-operatives so as to determine the extent to which they were democratic. It examined the relationship between the peasant co-operatives and state agencies and how this relationship affected democratisation. The Nigerian government's policy measures in the area of co-operative development were also evaluated to assess the quality of official commitment to the co-operative idea.

The study hypothesis was that in the 1990s peasant co-operatives had been successful in challenging the authority of central or local governments in ways that opened space for democratic practices and socio-economic policy change, rather than collaborate with government organs to perpetuate authoritarianism.

Several research questions were considered, including: what are the roles and activities of peasant-based co-operatives in Nigeria? What are their basic social, political and economic problems or constraints? What are the effective approaches to mobilise peasant farmers into viable co-operatives? In which way is democratisation initiated and sustained in the co-operative movement? To what extent do co-operatives serve as a force for the development of a participatory democratic culture? How democratic are the internal organisational structures and procedures of the peasant co-operatives? What is the quality of official commitment in Nigeria to the fundamental objectives of the co-operative movement?

Wider questions arose regarding the role of co-operatives in promoting agricultural transformation and the democratisation of economic, social and political relations in society. Among issues explored were: do peasant co-operatives collaborate with the state to foster and perpetuate authoritarianism? How far is it true that peasant co-operatives initiated, sustained and funded by the peasants themselves are more democratic than government-imposed and controlled ones? What special political, social and economic concessions and benefits have the peasant co-operatives gained from their interactions with state agencies? What is the class, gender and ethnic composition of the peasant co-operatives, and how does this impact on the patterns of leadership and internal governance of the co-operatives and the distribution of the benefits of their work?

Peasants, Peasant Organisations and Co-operatives

The term peasant or peasantry has been defined in various ways. As observed by Mintz (1973), one feature of peasant studies is the lack of consensus about the definition of peasantry. The 'peasantry' has usually been considered to be a social group, not necessarily a social class, falling between large landowners on the one hand and the landless on the other. However, especially in systems outside Europe, the non-existence of a large land-owning class has not prohibited the use of the term. It is still used to describe small-scale agricultural producers who either own their farm or have rights to the use of land; who farm largely with their own labour and that of their family; who own their own implements, and who purchase their own seeds, fertilisers, and so on (Olatunbosun 1975).

Other scholars, such as Redfield (1956) and Thorner (1962), define peasants simply as rural producers securing their livelihood through the use of family labour on family land. They view the peasants as backward and primitive, with for example Hyden (1980) identifying peasants with everything that appears to be technologically and culturally backward. For Redfield, peasants are those for whom 'agriculture is a way of life, not a business for a profit'. Those who carry on agriculture for 'reinvestment and business', looking on the land as capital and commodity, are not, in Redfield's reckoning, peasants but farmers. Yet other scholars, such as Wolf (1966), Bernstein (1979) and Williams (1976), are more particularly concerned with the age-long disadvantaged position of the peasantry in terms of access to production and marketing resources and subordination to and repression by a central authority. Thus Williams (1976) defines peasants by their relation to the market and state, while Wolf (1966) defines them as 'rural cultivators whose surpluses are transferred to a dominant group or rulers'.

Controversy has existed over the question of whether the term 'peasant' is applicable to African conditions (Mafeje 1996). In this respect Fallers (1961) argues that the African villager might be regarded as a peasant in both economic and political respects but not culturally. He is thus, according to Fallers, a 'proto-peasant' or an 'incipient peasant'. Hill (1986) prefers to use the uncomplicated term 'farmer' partly because of the built-in condescension and degree of rusticity attached to the word 'peasant'. Contrarily, Saul and Woods (1971) have justified the use of the term 'peasant', although they also conclude that a 'significant number of African cultivators have moved out of the peasant category and must be called capitalist farmers'.

Peasants in Nigeria can be defined as small-scale, petty rural agriculturists, producing predominantly for consumption but at times for exchange, and whose activities focus mainly on food crops although they may also practise cash cropping. They are in a disadvantaged position given the anti-democratic control over the use of agricultural production and marketing resources.

Various policy efforts in Nigeria to improve the agricultural productivity of the peasant farmer have rarely given adequate attention to the way peasants farm, the farm inputs they use, and their soil enrichment strategies. This neglect of the peasants has compelled the able-bodied to drift to the urban areas where per capita income levels are expected to be higher than obtainable in the rural areas.

But a significant area of concern in agricultural development is the lack of foresight in organising peasant farmers into viable co-operative associations.

The term 'co-operation' has been subjected to a multitude of interpretations. For example, Osuntogun (1983) defines co-operation as the system by which different people in any economy produce for each other and exchange goods and services with each other through the mechanisms of the market. In a similar sense, nations are said to co-operate when they trade with one another, negotiate treaties or form alliances. In general, any working together, joint action or concurrence in action meets the dictionary definition of co-operation.

Some scholars view co-operation as primarily a sociological concept (Taylor 1949), while others held the view that co-operation is Christianity applied to business.

The economic philosophies of co-operation can be categorised as reformistic or revolutionary philosophies and evolutionary philosophies. According to the former, a co-operative is regarded as a means of redistributing income and wealth, eliminating or harmonising conflicts of interests between labour and industrial management, and reducing conflicts between producers and consumers. In a simple and direct definition of a co-operative, Russell (1958) defines a co-operative society as an association of shareholders usually described as members, who as individuals have contributed a fund or capital to be employed under their collective direction for trading purposes in providing for their needs as producers or as consumers. Nasko (1988) defines a co-operative as a group of people who voluntarily come together to form an association with the aim of promoting a common economic interest in accordance with the laid-down co-operative principles. The rural or agricultural co-operatives are therefore associations formed by rural dwellers to cater for their common interests or needs.

The collective co-operative is a philosophy of common ownership, common sharing, and consumption and investment. It also extends to collective production and consumption culture. Emphasis is on service, human wellbeing and rounded development of the whole person and society rather than on profit. In its own contribution, the International Labour Organisation (1976) defines a co-operative as an association of persons, usually of limited means, who have voluntarily joined together to achieve a common economic goal through the formation of democratically controlled business organisation, the equitable contribution to the capital required and acceptance of a fair share of the risks and benefits of the undertaking.

Peasant organisations have been defined as forms of associative endeavour employed by small-scale holder agriculturists such as mutual self-support groups, voluntary associations and solidarity groups. For our purpose in this study, we define peasant-based co-operatives as associations formed by rural dwellers (specifically the peasants) to cater for their common interest or needs. We include in our concept of co-operatives both the formal and legal and the informal or 'pre-co-operatives', which often later transform into formal co-operatives.

Studies of peasant organisations and co-operatives have not been exhaustive enough to examine how the internal democratic dynamics of peasant co-operatives are conducive to promoting democratisation processes. As observed by Moyo (1995), most studies of peasant organisations and rural non-government organisations are imbued with technicist or developmentalist prescriptions concerning the activities of rural civil society, but tend to underestimate their potential to influence change in the political sphere. This study attempts to explore the potential of peasant-based co-operatives to promote democratisation.

Study Approach

Data for this study was obtained mainly from primary sources using questionnaire surveys, personal interviews and personal observations. These were supplemented by secondary sources in the form of government reports, gazettes and other public and private documents. The interview and questionnaire were the principal instruments of primary data gathering. These two instruments had been chosen not only because together they yield adequate data to support generalisation, but also because each could serve as means of checking the reliability of data generated by the other.

Data collection was concentrated on two states of the federation in the former Western Region of Nigeria. While this selection was judgmental, it was reasoned that the region had been in the forefront of co-operative development in Nigeria. Essentially, the sampling design was basically purposive and judgmental. However, an effort was made to try to systematise the procedure so that replication of the study by other researchers could be possible. We resorted to a directory of government officials handling co-operative matters in these areas. A random sample of the peasant-based co-operatives was selected, and in turn a random number of their officials and a random sample of the members interviewed through the questionnaire process. The fieldwork was carried out in the early months of 1996. A total of 120 co-operators were interviewed, while oral interviews were held with about 20 co-operative officials. Responses to each question, however, varied (see Appendix). The analysis involved the use of simple descriptive statistics with respect to data relating to the characteristics of peasant co-operators. The research design was essentially exploratory and the analysis was qualitative.

The Growth of the Nigerian Co-operative Movement

Co-operation is inherent in the tradition and culture of the people of Africa. The formation of co-operatives in Nigeria antedated colonialism and the incorporation of Nigeria into the capitalist system. The foundations of an agricultural co-operative organisation in Nigeria are generally thought to have been laid in the period between the

First and Second World Wars, and more especially in the mid-1930s, with the passing by the colonial administration of the Co-operative Societies Ordinance of 1935 (Beer 1976; Alubo 1988). However, prior to this there had been successful attempts, not only by the colonial government, but also by indigenous groups (aside from the traditional co-operative style such as the Esusu by the Yorubas; Etoto by the Ibibios, Adashe by the Hausas and Osusu by the Ibos) to form what might be regarded in some cases as 'pre-co-operatives'.

The colonial administration's motives in encouraging such societies were twofold. First, they served as useful links in the marketing chain between the European buyer and the peasant farmer. Secondly, they could be utilised to promote the dissemination of knowledge on agricultural improvements. The unco ordinated growth of co-operative-type societies in the Western Provinces, the exploitation of the peasant farmer by sub-buyers and the very growth of the cocoa growing industry itself led to efforts by the colonial authorities to regulate the societies under the law and to bring them under government influence and supervision. The Co-operative Ordinance of 1935 and the Regulations that followed in 1936 formed the basis for co-operative development in Nigeria until the present time by creating considerable executive powers over societies.

The influences and motives behind the development of the co-operatives in Western Nigeria were varied. The initial desire by the Department of Agriculture to improve the quality of the Nigerian cocoa crop led to the utilisation and promotion of producers' co-operative organisations, which were seen as a major tool for achieving this objective. The attraction of co-operatives was the higher rewards for high-quality products and the facility they offered for agricultural officers to meet groups of farmers. The co-operatives also fitted well into the indirect rule system and allowed agricultural and co-operative officers to influence the farmer through intermediaries acceptable to him as his representatives. The desire of farmers to use the co-operative system to displace the monopoly of the oligarchic export trade firms was reinforced in the 1930s by the government-sponsored co-operative movement. The emphasis was also on replacing the African middleman buyer, and on protecting the peasant farmer from indigenous capitalist exploitation. The Co-operative Department and the registrar, while maintaining their passionate opposition to the impoverishment of the cocoa farmer by the Nigerian trader, during the years of controls developed an equally strong opposition to the exporting firms.

Although the question of direct involvement of the co-operative movement in politics was to re-emerge from time to time, the political neutrality principle ultimately won the day. Official opinion during the 1950s persuaded co-operative leaders to refrain from declaring open electoral or party allegiances. Nevertheless, the registrar too was aware that the movement needed greater political strength and protection. To the department, the answer was to form a central union that would hold a watching brief on the movement's political interests, and serve as its political, propaganda and public relations organisation. Such a union would ally with no party but seek to influence all.

At the time, conditions for co-operative registration were stringent and deliberately tailored to the needs of the colonialists rather than the co-operatives. Accordingly, enormous and sweeping powers were vested in the Registrar of Co-operatives. He could, for example, dissolve any co-operative that in his opinion was not functioning in the interest of members. Furthermore, he derived his power and 'political support not (from) the co-operatives but from government' (Kohl 1969). He was essentially a paramount ruler of co-operatives. It is thus clear that even in conception the co-operative in Nigeria was an analogue to and a vehicle for the advancement of exploitation — a conception which contradicted the Rochdale pioneers on whose principles co-operatives are supposedly based. In this way, co-operative's anti-capitalist rhetoric during colonialism (such as 'co-operation is the salvation from economic strangulation') was bandied around as a matter of routine rather than persuasion.

Although co-operatives as a form of voluntary self-help organisation had been known in many parts of Nigeria for several decades, co-operative programmes in their own right never featured in the National Development Plan until the third plan (1975-80). This reflected both the slow progress of the movement in Nigeria and a lack of active government interest, despite the fact that co-operatives are potentially an important instrument of social transformation, especially in the rural areas. Before 1980, fewer than 1 per cent of the Nigerian population were members of the co-operative movement. In that year the Federal Government of Nigeria adopted a policy on rural co-operatives in its 'Green Revolution Programme'.

A number of capital-intensive co-operative projects were then undertaken. They included feed mills, storage depots, inshore fishing projects, food processing, transportation and inter-state co-operative marketing of food crops. The feed mill project was implemented to process farmers' raw produce such as maize, guinea corn, cottonseed and groundnut cake into animal feed. The establishment of storage depots was intended to collect and store food crops for eventual marketing and distribution, so as to promote the more regular distribution of foodstuffs, improve the price structure and ensure some stability of supply and pricing to both consumer and farmer.

It has been argued that co-operatives have continued to perform essentially the same function of facilitating rural exploitation and strengthening the agrarian base of the modern state (Beckman 1981). True, co-operatives are conceived as vehicles for rural 'development' (see Nigerian Third National Development Plan), which is itself an imposition of top-down programmes by developers. It is within this concept of imposed rather than collective development that co-operatives must be understood. Indeed, as the Third Development Plan clearly articulated, co-operatives are powerful instruments for increasing productivity and rapid rural transformation, hence their top-down character and the overbearing presence of the Registrar of Co-operatives.

To facilitate the achievement of these objectives, access to loans, fertilisers, herbicides, high-yielding varieties and similar inputs was contingent on the formation of co-operatives (King 1981; Adelakun 1981). As King has illustrated, this colonial conception has not been significantly questioned, let alone altered. In the main, co-operatives foster farmers' dependence on 'decisions beyond their control' and which erode systems of self-sufficient food production. This is how ordinary farmers are denied a say in a supposedly democratic organisation.

Rather than nurture democratic principles and practices, the state imposes district and prominent farmers as co-operative 'leaders' who are more accountable to the 'paramount ruler of co-operatives', the registrar.

The nature and primary objectives of co-operatives at this time largely explains why they were totally ineffective in protecting peasant cocoa producers from gross exploitation through price cuts by the marketing boards and other agents (see Helleiner 1966). Both in conception and operation co-operatives have not changed from the colonial mode when they were instruments that forged marketing and financial links with the government mainly to promote the availability of rural products for urban consumption and export (King 1981).

Organisational Structure of Co-operatives in Nigeria

Generally, co-operatives are organised in tiers. The most common structure is a three-tier or pyramidal system with the primary society as the first tier, the secondary society or union as the second and the apex as the third tier. The first tier is based at the village level where individual co-operators are members. The secondary level is a combination of primary societies whether of the same or different type but from the same geographical (local government) area. The apex is an amalgamation or several secondary societies either of the same or different type from the same geographical (state) area. Broadly, four types of co-operatives are represented in the country: agricultural co-operatives, consumer co-operatives, co-operatives of craftsmen and artisans, and co-operatives of thrift and loan societies.

Agricultural co-operatives are the most common and perhaps the most developed type of co-operative. They concentrate mainly on the collective marketing of agricultural commodities, particularly export crop like cocoa, cotton and groundnuts. They operate as 'licensed buying agents' to purchase produce on commission for the marketing boards in competition with private merchants, usually relying on the patronage of members who produce the commodity concerned. Co-operative marketing has not been extended into the area of domestic food crops like rice, maize or guinea corn, nor has co-operation been extended much into actual production by farmers who typically operate on a small scale or into the processing of farm output. These areas have the potential to benefit co-operative members and the country substantially.

In consumer societies, co-operators join together to undertake mostly retail trading for the benefit of their members. In the western part of the country consumer co-operatives are doing a flourishing wholesale and retail business, backed by an effective Co-operative Supply Association. Given good management, consumer co-operatives can provide a valuable means of stabilising the prices of consumer goods, especially since they aim at supplying goods to their members at modest prices.

Co-operatives of artisans and craftsmen exist, but are not yet widespread despite the abundance of self-employed artisans in such trades as tailoring, wood working and blacksmithing. Many individual operators in these trades cannot by themselves finance the quantity and quality of inputs required to operate on a viable scale, but in co-operation, it would be easier for them to raise the necessary resources. This type of co-operative is common in Kano state but has so far not been exploited adequately in other parts of Nigeria.

Finally, there are the co-operative thrift and loan societies which have long operated in various parts of the country. Members make a regular contribution to a fund, which is passed on to individual members of the association in rotation, usually for investment purposes. It thus operates as a kind of savings club. These societies are common among small traders and even urban wage earners, and constitute a valuable means of mobilising small savings for productive investment.

Co-operative Movement in the Oyo State of Nigeria

Oyo state was carved out of the former Western Region, a region that had been in the forefront of co-operative development in Nigeria. In fact, the state witnessed the beginning of the national co-operative movement with the formation of Gbedun Marketing Co-operative Society. Although several types of co-operative societies operate in the state, they fall under the umbrella of the Oyo state co-operative movement. Basically, there are two agencies set up to achieve the objectives of the co-operatives in Oyo state. One of these, although itself a co-operative apex organisation which serves as the mother to all types of co-operatives, is the Oyo State Co-operative Federation (OSCOFED), formerly the Oyo State Co-operative Union. Its major function is to foster the growth of all types of co-operative societies through an effective use of its public enlightenment machinery and arranging for the continuous education of co-operators throughout the state.

The organisational structure is a three-tier system. The apex organisations include OSCOFED, the Co-operative Supply Association (CSA), the Association of Nigerian Co-operative Transport and the Co-operative Press. There is also a central society that has drawn its membership from the co-operative consumers societies and the co-operative buying clubs generally called the Oyo State Consumer Union.

The secondary societies include the Farmers Multi-purpose Co-operative Union (FMCU), the Cocoa Produce Marketing Union (CPMU), the Co-operative Thrift and Credit Union (CTCU) and the Co-operative Investment and Credit Union (CICU). Below these are the primary societies, including co-operative crafts societies, co-operative consumer societies, co-operative buying clubs, co-operative produce and marketing societies, co-operative thrift and credit societies (CTCS), and others like the departmental societies which are mainly for government staff.

The agency called the Co-operative Services Division of Oyo State Ministry of Commerce and Industry, under the Director of Co-operative Services, is responsible for formulating and implementing the state government's policy on co-operative development through dissemination of co-operative ideology and through suitable extension methods and communication techniques.

The general supervision of the co-operative movement in Oyo state is focused on the structure and operations of the primary co-operative societies. These are determined not only by their co-operative form of organisation but also by the governing legislative framework. After formal recognition, the co-operatives remain under the supervision of co-operative inspectors (now co-operative service assistants), who ensure that rules guiding them are enforced. Reports are given by government-appointed auditors on each society for proper evaluation of its progress. For effective supervision of the co-operative societies, the field staff of the Co-operative Services Division are spread evenly all over the state. The Co-operative Services Division has zoned Oyo state into six zonal offices, namely Ibadan, Ogbomoso, Ilesa, Oyo, Saki and Osogbo zones. A chief co-operative services officer heads each of the zones. Within the zones there are area officers for each local government area, with the exception of Ibadan municipal government, which has four area offices owing to the large number of secondary and primary societies. A co-operative services officer supervises each area

At the village level co-operative service assistants, who are grassroots workers of the division, are stationed within easy reach of every co-operative society and prospective co-operators. The Co-operative Services Division handles audit services to all the societies and takes charge of extension services for membership and publicity.

The supervision of the various co-operative activities largely depends on the intellectual and moral qualities of the supervisors, hence formal co-operative education of the supervisors is a prerequisite. For this purpose, there is the Federal Co-operative College, Ibadan and the Oyo State Co-operative College in Oyo town. Training centres have been established in the six zonal headquarters, where co-operative secretaries and local managers of co-operative societies are trained. Committee members of primary and secondary societies also undergo refresher courses in these centres. As at 1990, Oyo state had 6,622 primary societies, 248 secondary societies and six apex organisations (see Table 6)

The governing body of each society is the general meeting of its members, which occurs at least once a quarter to review and direct the work of the society. The general meeting is the supreme body of the co-operative. Each member can attend and is entitled to one vote. The general meeting is empowered more particularly to approve the reports on the society's work and the balance sheet of the financial year, to decide how any surplus of the year's work should be used, to elect members of the committees, to direct the affairs of the society, and to approve internal rules and amendments to the by-laws.

The management committee elected by the general meeting of the co-operative carries out the policies decided by the general meeting. The committee is entrusted to conduct directly or indirectly the day-to-day business of the co-operative and to stimulate members' participation in its affairs.

The management committee is responsible for all the assets of the co-operative and has to ensure that the accounts are regularly kept and are in accordance with existing rules.

Since the time the government accepted the responsibility to promote the establishment of co-operatives in Nigeria, it has provided funds annually in budget grants as in aid and subsidies, and disbursed technical assistance from other countries. For example, during the Third National Development Plan of 1975-80 the Federal Government spent N129.3 million on various projects like co-operative education, the National Co-operatives Insurance Company, the Co-operative Financing Organisation and National Co-operatives Wholesale Organisation.

Towards Democratic Peasant-based Co-operatives

When co-operatives become vehicles for rural exploitation and attendant pauperism it is a travesty of their original conception as bastions against capitalist exploitation (Beer 1969; Olajuwon 1981). But rather than nurture these values, co-operatives in Nigeria have been incorporated into the existing system of exploitation. Indeed, as Olajuwon (1981) has illustrated, the ideal of self-help can (only) be given practical expression through the rejection of excessive governmental aid or interference in co-operative affairs, and the avoidance or total rejection of special privileges from government or any other agencies.

Even as the government, at least in principle, accepts this objective of self-help and democracy, it makes access to and utilisation of several benefits contingent on membership of co-operatives. Hence, the government's relation to co-operatives is a patronising and paternalistic one.

It is, however, possible to have a truly democratic co-operative that emerges from and reflects the grassroots. But for popular co-operatives to emerge, certain problems in their formation and relationship to the state must be delineated and resolved.

A co-operative must be a vehicle for the mutual and collective benefit of co-operators rather than those of external agencies. As we have seen, government interference is the most urgent problem faced by co-operatives in Nigeria. The current situation of the co-operatives and the role of government officials have led to the co-operatives being authoritarian rather than democratic in nature.

The general level of awareness of the population is also an important factor in the emergence of a popular and democratic co-operative movement. In a situation where general and political illiteracy is the norm, and the populace is overburdened with problems of survival, food crises, poll and other taxes, especially since the introduction of the Structural Adjustment Programme, the focus on problems of democracy by co-operatives has been limited.

The combined effects of these factors tend to breed apathy and resignation, the very antithesis of mobilisation. Furthermore, the problem of general illiteracy has meant that the few so-called foresighted and enlightened people who head co-operatives often perpetuate themselves in leadership. The political education programme of the government-appointed National Orientation Agency might help. However, the more crucial problem remains one of mass literacy of which political education is a part. There is need for a programme to promote effective grassroots mass literacy in the country.

Related to the foregoing is the crushing burden of class oppression, particularly in rural areas. In the Nigerian experience, rural poverty is correlated to the neglect of rural dwellers and their exposure to political manipulation and further exploitation. This growing oppression is perpetuated by the covert support of the state of massive land acquisition by the elite. Several state governments have been openly soliciting 'industrialists' with promises to freely handover peasants' land. This problem of land grabbing is further exacerbated by the activities of gentlemen farmers—the emerging kulaks comprised mostly of a coterie of retired and serving senior military officers and other highly placed Nigerians with powerful connections in government.

The government itself also acquires land for rural 'development' projects. The class contradictions in relation to land are particularly serious because whenever peasants protest this forcible—sometimes illegal—acquisition of their land, the state tends to intervene on the side of the land grabbers, the dire consequences of which the Bakolori massacre (Odeyemi 1981) illustrates. One can only expect suspicion and scepticism when the same state that intervenes on the side of a particular class turns around to 'mobilise' the very class it has been oppressing. In other words, until the class polarisation and attendant antagonism are resolved the Nigerian people will be distrustful of the 'Operation Food First' programme and the government-appointed National Orientation Agency in general.

The effectiveness of co-operatives depends much on effective leadership and administration, and such leadership should not be imposed from outside. Members should be encouraged to elect the leaders they trust. The ideal of such leaders would include qualities of honesty, selflessness, firmness and some degree of political consciousness. They would be firm but at the same time flexible enough to work with people. They would try to educate the people and at the same time learn from them. Such leaders would promote healthy and frank discussion and criticism, and hold meetings regularly to avoid cliquism and nepotism—attributes which easily sow distrust and destroy co-operative organisations. The leaders would try as much as possible to sustain a group spirit and promote democratic decision-making.

In addition, the leadership should receive at least minimal technical training in the use of production technology and accounting techniques.

As much as possible, these basic techniques should be diffused through publicity campaigns, demonstrations at village and district levels, and the operation of model stations. Further, exchange visits between co-operatives to share experiences should be promoted. Co-operatives need enormous assistance from the government, even though they are supposed to develop the spirit of self-reliance. Government, apart from providing technical assistance and credit facilities, must try to remove obstacles in the way of co-operatives.

This approach, it is believed, would help develop co-operatives in Nigeria into a weapon of the people, well organised and with dynamic leadership, founded on democratic principles and working to defend and advance the interests of the people. Organised and politically aware, they would confront the natural forces that oppress them by increasing production and overcoming the oppressive activities of feudal lords, usury and merchant capitalists, middleman, profiteers and any other forces that may militate against their human dignity.

Concluding Remarks and Suggestions

An attempt has been made in this chapter to examine the internal democratic dynamics and potential of peasant-based co-operatives in promoting democratisation in Nigeria. We have argued that by their present structure and operation, co-operatives are instruments for rural exploitation and pauperism—the same functions they fulfilled during colonialism. Against the backdrop of this colonially bequeathed role, we examined the prospects for genuinely popular democratic co-operatives. We argued it remains possible for co-operatives that will reflect and articulate the needs of members to emerge. The government-appointed National Orientation Agency—assuming it has honest intentions—can provide some assistance. However, the extent to which the NOA can challenge and/or contradict existing structures of oppression, in the final analysis, will determine the extent of the democratisation of Nigeria's co-operatives.

A highly sensitive problem of mobilising and democratising peasant movements is the nature of the leadership that may emerge. Hence attempts should be made to ensure that leaders with an interest in private capital accumulation rather than shared capitalisation are avoided. To achieve this, the concept of the 'working cultivator' with a relatively similar resource base and endowments to the members could provide the key to the constitution of the co-operative membership. Moreover, even where peasant-based co-operatives are organised through individual family farms or collective farms, new types of exploitation and abuse may well emerge. These may take the form of state extortion through pricing policy, taxation or the greed of new rural elites. Indeed, participatory peasant movements are important and necessary, but difficult to organise. Hence it is important to defend strategies which oppose state extortion, external manipulation and internal abuse, since they offer the only viable path towards political and social democracy.

For genuine rural development in Nigeria, there is a need to pay attention to the existing territorial organisation of communities with a view to transforming such structures into viable and conducive systems for mobilising rural communities for development and participatory democracy. Hitherto, in seeking to resolve this crucial issue of organisation, the rural development projects in the country have concentrated on spatial aspects of organisation through the provision of farm service centres at central places, rural access roads, rural markets and a network of extension services. However, farmers have remained relatively unorganised. To facilitate popular participation in the rural development process, therefore, there is clearly a need for a well-designed strategy of socio-spatial organisation of rural producers into viable and well-managed co-operatives.

Practical experience from countries such as China, India, Yugoslavia, South Korea, Indonesia, Zimbabwe, Tanzania and Cuba have clearly revealed that through effectively mobilised and organised farmers' co-operatives, immense benefits can be realised.

From a more practical point of view, co-operatives constitute an ideal method for enhancing the productivity and living standards of the rural population by encouraging the adoption of innovations by farmers, by improving the competitive power of the farmers vis-à-vis their actual or potential exploiters, and by encouraging grassroots democracy and the reduction of inequalities at the local level. In spite of the significance of co-operatives as highlighted above, there is at present a low level of co-operative formation in the country.

References

Alubo, S.O. (1988) 'Democratizing co-operatives in Nigeria', proceedings of a national conference on Mobilising Co-operatives for Self-Reliance in Nigeria, Directorate for Social Mobilisation (MAMSER), The Presidency, Abuja, Nigeria.

Beckman, B. (1981) 'The agrarian bases of the post-colonial state', in J. Heyer et al. (eds.) *Rural Development in Tropical Africa*, London, Macmillan.

—— (1985) 'Bakolori: Peasants versus state and capital', *Nigerian Journal of Political Science*, 4: 1-2.

Beer, C. E. (1976) *The Politics of Peasant Groups in Western Nigeria*, Ibadan, Ibadan University Press.

Bernstein, H. (1979) 'African peasantries: A theoretical framework', *Journal of Peasant Studies*, 8, 4: 430-1.

—— (1990) 'Taking the part of peasants?', in H. Bernstein et al. (eds.) *The Food Question: Profits Versus People?*, London, Earthscan.

Boserup, E. (1970) Women's Role in Economic Development, London, Allen & Unwin.

Brown, A. and Barrett, H. (1988) 'A women's work is never done', *Geographical Magazine,* July.

Dey, J. (1981) 'Gambian women: Unequal partners in rice development projects?' in Nici Nelson (ed.) *African Women in the Development Process*, London, Frank Cass.

Dunmoye, R.A. (1987) 'Agriculture and Economic Growth with Special reference to the Developing countries : A survry of relevant theories' in Okello Okuli (ed.) *Nigerian Alternatives*, Zasia, Nigeria ; Yaskiya Corporation.

Fallers, L. (1961) 'Are African cultivators to be called peasants' *Current Anthropology*, 2.

Forrest, T. (1981). 'Agricultural policies in Nigerai, 1900 - 1978' in Judith Heyer et al. (eds.) *Rural Development in Tropical Africa*, New York, St Martins.

Gana, J. (1988) 'Welcome Address', National Workshop on Mobilising Co-operatives for Food Self-Sufficiency, Jos, Nigeria, 6-9 September.

Griffin, K. (1975) *The Political Economy of Agrarian Change*, Cambridge, Mass., Harvard University Press.

Guyer, J. (1984) 'Women in the rural economy: Contemporary variations', in Margaret Hay and Sharon Stichter (eds.) *African Women South of the Sahara*, Harlow, Longman.

Helleiner, G. (1966) Peasant Agriculture: Government and Economic Growth in Nigeria, Homewood, Illinois, Irwin.

Hill, P. (1986) *Development Economics on Trial: The Anthropological Case for a Prosecution*, Cambridge, Cambridge University Press.

Hyden, G. (1980) *Beyond Ujamaa in Tanzania: Under-Development and an Uncaptured Peasantry*, London, Heinemann.

International Labour Office (1976), *Poverty and Landlessness in Rural Asia, A Study by the World Employment Programme*, ed. Keith Griffin and Azizur Rahman Khan, Geneva, ILO.

—— (1987) 'Women's participation in co-operatives', *Women at Work*, no. 1

Jacoby, E. and Jacoby, C. (1971) *Man and Land*, New York, Knopf.

King, R. (1981) 'Co-operative policy and village development in Northern Nigeria', in J. Heyer et al. (eds.) *Rural Development in Tropical Africa*, London, Macmillan.

Lappe, F.M., Collins, J., and Fooler, C. (1977), *Food First Beyond the myth of scarcity*. Boston, Houghton.

Mafeje, A. (1993) 'Peasant Organisations in Africa: A potential dialogue between economists and sociologists—some theoretical and methodological observations', *CODESRIA Bulletin*, 1: 14-17.

Mascarenhas, O. and Mbilinyi, M. (1983) *Women in Tanzania: An Analytical Bibliography,* UppsalaScandinavian Institute of African Studies.

Mintz, S. W. (1973) 'A note on the definition of peasantries', *Journal of Peasant Studies*, 1, 1: 91-3.

Moyo, S. (1995)'Peasant organisations and research issues on democratisation in Africa', paper presented to CODESRIA Multinational Working Group Methodology Workshop on Peasant Organisations in Africa, Dakar, 21-30 November.

Nasko, M. G. (1988) 'Keynote Address', workshop on Mobilising Co-operatives for Food Self-Sufficiency, Hill Station, Jos Plateau State, 7 September.

Nkom, S.A. (1982) 'Government policy, rural development and the peasantry in Nigeria: Mobilisation or marginalization?', in F. Odekunle (ed.) *Mobilisation of Human Resources for National Development*, Lagos, NEPA.

Odeyemi, I. (1981) 'Massacre of peasants at Bakalori', in B. Usma (ed.), *Political Repression in Nigeria*, Zaria, Gaskiya.

Olatunbosun, D. (1975) *Nigeria's Megliected Rural Mojority*, Ibandan, OUP.

Osuntogun, A. (1983) Co-operatives and Nigeria's Agricultural Development, Inaugural Lecture Series no. 61, Ife, University of Ife Press.

Rahmato, D. (1991) Peasant Organisation in Africa: Constraints and Potentials, Dakar, CODESRIA Working Paper 1/91.

—— (1993) 'On peasant studies —A reply to Archie Mafeje', *CODESRIA Bulletin,* no. 2.

Redfield, R. (1956) *Peasant Society and Culture*, Chicago, University of Chicago Press.

Rogers, B. (1980) The Domestication of Women, London, Tavistock.

Saul, J. S. and Woods, R. (1971) 'African peasantries', in T. Shanin (ed.) *Peasants and Peasant Societies*, Harmondsworth, Penguin.

Taylor, C. C. (1949) 'Objectives of farmer co-operatives: By a sociologist', *American Co-operation*.

Thorner, D. (1962) 'Peasant economy as a category in economic history', proceedings of the Second International Conference of Economic History.

Wallace, T. (1980) 'Agricultural projects and land in Northern Nigeria', *Review of African Political Economy,* no. 17.

Williams, G. (1976) 'Primitive accumulation: The way to progress', *Development and Change*, 18, 2.

Wolf, E. R. (1966) *Peasants*, Englewood Cliffs, N. J., Prentice-Hall.

Appendix

Table 1: Age Distribution of Co-operative Members in Oyo State, Nigeria

Age Group	Male (%)	Female (%)	Total (%)
Less than 20 years	-	-	-
20-29 years	10	-	10
30-39 years	28	12	40
40-49	19	11	30
50-59	8	4	12
60 and above	7	1	8
Total	72	28	100

Source: Field Survey, 1996

Table 2: Main Occupations of Co-operative Members in Oyo State, Nigeria

Occupation	Number of respondents	% of Total
Farming	68	72.1
Trading	12	13.0
Public Service	6	6.4
Others	8	8.5
Total	94	100.0

Source: Field Survey, 1996.

Table 3: Level of Education of Co-operative Members in Oyo State, Nigeria

Level of education	Number of respondents	% of Total
Primary school	34	34.0
Secondary school	13	13.0
Post secondary	2	2.0
None	51	51.0
Total	100	100.00

Source: Field Survey, 1996.

Table 4: Household Structure of Co-operative Members in Oyo State, Nigeria

No. of wives	No. of respondents	% of total	No. of children	No. of respondents	% of total
0	9	13.0	0-2	11	15.9
0	32	46.4	3-4	17	24.6
2	24	34.8	5-6	23	33.
3	4	5.8	7 and above	18	26.1
Total	69	100	Total	69	100.0

Source: Field Survey, 1996.

Table 5: Dependant Relatives of Co-operative Members in Oyo State, Nigeria

Number of dependent relatives	Number of Respondents	Percentage of total
0	10	14.5
1	14	20.3
2	26	27.7
3	10	14.5
4	7	10.1
5 and above	2	2.9
Total	69	100

Source: Field Survey, 1996

Table 6: Period of Membership of Co-operative Members in Oyo State, Nigeria

Range (Years)	Number of respondents	% of Total
0-5 years	08	8.4
6-10 years	19	20.0
11-15 years	18	19.0
16-20 years	25	26.3
21-25 years	13	13.7
26 and above	12	12.6
Total	95	100.0

Source: Field Survey, 1996.

Table 7: Benefits Derived by Co-operative Members in Oyo State, Nigeria

Benefits	Number of respondents	Percentage of total
Obtain financial assistance	23	23.0
Improved farming	12	12.0
Supply of materials and equipment	8	8.0
Promotion of business	14	14.0
Encourage group activities	3	3.00
Offered saving facilities	21	21.0
Encouragement of thrift	18	18.0
Others	1	1.0
Total	100	100.0

Source: Field Survey, 1996.

Table 8: Co-operative Unions and Societies in Oyo State, Nigeria

Types of societies/Unions	Number of societies in each zones						Total
	Ibadan	Ogbomoso	Ilesa	Oyo	Saki	Osogbo	
1. Co-operative Produce & Marketing Union (CPMU)	2	-	4	1	-	-	7
2. Co-operative Produce & Marketing Society (CPMS)	159	-	318	-	-	-	477
3. Farmers Multi-purpose Co-operative Union (FMCU)	8	4	4	-	1	1	18
4. Farmers Multi-purpose Co-operative Society (FMCS)	154	131	74	-	41	35	435
5. Co-operative Thrift and Credit Union (CTCU)	25	5	16	18	18	8	
6. Co-operative Thrift and Credit Society (CTCS)	1,865	350	772	850	791	352	
7. Co-operative Investment & Credit Union (CICU)	17	4	-	1	-	2	

Table 8: continued

Types of societies/Unions	Number of societies in each zones						Total
	Ibadan	Ogbomoso	Ilesa	Oyo	Saki	Osogbo	
8. Co-operative Consumers Societies	35	5	1	1	10	3	
9. Co-operative Buying Clubs	86	-	-	-	-	-	86
10. Co-operative Investment & Credit Society (CICS)	446	32	-	16	-	68	
11. Informal Co-operative Societies (Crafts Societies)	43	23	18	20	15	17	
Sub-total	2,840	554	1207	907	486	486	6,870

3

Peasant Organisations and Democracy in Nigeria: The Ayamelum Improvement Union

Onyebuchi Emmanuel Ezeani and Damian Ayichi

Introduction

> Social struggles against ethnic marginalisation, economic deprivation, religious and cultural oppression and exclusion, and human rights violations, can lead to the rise of social movements ... where state institutions fail to serve, or violate and contradict people's interest (Fuentes & Frank 1989: 180).

Social movements articulate demands that are not heard or satisfied through ordinary political channels or take action where community and state structures fail. The ultimate goal of most popular movements is to establish democracy. Democracy is the process and act by which government is made accountable by stakeholders through rules and institutions that allow for open competition and participation in government (Olukoshi 1995: 247; Bangura 1989: 3-7). Democracy is not just an object but a process of social emancipation. It entails struggles against economic exploitation, political domination, ideological alienation, ethnic marginalisation and the social disinheritance of the broad masses (Syahuka-Muhindo 1995: 493).

Over the years, Nigeria and other African countries have witnessed the emergence and growth of peasant organisations as social movements. The proliferation of a wide variety of peasant organisations in rural Africa has been attributed partly to the failure of the postcolonial state to achieve the much-expected improvement in living standards and general wellbeing of the people. Consequently peasants attempt to survive through associative endeavours (Rahmato 1991), because 'it may in fact be argued that unfavourable social conditions promote or enforce solidarity and indigenous structures'.

The role of peasants and peasant organisations in the establishment of democracy is a subject of growing interest among scholars (Rahmato 1991; Hough 1988: 185-206). Rahmato (1991), for instance, opines that because of their grassroots nature peasant organisations ought to be schools for popular democracy, platforms for the advocacy of issues and causes of significance to the people, and media for real and meaningful political participation. This historical role of the peasants in Africa in struggles against domination by the state, landlords or merchant capital, through a variety of organisational forms, associative activities and relationships is well documented (Beer 1976; Isaacman 1990; Mafeje 1993; Syahuka-Muhindo 1995; Smith 1989: 276-80). For instance, Amadiume (1995: 36, 50) explores the case of Nigerian patrilineage men, umunna, and patrilineage daughters, umuokpu or umuada, serving as agents of democratic struggle in the traditional social structure of Nnobi.[1]

The resentment of forced labour and mandatory taxes fuelled Ugandan resistance to colonial rule through local rural organisations and led to a major protest in 1919 (Syahuka-Muhindo 1995: 291-543). The formation of the Rwenzuru Movement by the Bakonzo of Uganda in 1919 and into the 1920s was a response to oppression and subordination by the colonial state. In the Tanzanian region of Kagera, the colonial state's 'Grow more coffee campaign' was strongly resisted by peasant farmers in the 1930s following the tight administrative controls and numerous regulations associated with the campaign (Smith 1989: 216-80).

Therefore, in addition to the traditional goals of defending, preserving and promoting peasant needs, interests and welfare, peasant organisations have been associated with the larger and more complex processes of people's participation, grassroots empowerment and democratisation. Despite these potentials of peasant organisations as agents of democratisation, few in-depth studies exist on peasant organisations and democratic struggles in general (Moyo 1995). While most previous studies focus mainly on the socio-economic roles of peasant organisations (Agu et al. 1990; Eboh 1993, among others), peasants continue to organise themselves to fight against all forms of socio-economic deprivations and to pressurise for democratisation, responsible governance and accountable leadership.

This chapter discusses the rise of the Ayamelum Improvement Union (AIU) as a social force that aims to improve standards of living and to empower the people of south-eastern Nigeria. Ayamelum is a clan in the present Ayamelum local government area of Anambra state of Nigeria.

Defining Peasant Organisations

Any discussion of peasant organisations should be based upon clear understanding of who the peasants are. Peasant studies tend to lack consensus about the appropriate definition of peasants and the peasantry (Mintz 1973: 91-3). The failure to address this particular question occasioned a major debate in CODESRIA based upon a critique of Rahmato's (1991) work by Mafeje (1993: 14-17).

Klein (1980) argues that there are generally two perspectives used in conceptualising peasants. One views peasants as non-enterprising rural producers[2] who are simple rural producers securing their livelihood through the use of family labour or family land. The peasant is unable to treat the family farm as an enterprise for reinvestment and business, and thus land is not regarded as a commodity or capital but as a social system where culture and tradition are dominant. 'Peasants' are in this vein regarded as backward and primitive people who are not keen to develop themselves. Thus Hyden (1980, 1983) identifies 'peasants' as everything which appears to be technologically and culturally backward. In his approach, 'peasant' is the semantic successor to 'native'.

The above definition of the peasantry is counterpoised by the political economy school, which contends that subordination to a central authority and dependence on a national market are the hallmark of a peasant society. Accordingly the state market, which is effectively a buyers' market, subjects the peasant to unequal exchange. Thus, according to Bernstein (1979: 421-2), while the first perspective above may describe the form of peasant production it does not have any theoretical content that is historically and socially specific. And, according to Williams (1976: 142), 'peasants' should be defined by their relation to the overlord, market and state, since there is no 'peasant economy' independently of its relation to commodity exchange.

Based on this understanding, Bernstein (1990: 72) contends that 'the starting-point must be to view peasants today as agrarian petty commodity producers within capitalism'. This definition retraces Wolf's (1966: 8-10):

> Peasants ... are rural cultivators whose surplus is transferred to a dominant group of rulers ... A peasantry always exists in more complex societies' social relations which are not symmetrical, but are based, in some form, upon the exercise of power ... the term 'peasantries' denotes no more than an asymmetrical relationship between producers of surplus and controllers.

Klein (1980: 11) saw the peasantry as mainly agriculturalists who control the land they work, either as tenants or smallholders. Landless labourers are not peasants but are organised largely in household units that meet most of their subsistence needs and are ruled by other classes who extract a surplus (rent) either directly or through control of state power (taxes). Their culture is distinct from but related to the culture of the dominant group. While this study shares the latter definition and attributes of the peasantry, it is important to note that the definition of the peasantry in terms of its asymmetrical relationships to external power does not imply that they form a homogeneous mass. The fact remains that peasantries are always and everywhere typified by internal differentiation along many lines.

The implication of the internal differentiation among the term 'peasant organisation' needs some clarification. Mafeje (1993) has distinguished between 'peasant organisation' and 'peasant organisations'. According to him, 'peasant organisation' refers to a generic mode of social organisation which is associated with the peasantry of classical antiquity. However, 'peasant organisations' refer to 'a variety of discrete associations formed by and involving peasants' (Ibid.). Thus although peasant associations occur in given agrarian settings, in their totality they do not constitute 'peasant society' but rather are its product. Peasant associations tend to be resistance movements or organisations against domination by the state, landlords, merchant capital, or men in the case of women's associations (Ibid.: 17).

It is important to mention that peasant organisations need not necessarily be formed by the peasants or composed exclusively of peasants. In raising this issue Sibanda (1997) argues that organisations are considered 'peasant' or peasant-oriented if they seek to advance the interests of those who survive on land, and they need to be organisations for peasants but not necessarily of peasants.

This study examines the extent to which the Ayamelum Improvement Union (AIU) has served as an agent of democratic struggle. The specific study objectives were to trace the origin, objective and organisational structure of the AIU and to determine the extent to which the AIU could be regarded as an agent of democracy. The study interrogates how the struggle for a separate local government for the Ayamelum clan and against displacement from their land by the state and capitalist farmers without adequate compensation, has contributed to local democratisation processes. The study also examines the extent to which and how the working of internal democracy within the AIU contributes to political development in the region. Since the AIU has largely been seen as a tool for socio-economic development of the Ayamelum clan, the study assesses the relationship between the state and the AIU so as to determine the problems militating against the potential of the AIU to become an agent of democratic change.

The Organisational Structure of the Ayamelum Improvement Union

The AIU was the principal peasant organisation, formed in the early 1950s by the people of the Ayamelum clan as an instrument of democratic struggle. Tangential to this political objective was its pursuit of various socio-economic and political rights, which had been denied over the years. The union grew out of the need to better the livelihood of the Ayamelum person, fight for the causes of the clan and engage in community development efforts. Initially formed by Ayamelum people who had gone to the 'towns' as businessmen, artisans or civil servants as an enlightened response to the individualism and anonymity of the urban milieu, the organisation soon spread to all the communities that made up the Ayamelum clan. Thus, as a body that derived a sizeable part of its membership from the peasants, it cannot be called an organisation of the peasants, for it was not originally formed by the peasants; rather, it is an organisation for the peasants.

The union had branches in the various cities and towns where Ayamelum natives resided, and each branch had its own executive, which comprised a chairman, vice-chairman, secretary, assistant secretary, treasurer, financial secretary, public relations officer and provost. The Ayamelum Peoples Assembly (APA) met at the home level, and its membership was on a representative basis throughout the branch unions. Traditional rulers of the various communities were also members of APA, but women still remained a minority. The AIU derived its financial resources from membership subscriptions.

The union was guided by democratic principles in the working of its structure and organisation. The selection of leaders was through an election process conducted bi-annually during a general meeting. Every eligible member, irrespective of sex, qualified for leadership positions in the various branches of the union. The principle of one-person-one-vote was also applicable. Decisions were taken through consensus, and divergent viewpoints on issues important to members were openly expressed, debated and consensus reached during general or emergency meetings. Membership of various working committees also provided opportunities for members to participate and to influence decision-making. It is important to mention that the educated elite had dominated the leadership of AIU over the years, and therefore education was an important prerequisite for leadership.

The Land Question and Democratic Struggle

The peasants in the Ayamelum clan, just like their counterparts in many parts of Africa, have been victims of displacement, exploitation and proletarianisation arising from capitalist agriculture that was and is still supported by the state. The AIU has championed the demand of the peasants both for adequate compensation for their expropriated land and for damaged economic trees. The alluvial soil of the area accounts for the rich agricultural potentials that make the area the food-basket of Anambra state, and the Ayamelum people are predominantly farmers. According to Fabiyi (1979), apart from labour, land is the major factor of production in traditional agriculture in Nigeria. Agricultural production was more constrained by availability of land than other production factors owing to the tenurial arrangements.

Before the promulgation of the Land Use Decree of 1978 by the then Federal Military Government of Nigeria, traditional or customary land tenure in the Ayamelum clan, as in most parts of south-eastern Nigeria, was based upon family and/or communal or community ownership. Land was seen as a common property managed by the family heads for the past, present and future generations. The sale of land was regarded as illegal and an abomination as land was sacred, and the most powerful deities in the area were usually those associated with it. The introduction of capitalist agriculture into the Ayamelum clan by 1970 brought a rapid change in production and land ownership relations in the area. The transformation of the peasantry from the presumed 'economy of affection' (Hyden 1980, 1983, 1986), which impeded the development process towards commercialised agriculture, underlay the changes in Ayamelum.

In this context, international aid agencies and liberal social scientists argued for orienting the mass of peasant producers towards the national and international capitalist market through a massive infusion of capital in the form of new farming methods, new seeds and better infrastructure into the peasant economy (Ibeanu 1993: 1). The Second National Development Plan (1970-74) of Nigeria had a negative view of peasant subsistence production, which saw no realistic need for changing the agrarian system based on a hoe and cutlass technique of cultivation. Thus, it was the view of the federal government that proper rural transformation would be achieved through public and private sector-operated large-scale production for livestock food and export crops. Improved productivity was thus externally imposed on the people of Ayamelum. In line with this conception of agrarian development the Nigerian government enacted the Land Use Decree of 1978 to free land from traditional holding systems which are said to be 'antithetical to an unfettered penetration of capital into agriculture' (Ibeanu 1993). The Land Use Decree was effectively employed by the capitalists to dispossess the peasants of their land resources in the Lower Anambra river basin where the Ayamelum clan constituted a significant number of inhabitants.

The inception of capitalist agriculture in the Ayamelum clan be traced to the early 1980s with the establishment of Lower Anambra Irrigation Project (LAIP). The Federal Government of Nigeria and the World Bank, supported by the state government-owned Ada Rice Project, established this project wherein about 15,000 hectares of land were acquired for various sub-projects. Compensation (though in most cases inadequate) was paid for economic property destroyed. Furthermore, the agencies promoted the project as a 'participatory farmer scheme' whereby indigenous farmers benefited from the irrigated lands developed by the agencies. Land committees were set up to handle all issues relating to land matters in the affected communities. But even with the above measures, there were many protests emanating from lack of compensation for lost lands (Ofuebe 1992).

By the mid-1980s, the then Federal Military Government mounted an aggressive agricultural and food self-sufficiency programme that was effectively supported by attractive incentives that the Nigerian capitalist class found irresistible. Consequently, the urge to own a farm increased and the Ayamelum clan region became a target because of its fertile land. The land committees and even some traditional rulers colluding with the local government authorities allocated and signed various lease and sale agreements with private commercial farmers, industrialists and land speculators without adequately compensating the original owners of land. According to Akor (1986: 1), 'thousands of peasants have been dispossessed of their land in the (Ayamelum) area by a number of companies and individuals'.

Table 1: Particulars of Land Sales in the Ayamelum Clan

S/No.	Name of Community	Name of Lessee/Buyer	Address of Lessee/Buyer	Hectarage	Purpose
1.	Omasi	SICO Farm Complex Ltd	2 E. Eloseh St, Aguda, Surulere, Lagos	480.782 ha.	Agricultural
2.	Umeje	Dr L.A. Omonye	31 Ugwunobamkpa Rd, Onitsha	259. 394ha	Agricultural
3.	Omasi	Chilian Farms Ltd	12 Okwuenu St, Onitsha	202. 317ha	Agricultural
4.	Anaku	Chief Rommy Amadi Ezeonwuka	Amaife-Akpo, Aguata LGA	40. 779ha	Agricultural
5.	Umeje	Offor-Offor	20 Park Lane, Nkpor	41163.946 sq.m.	Agricultural
6.	Umerum	Prince Mark Farm Ltd	30 Ifejika St, Onitsha	3297.241 sq.m.	Agricultural
7.	Omor	Emmanuel Enweh	CMS Church, Itchi	4453.759 sq.m.	Commercial
8.	Omasi	Rojel Farms Ltd	5A Aria Rd, Enugu	652.349ha	Agricultural
9.	Omasi	Alphonsus Akana	6 Ichida St, Onitsha	261.036ha	Agricultural
10.	Omor	Samuel Anataogu	18A Calabar Road, Ikom	1224.412 sq.m.	Commercial
11.	Omor	Leonard Ekeiyi	47 Calabar Road, Ikom	1690.033 sq.m	Commercial
12.	Omasi	Tempo Mills Ltd	PMB 1580, Onitsha	1142.349 sq.m.	Agricultural
13.	Umeje	Tempo Mills Ltd	PMB 1580, Onitsha	1019.581ha	Agricultural
14.	Anaku	Phili Ezeoba Mofus	20 Nwangene St, Onitsha	501.987ha	Agricultural
15.	Omasi	Olympic Farms Ltd	Abagana, Njikoka LGA	260.788ha	Agricultural
16.	Anaku	Ekene Dili Chukwu Motors Ltd	1 Ilodibe Road, Awada, Onitsha	43 649ha	Agricultural /Industrial
17.	Anaku	Ekene Dili Chukwu Motors Ltd	1 Ilodibe Road, Awada, Onitsha	90 236ha	Agricultural /Industrial
18.	Omor	Justin Akukalia	Box 342, Abagana	26891.156 sq.m	Residential
19.	Umeje	Ramac Farms & Agro Industries Ltd	43 Francis St, Onitsha	134.400ha	Agricultural
20.	Ifite-Ogwari	Ramac Farms & Agro Industries Ltd	43 Francis St, Onitsha	270.737ha	Agricultural
21.	Omasi	C.U. Uzo-Amobi	19 Aria Road, Enugu	80ha	Commercial /Residential
22.	Ifite-Ogwari	SESAIC Industries Ltd	1 Ukwa Street, Enugu	178.00 ha	Agricultural
23.	Umeje	Dalgo Ltd	48 Ibafon St, Apapa, Lagos	264.139ha	Agricultural
24.	Omor	GMO & Co. Ltd	1 Abimbola Shodipo St, Lagos	473.088ha	Agricultural
25.	Omor	Samuel E. Mbanefo	18 Chukwura Lane, Onitsha	3.015ha	Residential/ Agricultural
26.	Ifite-Ogwari	A.C. Ekpechi	Umuru Village, Ogidi	509.051ha	Agricultural
27.	Ifite-Ogwari	WINCO Foods & Agric Complex Ltd	P.O. Box 14, Isuofia, Aguata LGA	337ha	Agricultural

Source: Oyi Local Government Office documents.

Furthermore, there were conflicts among the peasant farmers resulting in the less privileged being evicted by force. By 1991, over a hundred individuals and private firms had acquired various sizes of farmlands for agricultural and industrial purposes. Table 1 shows the particulars of some land sales in the Ayamelum clan. One logical result of expropriating the land of the peasantry was their proletarisation, linked to the exploitation and alienation of their labour through paying wages below the statutory national minimum wage.

The outcome of the expropriation, proletarisation and exploitation of the peasants was a series of petitions and litigations to the government by the peasants through the AIU against the capitalist farmers and their collaborators. Individually and collectively, the leaders of the union wrote several petitions to the government on the activities of the land speculators and capitalist farmers. For example, in a letter sent to the then military governor of Anambra state dated 7 February 1989, some prominent leaders of AIU complained thus:

> ... a substantial proportion of the fertile agricultural land in Ayamelum clan has now been acquired by very rich individuals and groups who are mostly from Onitsha or Awka Senatorial Zone. Such wealthy individuals and groups have now become the super tenants (much more powerful than their landlords) in Ayamelum clan.

In another development, two prominent members of AIU and community leaders of Omasi, Simeon Anah and Simeon Okeji, in a petition to the state government dated 2 July 1986, decried the illegal sale of land in their community and expressed fears that they would 'be turned into beggars and wanderers in their own land if the situation was not arrested'.[3]

The Struggle for Local Autonomy of the Ayamelum Clan

As Oyovbaire (1974: 145) has rightly argued, the raison d'etre for agitation for political autonomy is basically rooted in 'the individual's and group's perception of their positions in the modernisation processes, and their relationship to others as regards questions of socio-economic welfare, employment, security, status, power and distributive justice'. The quest for communal autonomy of the Ayamelum clan is a typical case of response by a people to relative deprivation.

Ayamelum comprised eight autonomous communities, namely Anaku, Ifite-Ogwari, Igbankwu, Omasi, Omor, Umuerum, Umueje and Umumbo.[4] Historically, the eight towns that make up the Ayamelum clan had a common dialect, lifestyle, religious beliefs, titles, names, age grades, traditional administrative system, socio-cultural affinity and behavioural patterns (AIU 1995: 1). The clan lived in a riverine peninsula surrounded by the Anambra river on the west, the Ezu river on the south and the Obina river on the north. This structure of land in the area caused swampiness and drainage difficulties, experienced particularly at the four basin-bottom villages during the rainy season. Equally, the area was poorly accessible because of difficult topography. The consequence was that it remained untouched by the forces of modernisation until the early 1950s. For decades, the entire clan remained unpenetrated and inaccessible to the white colonial masters, missionaries (both European and African) and merchant sojourners from outside the clan. According to the National Association of Ayamelum Students (NAAS) (1987: 1), the result was a lack of educational infrastructure leading to illiteracy and inadequate representation in government

There was also the political dimension to the Ayamelum autonomy question. From the inception of the British colonial administration in Nigeria until 1949, the Ayamelum clan was under the then Onitsha Rural Areas Native Authority (ORANA). There are diverse opinions on the fate of the Ayamelum people under ORANA. While the leadership of the Ayamelum Improvement Union (1995: 1) contended that the clan had adequate political representation under ORANA, some other people within the clan believed the Ayamelum people had been totally neglected. However, in 1950-1, the colonial administration placed the clan under the then Nsukka province. The reason for the change of administrative area was what the then administration called 'administrative convenience' as the rivers mentioned earlier constituted a barrier to easy administration of the clan from Onitsha (Ibid.: 1).

The political problem of the Ayamelum people had its genesis in the transfer of ORANA to Nsukka as the relationship between the two was one of mistrust. Hence, under the Nsukka administration the people believed that discriminatory treatment was meted out to them (Ofuebe 1986: 80). Similarly, the Ayamelum Leaders of Thought (1974: 2) believed that the 15 years under Nsukka administrative system had contributed to more illiteracy among the youth as the administration opposed the siting of a secondary school in Ayamelum. As a result of the dissatisfaction expressed by Ayamelum leaders and under mounting pressure from the Ayamelum people, the then military administrator of the eastern group of provinces under Lt-Col. Odumegwu Ojukwu severed the clan from Nsukka division and merged it with Anambra division.

But the reunion was short-lived as the then East-Central state government created Uzo-Uwani division in 1973 and incorporated

Ayamelum clan in it without consultation.[5] In protest at this 'unholy wedlock' with the Uzo-Uwani people, the Joint Commission of the Communities in Ayamelum Clan, which met at Omor, sent a letter containing their resolutions to the then East-Central state administrator. In the letter, the joint commission noted that the creation of Uzo-Uwani division without consultation with the Ayamelum people amounted to a negation of the well-known and clearly declared policy of the East-Central state government — to bring the government nearer to the people, allow the people to take their destiny into their hands and develop at their own pace.

All the petitions of the Ayamelum people proved abortive. But they were persistent in their struggle for social justice and equity. Consequently, in 1975, during the local government reorganisation in the then Anambra state, the Ayamelum clan requested an Ayamelum Local Government Council. This marked the beginning of almost two decades of struggle by the Ayamelum clan for a separate local government area. The Ayamelum Improvement Union emerged as the major champion of the cause for political autonomy of the Ayamelum clan following the failure of the much older organisations, namely the Ayamelum Leaders of Thought and Federation of Ayamelum Students, to achieve this goal.

As a result of the failure of the government of the then East-Central state of Nigeria to grant an autonomous local government to the Ayamelum clan during the local government exercise of 1975, in 1982, the Ayamelum people through the AIU sent another request to the then civilian governor of Anambra state, Chief Jim Ifeanyichukwu Nwobodo. But, instead of granting the request for a separate local government council, the Ayamelum communities were split into two, with Omasi, Umueje and Ifite-Ogwari in the Igboda Local Government Area, and Igbankwu, Anaku, Omor, Umuerum, and Umumbo in the Uzo-Uwani Local Government Area. Consequently, the people of the Ayamelum clan became the minority in each of the two local government areas (AIU 1995: 1-2).

This development did not last long for the military administration that intervened in the country's politics in 1983 dissolved the local government areas created by its predecessor. Again, in 1989, the AIU spearheaded the request by the Ayamelum clan for an Ayamelum Local Government Area but were instead included in the present Oyi Local Government Area. 'The people of Ayamelum clan were, however, happy in Oyi Local Government Area as a better alternative to being in Uzo-Uwani Local Government Area' (AIU 1995: 2). Yet they did not relent in their efforts to obtain a separate local government area. Thus, in 1995, the AIU sent another request to the then Nigerian military head of state, General Sani Abacha, in which it stated:

> ... having gone through ... severe shifts, pull and pushes, divide-and-rule under different political administration, the quest for autonomy, political emancipation and self-determination of the people of Ayamelum clan remains paramount ... Conscious of the concern of the present administration for the welfare of the masses, justice and fair play to all, and the development of the underdeveloped areas, the people of Ayamelum clan comprising the eight autonomous communities of Anaku, Ifite-Ogwari, Igbaukwu, Omasi, Omor, Umerum, Umueje, and Umumbo, hereby respectfully request for the creation of Ayamelum Local Government Area with the Headquarters at Ifite-Ogwari (AIU 1995: 2).

As a result of the above-mentioned requests, which were spearheaded by the AIU, the Federal Military Government created the Ayamelum Local Government Area in 1997. However, the headquarters of the local government area was located at Anaku instead of Ifite-Ogwari as requested by the AIU. In its letter of appreciation to the president of the federal republic, shortly after the creation of the Ayamelum Local Government Area, the AIU said:

> It is with utmost sense of humility and responsibility that we hereby seize this ample opportunity to express our profound and immense gratitude to the Abacha Administration for their judicious and commendable decision in creating Ayamelum Local Government Area of Anambra State. This noble act of the Abacha Administration puts a final end to the denial of Ayamelum people of political autonomy, political emancipation and economic self-determination; and equally heralded the death knell for the politics of manipulation, exploitation and gerrymandering which have been the lot of Ayamelum people occasioned persistently by the invidious imperial pretentions of the extraneous elites of neighbouring communities.

Lobbying for Social Infrastructure

In spite of this positive result, however, the AIU had various weaknesses. Some of its leaders and prominent members, including some traditional rulers, were involved in the scandalous sale of farm lands to capitalist farmers, thereby thwarting the efforts of the AIU to check the arbitrary and illegal sale of land. Also the inability of some notable leaders of the AIU, such as Paul Obidigwe, Hon. Simeon Okafor (who were once chairmen of the union) and Hon. Raphael Osita, to make any significant impact on the campaign of the Ayamelum clan for local autonomy was blamed on their pursuit of personal interest to the neglect of the collective interest of the Ayamelum people (Ofuebe 1986: 117-27).

Intra-class conflict was also a major constraint on the democratic struggle of the AIU. A good example was the case of the two prominent leaders of the AIU mentioned above—Hon. Simeon Okafor and Hon. Raphael Osita. Both men were elected in 1979 as members of the Anambra State House of

Assembly. Unfortunately, rather than work harmoniously towards improving the lot of the Ayamelum people, the two men engaged each other in a tussle for supremacy. According to Mazi Okoye,

> As soon as they were elected into the House, they began to fight for who should be the leader of the other in that House. Consequently, they developed negative attitudes towards each other and could no longer see issues from the same angle. Their negative attitude adversely affected our struggle for local autonomy, for while one of them (Mr Okafor) was fighting side by side with Ayamelum patriots for an autonomous Ayamelum Local Government Area, the other (Hon. Osita) found solace in fighting on the opposite side, that is the side that was bent on the dismemberment of Ayamelum. Under this circumstance, it will be illusive (sic !) to expect a favourable result (quoted in Ofuebe 1986: 113).

In their obsession for personal aggrandisement, some leaders of the AIU teamed up with the government and other social forces opposed to the democratic ideals of the Ayamelum people. The AIU thus faced contradictions between the political and economic interests of its leaders and members.

The initial primary motive for the development of a popular peasant organisation must contain a substantial economic component, that is, a set of activities, services or reforms designed to improve the quality of life of its peasant constituency (Hough 1988: 191). These activities usually address the central problem of improving living conditions, principally through agriculture, the livelihood of the overwhelming mass of rural families. The range of socio-economic activities and issues that a peasant organisation may embrace as its raison d'être can vary greatly, from issues such as agrarian reform to access to services such as credit and production technologies, among others. The activities aimed at redressing the miserable economic conditions of the majority of the AIU membership were critical to both the struggle for autonomy and land compensation

The complementarity of economic interests and democratic political development can become integral to the development of peasant organisations. When these needs are serviced, the organisation is likely to grow and the possibilities of genuine democratic evolution are enhanced. When the needs are poorly serviced or neglected the opposite is the case, and the organisation's constituency tends to atrophy. The development and representation of economic and related interests can create considerable popular support that ultimately is the basis of political power and influence of an agrarian union or peasant movement. By effectively performing these functions, and by exploiting the complementary trade-offs and possibilities that exist, the movement can clearly enhance the importance of its own role within the body politic (Hough 1988: 193).

The role of AIU in the socio-economic development of the Ayamelum clan was mainly in the form of sensitising the government (local, state and federal) to the underdevelopment problems of the area. Through a series of letters and personal contacts the AIU was able to draw the attention of the government to the socio-economic problems of the Ayamelum clan. According to Chief A. Ofuebe (a prominent member of the executive of the AIU), the Ayamelum clan was inaccessible both by land and river to other parts of Anambra state up to the early 1980s. This was because there were no roads and big boats to link the area to other parts of the state. It was through pressure mounted on the government by the AIU that in 1982 a big boat was donated to the area by the state government thereby enabling access by sea. Also in the same year, the civilian government of Alhaji Shehu Shagari awarded a road contract linking the area to Onitsha by road through Otuocha and Adani. Raphael Machi (an ex-public relations officer of the AIU) attributed this AIU success to the fact that it had branches in every community of the clan as well as in the various town unions.

The AIU did not embark on any specific development projects, but through persistent advocacy in the form of letters, publication in newspapers and personal contacts with government officials at various levels, had drawn the notice of the government to the plight of the people. Most of the government projects in the area were sited as a result of the efforts of the AIU. The union therefore had a positive impact on the socio-economic lives of the people through drawing government attention to the area.

The Ayamelum Improvement Union and Democratisation

From its inception, the AIU was confronted with many problems which hindered or constrained the rapid attainment of its goals. One of its main problems was the inability of the umbrella union to co-ordinate the efforts of its various branches towards achieving a set of common goals.[6] The organisation was hamstrung by this lack of co-ordination as it relied on the small financial contributions from its various branches of poor peasant communities who found it difficult to pay their dues. The class and personal interests of certain prominent members of the union (including some members of the executive) had also in some instances been allowed to gain precedence over the goals of the union. Within the top leadership structure of the AIU, some politicians' interests and ambitions in national politics had been contradictory to the union's interests

The membership of the AIU provided a great opportunity for participation in politics in the wider Nigerian society, especially at the local and state level. As Table 2 shows, many prominent members of the AIU played an active role in local and state politics as members of the legislative arm of the government. The AIU usually offered moral support to political aspirants. However, an active role in the affairs of the AIU, especially at leadership level, provided tremendous exposure to people. The result was that by the time AIU leaders went out to contest wider political positions, they were already popular with the people of the clan.

This linkage of the AIU leadership to national politics influenced the development of the AIU as a peasant organisation and the nature of its relationship with the state. In their daily operations and in attempting to achieve their objectives, peasant organisations related to and made demands on the state. The state, on the other hand, sometimes tried to control peasant organisations, especially if it had been instrumental in their formation or origin.[7]

Table 2: Political Positions Held by Prominent Members of the AIU

Name	AIU	Government Institutions
Vincent Nwalieji	Active member	(1) Federal Parliamentarian
		(2) Chairman, Uzo-Uwani
S. N. Okafor	Chairman	(1) Supervisory Councillor, Uzo-Uwani
		(2) Hon. Member, Anambra State House of Assembly, Enugu
P. Obidigwe	Chairman	Hon. Member, Anambra State House of Assembly, Enugu
R. Osita	Active member	Hon. Member, Anambra State House of Assembly, Enugu
Anthony Nwuneli	Active member	Supervisory Councillor, Uzo-Uwani
Fabian Nnalue	Active member	Supervisory Councillor, Uzo-Uwani
G. O. C. Okoye	Active member	Chairman, Igbo-Oda L.G. Council
John Nweke	Active member	Supervisory Councillor, Igbo-Oda

Source: Memorandum by the National Association of Ayamelum Students (NAAS), 1995.

In its attempts to deal with the Ayamelum demand for local autonomy and for adequate compensation for displaced people and damaged economic trees, the AIU frequently came into contact with different state institutions regardless of the government in power. The relevant state institutions contacted included the legislature, the bureaucracy, the executive arm of government and many commissions and panels of investigation. In all its contacts with state organs, the AIU leadership put forward specific demands. The strategies and tactics used by AIU to pursue its demaand included drafting of petitions to the legislative and executive arm of the government, sending of memoranda to commissions and panels, the use of radio, television, and the press to raise a discussion on Ayamelum-specific problems, rallies/ demonstrations, and the sending of delegations and lobbying groups. In all these, the most frequently used approach was the petition.

It is important to mention that although the emergence of the AIU, unlike some other peasant organisations, was not initiated by the state, there were attempts by the latter to thwart or stifle its democratic struggles — especially the demand for local autonomy. Thus it was suggested that the state and its agent tried to achieve control of the AIU through blackmail of the leadership and prominent members, material inducements, campaigns of calumny, influencing committee reports, the threat of termination of appointments and non-promotion of such civil servants (Ofuebe 1986: 69).[8]

Conclusion

This study has attempted to ascertain the extent to which a particular peasant organisation — the Ayamelum Improvement Union (AIU) — can be described as an agent of democracy or democratic struggle. From our study it is seen that the AIU had come to represent organised political action arising from the Ayamelum autonomy question. The union had been organised and managed among the Ayamelum clan as an oppressed group within the Anambra state of Nigeria. For a good period of its existence and struggle, the AIU had been articulating the socio-historical grievances of the Ayamelum clan, which ranged from socio-economic marginalisation to political domination. The AIU played a prominent role in the achievement of political autonomy for the Ayamelum clan. It also joined forces with the peasant farmers in their struggle against capitalist farmers and land speculators who appropriated their lands without adequate and in some cases no compensation.

This study shows that the AIU had over the years served as a training ground for future politicians. Many prominent members of the AIU had contested and won elections as legislators at federal, state and local level. It is also important to mention that the AIU was found to be democratic in its internal operation despite the monopoly of leadership positions by educated elites. The study further revealed that class and personal interests often manifested themselves in peasant organisations. There were instances when prominent members of the AIU (including some members of the executive) in pursuit of their personal interests worked against the cause of the union. Some of them undermined the demand for local autonomy and colluded with land speculators and capitalist farmers in dispossessing peasant farmers of their land without compensation.

Despite the potential of the AIU to be an agent of democracy or democratic struggle, it had many shortcomings in the pursuit of its goals. These included among others: financial constraints, poor co-ordination of activities between the various branches and the pursuit of personal aggrandisement to the detriment of the collective good.

Notes

1 She also reported the cases of the dancing women's movement in 1925 and the spirit movement in 1927 as examples of peasant resistance to colonialism in south-eastern Nigeria.

2 The first was pioneered by Redfield (1956), followed by Floster (1967) and Hyden (1980, 1983, 1986, 1987).

3 Despite the limitations of these petitions, they contributed much to checking the excesses of land speculators and capitalist farmers. Many compensation issues were resolved following these petitions.

4 It had a total population of 121,355 people, according to the national census of 1963.

5 The decision to include Ayamelum clan in the Uzo-Uwani division was a sequel to petitions spearheaded by certain notable politicians from the clan who on personal political grounds objected to the transfer of the Ayamelum clan from Nsukka province to Anambra division.

6 A typical example was cited by Ofuebe (1986: 134). On 10 November 1981, when the AIU went to Enugu to protest against the splitting of the clan into two and putting the parts under a different local government, the Enugu branch was said to have not been properly briefed on its responsibilities to ensure the success of the demonstration. This resulted in the failure of the venture as the police had to disperse the Ayamelum demonstrators because, according to the police, no permission had been granted for such a demonstration.

7 The manner or strategies adopted and the level of success vary from one organisation to another and from one regime to another.

8 These threats were rampant between 1979 and 1983 because the civilian governor of the then Anambra state, Chief Jim Ifeanyichukwu Nwobodo, was from the part of the state (Anambra North) that was most opposed to the demand of the Ayamelum people for local autonomy.

References

Akor, J. (1986) 'Row over land grab: Community cries out to Governor Omeruah', *National Concord* 24 September.

Amadiume, I. (1995) 'Gender, political system and social movement: A West African experience', in M. Mandani and E. Wamba dia Wambe (eds.) *African Studies in Social Movements and Democracy,* Dakar, CODESRIA.

Ayamelum Improvement Union (1975) 'Request for Creation of Ayamelum Local Government out of the Present Oyi Local Government Area of Anambra State', Request to Nigerian Head of State, 24 July.

—— (1989) 'A Memorandum by the People of Ayamelum Clan in Uzo-Uwani Local Government Area Re-Affirming Our Time-Honoured Decision to Remain in Uzo-Uwani L.G.A., in Nsukka Senatorial Zone', 7 February.

—— (1996) 'Our Request for the Creation of Ayamelum Local Government Area (LGA) out of the Present Oyi LGA in Anambra State with Headquarters at Ifite-Ogwari', petition to the chairman, States Creation, Local Government and Boundary Adjustment Committees, Abuja, 4 January.

—— (1997) 'Request for the Relocation of the Headquarters of Ayamelum Local Government to Ifite-Ogwari', memorandum to the Head of State of Nigeria, 4 - February.

Ayamelum Leaders of Thought (1993) 'Special Appeal by Ayamelum Clan of Former Uzo-Uwani L.G.A. to Mr. President for the Creation of Ayamelum Clan into a Local Government Area in Enugu State', memorandum to the Nigerian Head of State, 27 April.

Beer, C. (1976). *The Politics of Peasant Groups in Western Nigerian*, Ibadan Univ. Press, Ibadan

Bangura, Y. (1989) 'Democracy and development in the Third World: A conference proposal', prepared for UNRISD, Geneva (mimeo).

Bernstein, H. (1979) 'African peasantries: A theoretical framework, *Journal of Peasant Studies* 8, 4: 430-1.

—— (1990) 'Taking the part of peasants?', in H. Bernstein et al. (eds.), *The Food Question: Profits Versus People?*, London, Earthscan.

Eboh, E. C. (1993) 'Analysis of indigenous systems of group saving and capital formation among Nigerian rural households', Report of the research sponsored by the African Development Foundation, Washington, D. C.

Fuentes, M. and Frank, A. G. (1989) 'Ten theses on social movements', *World Development* 17 (2), pp.179 - 91 .

Hough, R. (1988) 'Peasant organizations in democratic development', in R. M. Goldman and W. A. Douglas (eds.), *Promoting Democracy: Opportunities and Issues,* New York, Praeger Publications.

Hyden, G. (1980) *Beyond Ujamaa in Tanzania: Underdevelopment and an Uncaptured Peasantry,* London, Heinemann.

—— (1983) *No Short-Cuts to Progress: African Development Management in Perspective,* London, Heinemann.

—— (1986) 'Discussion: The anomaly of African peasantry', *Development and Change* 17, 4.

—— (1987) 'Final rejoinder', *Development and Change* 18, 4.

Ibeanu, O. (1993a) 'The deteriorating condition of the Nigerian peasantry', in O. Nnoli (ed.) *Dead-End to Nigerian Development,* Dakar, CODESRIA.

—— (1993b) 'The state and the market: Reflections on Ake's analysis of the state in the periphery', *Africa Development*, XVIII, N° 2, pp. 111-122.

Isaacman, A. (1990), 'Peasants and Rural social protest in Africa', *African Studies Review*, vol. 33, n° 2.

Klein, M. A. (1980) 'Introduction', in M. A. Klein (ed.) *Peasants in Africa,* Beverly Hills, Sage Publications.

Mafeje, A. (1993) 'Peasant organisations in Africa: A potential dialogue between economists and sociologists—some theoretical and methodological observations', *CODESRIA Bulletin,* 1: 14-17.

Mintz, S. W. (1973) 'A note on the definition of peasantries', *Journal of Peasant Studies* 1, 1: 91-3.

Moyo, S. (1995) 'Peasant organisations and research issues on democratisation in Africa', paper presented to the CODESRIA Multinational Working Group Methodology Workshop on Peasant Organisations in Africa, 21-30 November, Dakar.

National Association of Ayamelum Students (1987) 'Request for Creation of Ayamelum Local Government out of the Present Oyi Local Government Area of Anambra State', memorandum to the Head of State of Nigeria.

Ofuebe, C. (1986) 'The politics of relative deprivation and goal attainment: The quest for communal autonomy in Ayamelum Clan', unpublished M.Sc. thesis, Department of Political Science, University of Nigeria, Nsukka.

—— (1992) 'The state, peasants and land acquisition in Nigeria: The Lower Anambra Irrigation Project and the transformation of Ayamelum Clan', unpublished Ph.D. thesis, Department of Political Science, University of Nigeria, Nsukka.

Olukoshi, A.O. (1995) 'Bourgeois social movements and the struggle for democracy in Nigeria: An inquiry into the "Kaduna mafia"', in M. Mamdani and E. Wamba dia Wamba (eds.) *African Studies in Social Movements and Democracy,* Dakar, CODESRIA.

Rahmato, D. (1991) 'Peasant Organisations in Africa: Constraints and Potentials', Dakar, CODESRIA Working Paper 1/1991.

Redfield, R. (1956) *Peasant Society and Culture,* Chicago, University of Chicago Press.

Sibanda, E. A. (1997) 'Voicing a peasant alternative: A study of Rural Association for Progress (ORAP) in Zambia', final seminar report for CODESRIA Multinational Working Group Methodology Workshop on Peasant Organisations in Africa, Dakar.

Smith, C. D. (1989), quoted in K. E. Amin (1997) 'Peasant organisation, participation and the state: Two case studies from rural Sudan', final seminar report for CODESRIA Multinational Working Group Methodology Workshop on Peasant Organisations in Africa, Dakar.

Syahuka-Muhindo, A. (1995), 'The Rwenzuru Movement and the democratic struggle', in M. Mamdani and E. Wamba dia Wamba (eds.) *African Studies in Social Movements and Democracy,* Dakar, CODESRIA.

Williams, G. (1976) 'Primitive accumulation: The way to progress', *Development and Change* 18, 2.

Wolf, E. R. (1955) 'Type of Latin American peasantry: A preliminary discussion', *American Anthropologist* 57, 3.

—— (1966) *Peasants,* Englewood Cliffs, N.J., Prentice-Hall.

4

The Union Tunisienne de l'Agriculture et de la Pêche and Its Role in Strengthening Civil Society in Tunisia

Gilbert Naccache

Introduction

To analyse the l'Union tunisienne de l'agriculture et de la pêche (the Tunisian Union of Agriculture and Fishing), we need to question whether, to most researchers, democracy is not regarded essentially as an urban problem, which is concerned with changing the form of the central power, and which therefore concerns only a few rural actors.[1] A review of contemporary thought on society shows that peasants, who used to occupy an important place in analyses concerned with economic development or the beginning of the industrial era, are now rarely mentioned, and researchers no longer give much attention to them.

Perhaps in spite of being defined in a less restrictive way, democracy is not yet generally regarded as a means of becoming self-reliant, but rather as a transformation in the status of citizens; a transformation that gives them access to a certain number of rights they had earlier been denied, without the use of these rights being made a condition of the transformation.

So far as the present study is concerned, it should be noted that we had the greatest difficulty in obtaining information on the peasants organizations in Tunisia. The reticence of our interlocutors, even those who were most encouraging about our research, was a foretaste of the conclusions we could reach. We are still far from a climate of transparency and democracy in the organized rural world of Tunisia, and it was impossible to carry out the inquiries that had been planned, because of our inability to gain access to those concerned.

Definition of the research project and its context

It is remarkable that peasants organisations are so rarely mentioned in studies on the rural world, even in Hafedh Sethom's remarkable work, *Pouvoir urbain et paysannerie en Tunisie* (1992), which reviews the various ways in which the urban space dominates agriculture, excluding its organisational and political aspects. The bibliography at the end of this work does not even mention the term "organisation".[2]

From the point of view of the demand for democratisation and the emergence of civil society, which is both cause and effect of this democratisation, the question was to know whether farmers were really as far removed from this process as it seemed, and whether this absence was their own fault or was due to more general causes. We had first to inquire into the degree to which the farmers were organised and were active within their organisations.

Secondly, a study of the Union tunisienne de l'agriculture et de la pêche (UTAP), a 'national', that is, official, organisation[3] of peasants (and fishermen) in Tunisia is important in order to determine whether it is capable of playing a role in peasants' participation and mobilisation in the country's politics, or whether, being supported and controlled by the government, its goal is to keep the peasant world within the limits set for it by a central, essentially urban power.

We must be rigorous in drawing an explicit distinction between "farmers" and "peasants". Both of them cultivate the land in order to produce crops. In a country like Tunisia, peasants are to farmers what the craftsmen are to industrialists. Their processes of production are different, and so is the organisation of their production, the place their products have in the market, and the ways in which they are sold. Many modern elements may enter the peasant's life without changing his status, but a time comes when the peasant's entry into the circuits of the market makes him a farmer of some substance. This is the objective, avowed or implicit, of all the policies which governments promote to develop and modernise agriculture.

In Tunisia's case, to belong to the UTAP is in a way an act that may determine access to civil society. The individual producers are no more facing the market alone, or simply following structures directly linked to the market (co-operatives, or marketing boards, etc.) with as members of an organisation which is capable of speaking in their name and reflecting their demands, hopes, and fears.

Research Objectives

The present study is situated within the general problematic concerning the peasant world's ability to achieve autonomy and its degree of organisation within its own structures. In general, we are questioning whether there are any organisations, among those that can exist in the world, that might be capable of changing the common attitudes of farmers, particularly their political stance. In other words, given that the structures of the Tunisian state have long been those of a totalitarian state based on a single party, with all its satellite organisations, particularly in the area of agriculture, is there today— with the assertion that the single-party system is being abandoned— signs of deep transformations in these organizations, or have they remained basically the same, and are they still playing a role similar to the one they played before?

The organisations of the peasant world are often concerned with a precise goal or a particular product: collective interest associations (Associations d'intérêt collectif—AIC), agricultural service co-operatives, the Northern Cereal Producers' Association, the Olive Growers' Association, the Sheep Raisers' Association. There are also vertical organisations, which bring together all those involved in a given area of production (producers, exporters, processors) and professional organisations operating at the regional or national level: the Central Co-operatives, which theoretically bring together the basic agricultural service co-operatives for collecting and marketing certain products, and the Chambers of Agriculture, which under the supervision of the Ministry of Agriculture, have as their role to "provide links between farmers and government offices". The Association of Technicians Recipients of State-owned Lands, created recently, has been complemented by a Federation of Technicians Recipients of Lands, which is a member of the UTAP.

The UTAP, which looks like a union and has the characteristics of a corporate organisation, is the subject of the present study. Its brings together farmers on the basis of their locality, the base level unions. These are then united together in local unions, regional unions, and national federations, but still on a sectoral basis. We also need to consider the co-operatives and associations, which collectively belong to the UTAP. These are an important source of financing for the UTAP headquarters. The Federation of

Associations for Improvement and Development (La Fédération des sociétés de mise en valeur et de développement, SMVDA), a new form of agricultural organisation that has been recently recognised, and the Federation of Women Farmers, which also belong to the UTAP, differ completely in their nature and goals from the other federations. In addition, there is a federation of associations guaranteeing agricultural credits (these associations exist on the national and local levels). All this makes the UTAP a complex organisation that includes very different and sometimes competing social categories.

To begin with, we attempted to make comparisons, by seeking information on the role elsewhere of organisations similar to UTAP. But in the literature we were able to consult, we found nothing that might shed light on these aspects of the problem. We therefore abandoned the attempt to make such comparisons.

We were also unable to work out the number of farmers by production sector or by region of the country, nor attempt any sociological description of them, because of our lack of access to documents and to UTAP's archives. We had planned a history of UTAP's relationships with the government and the party in power, to determine how far possible reservations might have been expressed. We also raised the question whether UTAP's operations could lead towards a better or broader representation of farmers, or if one should consider this organisation, from the point of view of the emergence of civil society, to be an obstacle in the way of the development of other forms of rural organisations.

This question is linked to another, more general question: can the UTAP's functioning be democratic and accelerate democratisation in the countryside before the end of the single party regime? As a corollary, is it conceivable that if the single party, and its theoretical base, were to disappear completely, the UTAP might survive and transform itself into an autonomous farmers' organisation? This question, in fact, goes far beyond the simple case of one organisation, and concerns the general process of the transition from the government of "development" to a democratic state based on the rule of law.

In this study, we have ignored issues specific to fishermen who are also covered by the UTAP, but who raise special problems that do not concern our main subject.

The relevant literature and the concepts used

Bibliographical references

While there have been no analyses bearing directly or indirectly on the UTAP or on its relations with other expressions of representation or power in Tunisia, the context in which this organisation developed has been described in a number of general studies on the evolution of politics in Tunisia, on Tunisian agriculture, or social relations, which have been particularly analysed in relation to various crises.

The literature also offers ways of defining the role and place that professional peasant organisations can play, but these generally relate to democratic countries such as France, or to organisations that play a role in the revolutionary mobilisation of the peasantry (China, Latin America). In the appendix, we have reproduced extracts from a study that we have carried out that may clarify the place and role of the single party in "developing" states. However, some analysis remains to be done, and so it cannot be regarded as in any way complete or exhaustive.

The concepts used and the plan proposed

We must first explain the meaning of our analytical concepts. To do this, we offer brief definitions of the terms "democracy", "democratization", 'civil society', and "empowerment", putting the emphasis on the Gramscian approach to civil society.

Democracy, which has been so broadly defined that its meaning seems obvious, is first of all an active and conscious participation by all citizens in decisions affecting their lives in common, or 'politics', in the etymological sense of the term, or in other words, the life of the city. Without going into detail concerning the way it functions concretely in different democratic systems, we observe that it implies a certain number of public and private freedoms, chiefly the freedoms of thought, expression and organisation, and freedom to have correct information, and to participate in decision-making.

It is easy to contrast democracy with dictatorship—the various ways in which systems vesting absolute power in a single person, a family, or a clan, reject the exercise of democracy—in order to illuminate democracy further, and the process that may lead, in a continuous or discontinuous, or in a total or partial way, to the transition from one to the other, to democratisation. This latter clearly consists in according or allowing a partial enjoyment of a certain number of public freedoms that did not previously exist. This is the act through which dictatorship recognises itself for what it is, by beginning to deny itself and end up ceding some of its prerogatives to others. Democratisation implies the clear lack of complete democracy. If the term seems to indicate a process, it also implies its own limits. Democratisation of public life does not conclude with democracy, but with the end of overt dictatorships, something that is slightly different.

Civil society is sometimes defined as the part of society that does not exercise state power—a sort of "non-governmental society"—and sometimes as the part that does not participate directly in political life—thus "civil society" is opposed to "political society"—or again as the place where citizens express themselves, in an organised way about the problems they are experiencing. However "civil society" may be defined, the notion of organisation is particularly important. Civil society only exists in relation to ways of organising citizens, forms of organisation that are not those of the family or of production. In other words, modern civil society emerges when citizens organise themselves and step back to consider their everyday acts. It presupposes something additional on the part of the citizen, concerns that go beyond his "raw" status as a member of a family or a profession. According to Gramsci, it is the place where state power, conceived as legal-military-police power over society, becomes an ideological hegemony. In organising themselves into a civil society, the citizens adopt the ideological, moral, religious principles, etc., that dominate society, and contribute to the continuation of the social organisation. Through their organised action, citizens make the functioning of society less contradictory and reduce a certain number of abuses. They "moralise" its functioning, as it were, but do not challenge its class structure. Civil society goes along with democracy. When one is strong, the other is strengthened as well.

The process of empowerment, which means access to power, necessarily proceeds from societies in which democracy is weak or non-existent, or concerns populations, strata, or segments that do not have genuine access to democracy. The path of empowerment may involve setting up a genuine civil society, which incorporates within itself not only part of the power, but above all the ideology that gives the overall power its form, in the framework of a democratic system, that is, a system in which the majority of the citizens approve, directly or indirectly, the decisions that affect them.

We will not pursue this line of thought any further, apart from saying that it has allowed us to understand better the necessity of considering peasants organisations not only as structures for organising and acting in the rural world, but also as vehicles for consciousness, and not only for demands. They are among the places where the assertion of this consciousness and the process of self-determination are revealed, with all the hesitations, contradictions, and conflicts that have to be interpreted in order to describe a part of the real dynamics of the world.

The approach includes:

- a brief history of Tunisian agriculture and the UTAP, in order to understand some of the working conditions and limits of this organisation;
- consideration of aspects of the organisation;
- its reactions to the various crises in agriculture;
- statutory and regulatory aspects;
- the meaning of the creation of new federations;
- the current content of the UTAP;
- how the organisation is financed.

I - The evolution of Tunisian agriculture

1.1 - The situation at the time of independence

At the end of the colonial period, two kinds of agriculture existed side by side in Tunisia, one oriented toward the market that used modern techniques (colonists and large Tunisian landowners), and traditional agriculture, untouched by the market, and based on small farms, whose purpose was to provide a subsistence that had to be eked out by paid labour on the large modern farms. In the central and southern parts of the country, there were immense areas of pastureland, exploited extensively as "collective lands" by tribes or groups of tribes.

The new nation-state immediately attempted to "modernise" this agriculture, to give it a clear legal status under state control. This was supposed to make it possible to develop it, to increase its production, to alleviate the problems caused by increased population, and to enable it to penetrate the national market, while at the same time opening the whole of the rural world to the control of the state over individuals.

1.2 - The development of the government's agricultural lands

The state soon found itself in charge of a very large area of land, as a result of political measures aimed at breaking the power of the religious congregations, which managed large holdings of buildings and land. These measures involved the suppression of *habous*[4] (31 May 1956, or about 40 days after the declaration of independence), and the power of the bey,[5] who was overthrown by the proclamation of the Republic (25 July 1957), the nationalisation of the bey's property and the recovery of former government lands, and also the repurchase of some of the colonists' farms, in conformity with the independence agreements (some of these lands had been sold directly to Tunisian owners, but most were the object of public purchase).

In order to manage all this land, several offices charged with developing unused or poorly used lands were set up between 1956 and 1958. Other development offices undertook public works, which enabled labour to be employed on "unemployment projects". The Office of Government Lands confined itself to the direct cultivation or the renting to farmers of rich lands bought back from colonists or confiscated from the ex-bey.

On 12 May 1964, the Tunisian government proceeded to nationalise, directly or through state enterprises, all agricultural land still held by foreigners.. Farming by private companies was banned. Only individual farmers could own land.

The state's holding was increased by vast areas, including some of the most productive land.

1.3 - The "socialisation" of the countryside

In September 1962, the creation of "co-operative units of production" began, starting with the area of large-scale farms in the northern part of the country. To a "kernel" of state lands, roughly the same amount was added from the peasants' lands surrounding it. For each parcel thus added, one took a single claimant. The number of "co-operators" thus chosen per farm (and thus the total area) was determined by calculating the potential revenue: the goal was to realise an annual revenue of 50 dinars[6] for each co-operator. The general objective was to complete the destruction of pre-capitalist structures in the countryside and to integrate the rural world as a whole into the national capitalist market.

This process was extended to the zones of family tree-growing in Cap Bon and the Sahel, which led to resistance and disturbances, particularly in the Sahel, a fief of the central power.

The state's lands had been limited, and the nationalisation of lands in 1964 came at the right time to increase the area for the operation of establishing co-operatives. As early as 1966, it was clear that this would not be enough, and that sooner or later it would be necessary to include private property[7] in order to pursue the operation fully. A state of conflict with large landowners, most of whom had become farmers for fear of seeing their lands confiscated[8] in the name of the principle that "property is a social function", soon progressed from the stage of quiet grumbling to protest and confrontation.

The decision in 1969 to extend the co-operative system to all land also involved the takeover of all the sheep and cattle herds. This led to a general revolt among the peasantry, carried out everywhere by the systematic slaughtering of livestock, for sale at very low prices. Numerous incidents culminated in the bloody repression of peasants in Ouerdanine, a village famous for its devotion to Bourguiba. In September 1969, Bourguiba accepted that this policy had failed, and put an end to it.

1.4 - The results of the "socialist" period

After the abandonment of "socialism", came an era of trade liberalisation, especially trade in the principal agricultural products, which had been handled by state monopolies during the preceding period. The selling and buying of land was liberalised, and the renting of state lands to private farmers was authorised—a return to the procedure of renting out their land to small peasants.[9] The large landowners were naturally in the best position to benefit from this new policy, and also profited both from direct access to markets and from measures taken by the state to "encourage" agriculture.

The end of the "socialist" period coincided with the final completion of the capitalist transformation in the countryside, still within the framework of a single-party system.

The model of economic growth pursued from 1970 onwards is a model of "extraverted accumulation" based on exports and foreign financial aid, and especially on a large levy on the agricultural surplus. Ben Romdhane[10] has shown that for 1975, this levy amounted to 16.2% of the agricultural surplus, 15.2% of the country's total investment, and 130% of net agricultural investment. This model of accumulation has since accentuated its extraverted character. It can be said that the process has generally continued, and that the transfer of wealth from agriculture has persisted, even if the transfers were not the same for all regions and products. The large quantities of agricultural products imported made it possible to control on prices.

One might ask whether every category of farmer made a similar contribution to these transfers. This seems unlikely: small peasants made bigger contributions than their wealthier neighbours, who could obtain all the advantages, subsidies, loans at low interest rates, and the possibility to make reimbursements by instalments, which enabled them to amortise their loss of earnings.

We can thus understand that these large farmers play a role in the structures representing agriculture and in all the accompanying service organisations—co-operatives and others. For these reasons, they could more easily deal with state power and defend themselves by diversifying their activities (investments in commerce, transportation, the distribution and even the processing of products, and, of course, in urban real estate), without having to go in for protest campaigns.

1.5 - Policies on government lands

During the 1990s, the state decided to broaden the renting out of government land (and even transfer of ownership for technicians) in particular to firms engaged in improving and developing agricultural production. These firms were supposed to pump large quantities of capital into the sector and accelerate its modernisation. They were open to foreign capital, on condition that investment was in collaboration with Tunisians. They were supposed to be able to facilitate exports and bring foreign capital into other sectors as well.

The "levelling" of agriculture, to use the term currently fashionable in Tunisia, is carried out by these firms and, to a lesser degree, by the technicians who have been granted state land, and who may in future be in charge of the sector and its organisations. But will these new farmers bring in modernity and new techniques at the risk of spending a great deal of time waiting for results, or will they adopt the methods of farmers in the area—the largest farmers, naturally—and simply continue the present system?

II - The history of the UTAP

2.1 - The situation at independence

During the years between the achievement of internal autonomy and the declaration of independence (1954-1956), Tunisia was torn apart by a conflict that almost took on the character of a civil war. The Youssefist crisis, as it was known, was between the supporters of Bourguiba, the President of the Neo-Destourian Party, and those of Salah Ben Youssef, the Secretary General of the same party. It ended with Bourguiba's victory, largely because of the active support of the larger trade union group, UGTT (Union générale des travailleurs tunisiens).

The General Union of Tunisian Agriculture (Union générale de l'agriculture tunisienne, UGAT) was created in 1949 by farmers who were members of the nationalist political parties.[11] It included the large Tunisian farmers, who were close to the French colonists in their methods of production as well as in the markets for their products. UGAT supported Ben Youssef in the conflict with Bourguiba. This led the latter's supporters to create a new organisation, the Union

Nationale des Agriculteurs de Tunisie—UNAT (Tunisian Farmers Union) headed by Tahar Ben Ammar, a former prime minister and large landowner. This group included militant nationalists who were not all farmers, but it quickly won over most of UGAT's members and led to the disappearance of UGAT. According to Clement Moore,[12] UGAT, henceforth composed mainly of large landowners, did not feel capable of standing up to an alliance between Destour and the UGTT (for fear of strikes that the union might incite among their own employees!), gave up the fight and agreed to remain silent.

2.2 - Relations between the farmers' organisation and the government

The Neo-Destourian Party, which later became the Destourian Socialist Party, was created by and for the party in power. It went without saying that UNAT had to uphold it and support its policies at all times. In return, it received substantial rewards. We will discuss later on the period when it proved impossible to continue this support as the interests of the farmers, which the labour confederation somehow had to express, had become completely opposed to the policy implemented by the Minister of Planning and the National Economy, a former union leader who wanted to collectivise the countryside.

That is why, except during this period, and until 1987,[13] the relations between the farmers' organisation and the government were generally carried on within the framework of relations with "national organisations,"[14] based on the theory of National Union, according to which all organisations had to support the government that was working to develop the country. These relations had the following notable aspects:

The leaders of the agricultural organisation were in the decision-making bodies of the party in power (the political bureau and the central committee) and on the Destourian electoral banner (and therefore in the country's elected institutions), as well as in the institutions that consulted and defined the main political lines (for example, the Conseil supérieur du pacte national de 1989), which theoretically leads to adherence to the government's decisions and complete solidarity with all official goals.

- Government intervention in the constitution of the leadership organs of the organisation, in particular in the choice of a president from outside the world of productive agriculture, with another consequence on its level of functioning: the leadership nominated the officials of the sector federations. Elections in the other structures were essentially formalities (the future leaders were "recommended" to the delegates). See, in the appendix, the list of the presidents of the UNAT, then the

UNA, and finally the UTAP, which shows to what extent they were government personnel.

- Changes in name show that the confederation was transforming its role, its attributions, and its politics, at the same time as the country's politics were undergoing an important change. Firstly, after the collectivist period, there was a "constituent assembly" of a farmers' organisation called UNA in 1970: this was, theoretically, to designate farmers as subjects. The necessity of "creating" a "new" union probably meant accepting that the former UNAT was incapable of defending farmers against "co-operatives," but most of the former leaders remained in the supervisory structures, especially at the local and regional levels.

  Secondly at the congress in May 1990, under the framework of adapting to the "change of 7 November 1987," there was another change of name: the Union Nationale des Agriculteurs (National Farmer's Union) became the Union Tunisienne de l'Agriculture et de la Pêche.

2.3 - Positions taken by the UNA during crises (1961-1969)

After expressing reservations in 1961, at the beginning of "socialism,"[15] the UNAT fell into complete public silence. In 1965, it agreed to create a federation of cooperators. This proved ephemeral, and was seen as being dissolved de facto in 1967, with the constituent assembly of the National Union of Cooperation (Union Nationale de la Coopération).[16] No one doubted the hostility of the large farmers towards the way the country was developing. Certain rumours accused them, though without any proof, of being behind some of the peasant resistance.

After September 1969, the Political Bureau of the PSD decided to "create" the "National Farmers' Union" ("Union Nationale des Agriculteurs," UNA), and put Moamed Bellamine, who had been a Governor (a post equivalent to that of a regional prefect) in charge of its general secretariat. The Union's provisional committee was also named by the PSD.

In their delight at recovering their land, and apparently their power, the farmers started to demand for state support to a greater extent than ever before. The new Prime Minister, Hédi Nouira, put an end to that by requiring farmers to count more on themselves and help their more deprived fellows.[17] The Union kept quiet, and returned to its role as a supporter of government policies.

In 1977, with its new president, M. Mohamed Ghedira, who had been Minister of Agriculture, the Union strengthened its technical potential by creating a research office. It challenged indirectly and discreetly—its report remained confidential—the government's official conclusions regarding agricultural costs and prices. The research office (a structure conceived chiefly to "train the UNA leaders", according to former members of the organisation, by giving them an overview of the problems) was directed from 1981 on by M. Adelbaki Bacha, who first became Governor and then returned to the UTAP as President.

In the other crises that afflicted the country, the UNA provided unfailing support for the government and its decisions, although it maintained certain demands, the most pressing of which always concerned privatising state lands by sale or cession to private farmers.

This trend was later on confirmed by the extension of the process of renting certain state farms to agricultural technicians, and then by the authorisation, in 1993, of the creation of agricultural development firms, which could rent large state farms by presenting dossiers to an official commission, where political support may play a significant role.[18]

Nevertheless, the setting up of Chambers of Agriculture in 1988 brought about a test of strength. The government decided to create three Chambers of Agriculture to cover the entire country. The definition of the scope of these chambers antagonised the UNA, which saw in them a restriction of its own sphere. The conflict sometimes took on the air of a confrontation between the representatives of organised farmers and agents of local authorities. A compromise was found which allowed UNA headquarters to supervise these chambers, which were, moreover, poorly funded.[19]

2.4 - The Evolution of UTAP

The Tenth Congress of 1990[20]

It was ten and a half years after the "change of 7 November 1987" before the farmer's organisation came in line with "current tastes" and made democracy its official mode of operation. Only official, however, for all the measures announced in advance were adopted without change by the congress, and its president Taoufik Essid, who had been appointed under Bourguiba, in 1966, was re-appointed to his office by the "perestroika conference".

The change in the organisation's name,[21] which was henceforth to be called Union Tunisienne de l'Agriculture et de la Pêche ("Tunisian Union of Agriculture and Fishing"), indicated that it would include fishermen in its structures, in order to have "a wider scope." This also allowed it to legalise the membership in the central union of co-operatives and associations of producers and service providers (it became a Union of Agriculture, and no longer of Farmers).

The next measure concerned the extension of the prerogatives of sectorial federations, which were only consultative. Henceforth, they were represented in the leadership organs and were able to elect their officials at all levels, whereas they had previously been appointed. While this measure seems to provide a certain democratisation, it must be added that the broadening of these federations' powers does not seem to have been truly significant: the sectorial structures included members who were already organised in other federations (in which appointments were made by election). There was every chance that interventions by the sector federations would be reflected and supported by these other federations (which explains their consultative character). In any case, the representatives of these sector federations closely resembled those of other federations, for reasons relating to their real functions (which we shall discuss a little later on)—the importance of clientelism, and the favours that could be bestowed by the leadership.

The most important change was the creation of a Central Council for the organisation, which was intended to replace the Central Council and the administrative committee. The latter, elected by the congress, managed the headquarters, and from time to time called the Central Council, made up from presidents of regional unions and secretaries general of national federations, into session, in order to secure its approval of important decisions. One can see that there were two decision-making levels, one of which would allow the representatives in closest contact with the grass roots to have a voice in, or at least to exert an influence on, the final decisions, after having consulted their members. Henceforth, the Central Council had 50 members, including the presidents of regional unions and the secretaries of national federations, as well the president of the consultative committee (consisting of former presidents of the UNA). There was only a single decision-making authority. Grassroots representatives were involved in these decisions and no longer simply had to defend them to their members. Formal democracy seemed to be limited in this change, which was more like a strengthening of political control over the organisation.

Another measure, presented as a way of preventing the dispersal of effort, increased from 50 to 100 the minimum number of members necessary to create a union. This decision was to "to avoid inflating structures and to give priority to efficiency". The increase in the number of members required to form a union, as we know, was more intended to prevent the uncontrolled creation of new unions and to make more difficult the organised expression of demands not approved by the leaders. It was easy to assemble 100 persons when one had power, or even to add to that hundred (in order to ensure a majority), whereas it was all much more difficult when one was challenging individuals or methods.

In order to give an idea of the transformation of the role that this new leadership of the UTAP assigned for itself, we have only to quote the beginning of the congress's general motion: "Conscious of the role of farmers in realising the government's objectives to promote agriculture and to achieve a secure food supply..."

We can see that the concept of democracy and the role of civil society is hardly an original one.

III - The Organisation

3.1 - The members of the UTAP: an omnibus organisation

The statute, published after the eleventh congress in May 1995, is quite clear about the Union's objectives and its aim not to let any sector remotely connected with agriculture or fishing remain outside its structures.

As a "national trades union and professional" organisation, the UTAP accepts as members

> all those who are involved in agricultural activities, whether on land or on water, whether these an individual or legal entity, that is, private producers, whether landowners or not, development companies, family companies, producer co-operatives, service co-operatives, or others:
> - agricultural service co-operatives of every kind;
> - olive growers associations;
> - irrigation water users associations;
> - agricultural mutual insurance associations;
> - farmers professional groups;
> - fishmongers; and
> - generally, any organisation established by those involved in agriculture, whether on land or on sea; in the same way, the children of those involved in agriculture and their auxiliaries, on the condition that they are involved in agricultural activities.

The UTAP also has a "Fédération Nationale des Agricultrices" (National Federation of Women Farmers), with all the associated regional and local structures. The creation of this women's organisation met the recommendations of international organisations, and complemented the launching of the Chambre des Femmes Chefs d'Entreprise (the Chamber of Women Business Executives), a structure of the same kind within the employers industrial and trade organisation, UTICA. A Ministry of the Family and the Promotion of Women was formed around the same period, and within all the Ministries a special office for dealing with women's problems was created. Other official structures were established in addition, such as CREDIF, the Centre de Recherche, de Documentation et d'Information sur la Femme (Centre for Research, Documentation, and Information on

Women). Similar structures were created within political parties and other organisations.

The UTAP also provides training and educational activities for women involved in agriculture, and in particular organises conferences, excursions to research centres, and training sessions in specific topics in agriculture. It supervises certain rural development projects, many of which are intended to "improve the lot of the rural woman" by giving her a couple of milk cows or ewes (and a voice in the distribution of government grants or subsidies), but up to now, it has never really been active on other levels, or developed a specific approach to the problems of rural women.

The UTAP's desire to include absolutely everything within its sphere of action might be connected with the way this organisation came into being. It was conceived, run and still runs as a professional arm of the party in power. Its most important function is to maintain the most complete control possible over the country, by leaving no room for any rival organisation. In return, the organisation's executives, whether they were originally involved in agriculture or not (an increasing number of technicians are being brought into the Union apparatus), enjoy all the advantages connected with participating in the structures of power, moral advantages, but also direct financial benefits. And given that it is stipulated that belonging to the Union implies in particular that one is "committed to following all its decisions" (Article 5 of the statutes), it is easy to silence any possible challengers.

3.2 - The aims of the UTAP

The thirteen aims defined by the UTAP's statute cover activities of very different kinds, from the defence and promotion of farmers to the organisation of the rural world through the intermediary of structures of production or of service, and the initiation of projects for production or for participating in the management of operations of general interest in relation to the sector, such as irrigation water users and olive growers' associations, etc. (see appendix III)

3.3 - The running of the UTAP

The UTAP has an extremely centralised structure, the very model of a pyramidal structure, dominated at the top by the president, assisted by a consultative council composed of former central presidents of the Union, and presided over by Mohamed Ghedira (the most official of them all), and by the Central Council.

The structures function like those of any centralised organisation. One difference results from the fact that since many professional organs and associations are members of the UTAP, arrangements have been made for them to be represented according to their presence at a given level, local or regional, at the time of the congress on the structures involved. For example, if in a locality (a delegation, equivalent to a sub-prefecture), there is irrigation water users' association, it will be represented by a delegate to the conference of the local UTAP union, which is theoretically held every three years. The same applies at the regional and national levels.

Internal democracy is very relative. Centralisation takes priority over it. Thus, for example, the local congress "is obliged to make decisions in accord with the aims defined by the Union's executive bureau at the national level."

Candidacies for the various elective offices of the local or regional unions must be submitted a month before the congress meets, and published and accessible to all. This provision, which might be understandable for the national conference, seems to indicate a desire at the top to control candidacies in these intermediate structures, and possibly to promote others, if a given candidate's profile does not seem satisfactory.

Another point is rather puzzling: the varying periodicity of the congresses of grassroots unions, the local unions, the regional unions, and the national congress, which occur every two, three, four and five years, respectively. The delegates to the local union conference are "the members of the current executive committee, the committees of the grassroots unions (between 7 and 11 persons per union, depending on the number of members, according to article 11 of the statutes), one representative for each local sector structure, one representative elected by the grassroots unions for each 50 members of these unions, one representative for the professional structures..." One sees that it is highly likely that most of the members of this congress will have been selected more than a year ago, which also leaves a lot a room for intervention from the top of the Union, in case of disagreement. Things work the same way in the congresses of the regional unions, given that the representatives of the local unions are delegates to these congresses, as is the former executive committee of the regional union.

The national congress is composed somewhat differently. Apart from the former central council (composed in part of presidents of federations elected one or more years before the congress), it consists of representatives of the grassroots unions and the local sector federations, elected according to a scale determined by the Executive Bureau before the convocation of the congress. The omnipotence of the Executive Bureau is combined with very great influence on the Consultative Council, whose members have direct access to the structures of national political power.

We have been able to obtain only a few figures from the UTAP about the organisation:

- In 1993, there were 1,590 grassroots unions or one union per *imada* or locality. The number of 100 members per union seems to be a rather low average. It would give about 160,000 members. Since there are about 325,000 agricultural units, and several thousand fishermen, a membership percentage of less than 50%, seems an underestimation. We will nevertheless use this figure, for lack of more reliable statistics, and because keeping up membership of the UTAP is not done systematically by all farmers, and they do not necessarily renew it every year.
- The officers of these grassroots level unions numbered 14,310, that is, 9 per union. We have some idea about the level of education of these officers: 68.18% have an elementary school education, 28.4% a high-school education, and 4.64% have attended an institution of higher learning. They tend to be young: 32% are less than 40 years old, 31% between 40 and 50, while those over 50 constitute 39%, of which only 13.5% are over 60.

There were 206 local unions, with 1,850 officials or nearly 9 per union.

Unfortunately, it is very difficult to talk about how well represented are the different categories of peasants, since it is not possible to determine very precisely the composition and geographical origin of the mass membership or the leadership of the UTAP.

3.4 The UTAP's main activities

In addition to its activities of popularising its decisions and proposals, and of raising the consciousness of members about particular measures —in a word its activities devoted to training the rural world—the UTAP either gives direct support or sponsorship to producers in their dealings with the administration, or plays an indirect role by participating in official bodies concerned with supplying the countryside or in marketing its products. It can, if necessary, give an expert view on government projects and provide technical assistance to farmers. The federation of institutions guaranteeing agricultural credits also provides members of the Union with a number of financial services. All these services have, of course, to be paid for.

For many activities—authorisations for importing equipment or exporting products, obtaining bank loans, acquiring tax-free products for farmers and fishermen, etc. —the UTAP's role overlaps with that of the administration. It is authorised to provide attestations that a given claimant is a farmer or a fisherman. This activity helps to promote recruitment, since attestations are given only to those who are members. It must be emphasised that all these attestations must be paid for.

The UTAP is a member of the supervisory boards of many enterprises connected with agricultural and para-agricultural activities, as well as of the committees of certain co-operatives and marketing organisations. It is thus able to receive part of these institutions' profits, dividends, or rebates. It also has a say in the appointment of officials. Such appointments are not made at random, and are part of the power nexus of political and other deals. This gives its central organisation a rather different complexion from that of a simple trades union or corporate body.

In addition, the co-operatives or service enterprises created by the UTAP, or those which belong to it, also play an important role in financing the Union and in its activities. This provides more opportunities to find paid administrative jobs for members chosen by the most influential leaders.

These activities develop a clientelism, which enables the leaders to gain support at elections, for renewing appointments and even for services rendered through networks of power. Above all, they make possible the spread of a network of privileged people, not just those directly employed by the UTAP and its associated organisations, but especially the members of the union apparatus (15 for every 1,000 members), who are appointed, appoint or arrange for officers to be appointed in institutions where there are easy opportunities to make some money.

3.5 - The financing of the central organisation

The chief sources of finance are membership dues, gifts, payments for services directly rendered by the organisation, and income from activities controlled by the UTAP. Membership dues, according to various estimates, cover about 160,000 to 170,000 individuals and 2,000 to 2,500 bodies. This would produce an income of about 3,600,000 dinars per year.

One cannot give a figure for the gifts made by members, usually legal entities, but they are probably about as large as the membership dues, and they may have made a big contribution in the UTAP's role in the "26.26". Post-office account affair.[22]

When considering the services provided by the UTAP, we have to remember that every farmer who has dealings with a bank or with the administration has to go through the central organisation to get the attestations that he needs. Some subventions and government grants are directly administered by the UTAP. All these documents have to be paid for. They cost a minimum of 12 to 25 dinars apiece. It can be estimated that the central organisation issues between 150,000 and 200,000 documents per year, which would yield a minimal annual revenue of about 3,500,000 dinars. If we also count the rebates given by member co-operatives, its share of the profits of the service enterprises in which it holds shares, etc., we can understand that this organisation has ample resources to support its administration, pay its rent and construction loans, and even, if necessary, to provide its officers with important bonuses. Even more so, when we remember the Union's direct wealth conceals its "indirect" wealth, its ability to obtain advantages, privileges and even financial benefits for those under its protective wing.

This apparent financial autonomy nevertheless depends on its ability to maintain good relations with the representatives of the government. This is a further reason to doubt whether its policies and decision-making are completely autonomous.

3.6 - The organisation's press

The Farmers Union has published two periodicals in Arabic: *El Umma* (*The Nation*), and *El Fellah* (*The Peasant*). These periodicals come out very irregularly, with apparently small circulations (the number of copies printed was not mentioned in the magazine), and do not seem to have been concerned primarily with farmers' real conditions: the few issues we were able to find[23] set out political points of view supporting the government, some of them not related to agriculture. Cursory "inquiries" into a region in the country, one or two articles on agricultural awareness, and some "information of general interest" make up the rest of these numbers.

IV - The Single Party and National Organisations: the UTAP's and UGTT's and Degree of Autonomy

4.1 - Reminders about the single party[24]

One party regimes, *de facto* and *de jure*, no matter what label they attach to themselves, are based on the double hegemony of an all-powerful centralising state, which owns most of the country's means of production, and that of a single party, which reinforces and refines the domination of the population exercised by the higher levels of the state apparatus. .

In such societies, the single party—a deviation historically founded in the then USSR on the theory of the dictatorship of the proletariat—

monopolises the institutions of the state and shapes them in such a way that they can function only when they lean, so to speak, on this single party. This gives the party a legitimacy and a depth which it would not otherwise have.

Reinforcing, complementing, making profitable, and stimulating the action of the single party, "national organisations" (sometimes "mass national organisations" or with other names) carry to the point of caricature the party's main functions. These are mainly those that have to do with recruiting personnel and raising funds to remunerate this personnel, in order to "motivate" in a more substantial way those whose position allows the party to be everywhere, and to know at any time who are its "friends" and "enemies".

These organisations can have a certain degree of freedom, an apparent separation from the government or party, but they are not allowed to set their own goals. In return, no notice is taken of corruption, misappropriation of government funds, abuses of power, and all the other forms of denying justice, of which their leaders may be guilty, thanks in particular to their economic role in "co-operatives," which depend on these organisations or where senior appointments can be obtained only with their support.

That is why the myth of the necessary comprehensiveness of each of these organisations is developed and accepted even by those who are just ordinary members of them.

The single party is not a simple political party that happens to be in power at the moment. It is one of the basic mechanisms of domination by the group that controls society. To paraphrase Gramsci, in these societies where civil society does not exist, the state is the political-legal-military apparatus and the party is the apparatus of hegemony over society. This hegemony requires not only ideological action, but also a decisive influence over the everyday life of the people. The question of democracy and political pluralism very quickly raises the question of how to maintain this form of political domination by the ruling groups, and whether the single party should disappear or survive.

4.2 - The reactions of the two national organisations in times of crisis: The UGTT and the UTAP

"National organisations" in Tunisia were supposed to embrace and represent the various social categories. Two of the latter, the working class (and more generally employees) and farmers, have often complained about government policies. We can legitimately ask why the reactions of their organisations have been so different.

4.2.1 - The workers' union and state power

Let us first recall the facts regarding the trades union congress. At its congress in 1956, it adopted a programme judged to be unacceptable by the party in power. The party organised a split in the congress, by putting pressure on those union leaders who were closest to the party. The authors of the programme, for the most part teachers, had in fact outlined a programme for creating and consolidating a future "state bourgeoisie," as the members of GEAST[25] (who had published the journal *Perspectives* and had been severely repressed since 1968[26]) emphasised in 1972. The existence of recognised and relatively independent trades unionists, reflected in their criticism of the fraction of the leadership composed of 'civil servants',"[27] allowed the eviction of all those who had aspired to political power.[28] Later on, there was a reunification on the basis of a purely trade union programme, but still.[29] The government intervened on several occasions, to help one group to overcome another, as soon as the latter seemed to be demanding too much. But the crisis of 1964-1965 ended up unifying previously feuding trades unionists in an attempt to save their union, which was threatened by decisions made by the Destourian party,[30] under the influence of Ben Salah, the undisputed ideologist of "Destourian socialism": the creation of party "professional cells" in which the UGTT's structures were to be merged. Confronted by this threat to the organisation, and putting themselves forward for the first time in years as the defenders of the material interests of the working class, the leaders of the UGTT began to resist fiercely, supported by the CISL, of which Ahmed Tlili[31] was the vice-president. The government used a minor accident on a ferry boat owned by a development firm, of which Habib Achour was president, as a pretext for arresting and sentencing him, for seizing all the property and businesses belonging to the UGTT, and for bringing in a new secretary-general, Béchir Bellagha, who was then Governor of Jendouba. In silence, but without leaving the organisation, most trades unionists waited to see how things developed.

A third and still more serious crisis occurred in 1977-78. This led to a general strike and street demonstrations, in the course of which the army fired on the demonstrators, killing several hundred of them (the exact number of those killed was never known. According to different sources, it was from 120 to more than 600). There were thousands of arrests, followed by torture in custody and imprisonment. The leaders, headed by Habib Achour, attracted prison sentences of dozens of years. This conflict once again bore on the role of the union in political decision-making. The leaders had moved toward a kind of labour politics, which remained ill-defined, but called for the workers not to be once again the victims of a frenzied race for riches, tainted by scandal and corruption, which was beginning to be seen among those close to the centres of power.

It was noted that even if "socialism" had been abandoned in favour of liberalism, the single party continued to dominate political life, whether through the "state bureaucracy," which continued to hold a monopoly on power, or through the lack or weakness of middle classes who depended on market forces and on being at least tolerated by the state. The assertion of a shift to liberalism was thus not enough.

In this crisis, revealed by the events of 26 January 1978, the UGTT's bureaucracy, weakened as it had been after the 1965 crisis, could not make a deal with the government. In 1965, it had been amply shown that only leaders accepted by the workers could remain at the head of the union. Attempts to give some kind of legitimacy to the puppets installed after the 1978 crisis were a miserable failure.

4.2.2 - The farmers' organisation and the crises

We have dwelt briefly on the crises in relations between the state and the UGTT. However, we should also note that relations with the farmers' organisation never took on such a serious character, and never developed into an open public conflict.

The sole case worthy of note was the organisation's complete public silence, during the period of "co-operatives", and putting the issue on the back burner after its two public statements in October and November 1961, which we have already discussed, expressing reservations about the projected planned economy. Union officials only communicated their fears indirectly to the government. Most of the time, the "problems" with the union can be discerned in certain decisions made by the government (visits made to certain rural Governors' offices, increases in the prices of certain agricultural products, encouragement of certain forms of production and certain types of farming, etc.) or in the changes in leadership personnel, which did not follow congresses but general assemblies of the leadership organs. But in these cases, it was a matter of nominating people from the administration to replace other nominated people.

More than expressing grass roots demands, and because of the various advantages they derived from belonging to the supervisory structures, the UNA's leaders were in bed with one or another faction within the party in power. Any changes of direction that might have occurred, always occurred in a discreet manner, and were more a function of power relationships within the party in power than of any needs expressed by the farmers. Moreover, many important farmers were also members of this party, and played a considerable role as leading party members and also as informants providing the leadership with information on the situation in their area.

4.2.3 - Consequences

Given differing attitudes, one might think that the way in which their officials were appointed, and therefore the degree of autonomy enjoyed by the two organisations largely explains the differences between them during the crises. The UGTT and its leaders were not produced by a decision of the Destour Party. They had a separate existence from that of the party, and a reality connected with the working class's consciousness of having to defend its interests. They always showed that they could not be replaced. Efforts to do so always had to be given up in the end, particularly because sooner or later, the authorities had to be able to negotiate with accepted workers' representatives. Thus the intrinsic power of the working class remained significant.

This is not so with the UTAP. Farmers do not see it as a place to express and defend themselves. For them, it is something they have to deal with for certain procedures, a semi-official organization, where they can hope to find influential persons who might be able to intervene on their behalf, and help them get along in the complicated world of the market and of official procedures—rather than being just their representative.

We need to add to these points the more general problem of relations between the authorities and the cities in Tunisia. It goes without saying that the authorities and the working class have occasion to encounter each other in the cities, and even to confront each other there. But the countryside does not play a very obvious role in the mechanisms of power. UTAP leaders at all the levels are among the accomplices, and even the agents, of the authorities' dispossession of the countryside, in particular because of the various advantages, especially financial, which they can derive from their positions. Greater local power or more effective control of their management does not seem to them to be a high priority.

4.3 - The monopoly of representation

Speaking in the name of farmers who have no real place in the mechanisms of power, the UTAP acts more through "dialogue," "intervention," and "secret negotiations" than by mounting campaigns to bring pressure on the authorities, by mobilising its members, or by informing public opinion. It seems in fact to be completely integrated into the functioning of the one party state.

The Association of Technicians who have been granted state-owned lands[32] is an excellent example of the way in which the UTAP tries to "take over" all organisational initiatives, since it cannot prevent them, as it used to do before democracy had so strongly asserted itself. It tries to absorb into itself all those who want to do something, by creating parallel structures which deprive them of their members; all in accordance with the principle of "cutting the grass from under their feet." They may in fact be able to negotiate certain advantages from the government, which prefers to give benefits through the intermediary of the UTAP, rather than negotiating rights with autonomous partners. In short, this organisation acts as an "omnibus" organisation and monopolises the representation of the countryside.

We should note that this does not always work, and in any case produces a degree of resistance. Some grassroots co-operatives (perhaps about fifty) refuse to this day to join the UTAP, despite all the advantages dangled before them. Their leaders are not convinced that they would really gain anything by joining. They prefer to trust to themselves to defend their interests. This is simple, unorganised resistance at present, but this attitude is a sign that things could change and that there is some hope that other sectors of agriculture might organise together one day, outside the UTAP.

4.4 - The organisation and its capacity for autonomy

In its way of operating, by taking over the powers of civil society in the rural world, in the name of lawfully representing them, in the place given it by the government, the UTAP is much more an auxiliary of the single party in power than a representative of those in whose name it speaks. So that far from promoting associative life and the development of civil society, the UTAP acts as an agent of a state that would like to see all its power and its monopolies preserved.

Moreover, it is the place where all those who are going to direct the farmers' "autonomous" structures are appointed: co-operatives (local, regional, and central) marketing agricultural products, agricultural supplies, and other services of the same kind. These opportunities for obtaining "places" has enabled many people from agriculture and members of the UTAP's organisation (or people supported by them) to accumulate wealth, which has made them full members of the ruling class. To return to the problem we raised regarding transfers of wealth from the countryside to the city, the

UTAP is also the place where some farmers can arrange for this transfer not to inconvenience them, but even bring them some benefits, by laying hold of part of the income of small farmers. The total absence of transparency in all the appointments, the lack of clarity in the management of the "collective" organs that supervise agriculture—none of this makes the UTAP a place where real powers are conferred on civil society, a place where rural people are empowered. Some even accuse it of promoting the abuse of power over the countryside: enormous profit margins for the activities of intermediaries, for example, which disappear from the accounts. The difference between the "advances" paid to farmers and the market prices of products is often far more than the "rebates" granted, and is not justified by the management needs of the service organs involved.

V - Conclusion

The central question that arises here is the transition from totalitarian societies to democratic societies, to the rule of law, the existence of the single party and its accessories in various sectors, the national organisations, which openly share with it the ideology of national unity and the priority to be given to a so-called development, which never seems to get going.

If the transformation of the structures of government is called for by the increased need for democracy and justice, and also by an increasing integration into the world market, the transition from a widespread but poorly defined call for democracy to an organised struggle for this objective involves a recognition of the driving elements of the various social strata concerned in this change.

The transition to new state structures excludes the party's monopoly of power, and also excludes the rise of a new ruling social group that is outside of and different from the one which depends on the single party. Thus a 'bourgeois liberal', or democratic group that is interested in this change must emerge. As Gramsci emphasised, "a social group can and must be a ruling group, even before it takes over government power (this is one of the main conditions for taking over power); afterwards, when it exercises power, and even if it keeps it tightly under its control, it becomes dominant, but must also continue to be 'ruling'. In Tunisia today, liberal, democratic discourse is practised by the upper levels of a power structure, which is openly based on the hegemony of a *de facto* single party (always the same policy of "cutting the ground from under their feet"!)—that is, the "dominant group" is not the "ruling group." It develops ideological points of view that are not its own, so that it is in a false position and its utterances are far removed from reality.

Perhaps the delay in the emergence of this kind of new social group results from the fact that private entrepreneurs are still enormously dependent on government markets, and generally on the government's good will, especially because the main characteristic of the current development is lack of any capacity to innovate. This seems an insurmountable obstacle to creating a forward-looking industry.[33]

This does not mean that there can be no avant-garde groups, holding this ideology, who could try to transform themselves into a "ruling group". If there were any such groups, their task would be particularly difficult. The areas where rights can be freely exercised are limited and closely watched. In any case, it does not seem that the UTAP could be a place where the people could recover the powers, which they still find it very difficult to obtain.

This process seems to need the creation of new areas for organising farmers, new associations that could adopt different methods of action, and that would not surrender when confronted by the UTAP's efforts to "cut the grass from under their feet". We have already said above that this has already started to happen.

The outlines for development during the 1970s have been emphasised the application of the Structural Adjustment Programme imposed by the World Bank. This does not seem to be moving in the direction of decreasing the burdens on agriculture. Just the opposite, in fact. The multiplication of agricultural development agencies managed by, or with, large-scale farmers, and the arrival of management technicians to whom government lands have been granted, have aggravated social and cultural distinctions in the countryside, and show clearly that small farmers can look forward only to a rather limited future, all the more so because the resources from outside agriculture, which they used to get from emigrant workers, seem to be drying up. The steadily increasing role of "educated" persons (technicians who have come from the offices of Ministry of Agriculture or who have become farmers, the sons of farmers who hold university degrees) in the various agricultural organisations threatens to exclude small farmers even more from decision-making structures. It could also lead to a trade unionism of a different kind, or at a different level, led by men, often trained technicians, who have access to more information and better communications, and who can therefore get associative life going again on a new basis. The preconditions for these changes in the structure of agriculture and in the form of action in the sector probably already exist. The question is whether this sector, or others that have the same limitations, can in practice transform the relationships of power between civil society and the state, without the problem of the single party being radically resolved. Once again, it is politics that seems to decide the future form of society.

Notes

1 Thus the collective work published under the direction of Camille and Yves Lacoste, *L'état du Maghreb* (Tunis: CERES Production, 1994 ; "l'édition française en Tunisie"), which accords special importance to the problems of democracy, devotes only four pages to agriculture in a book of 572 pages!

2 In addition, during our research we looked at all theses from the Institute National Agronomique de Tunis, and found out that not a single thesis was devoted to the question of the organisation of agriculturalists, or even only to their collective and organised interventions.

3 In Tunisia, "national organisations" are ones that have been created or directly supported by the Neo-Destourian Party, which has been in power since 1956, which include, in addition to the UTAP, the General Union of Tunisian Workers (Union générale des travailleurs tunisiens, UGTT), a labour union whose relations with the government have had their ups and downs and that raises the question of the degree of autonomy even one of the most "official" organisations can sometimes assume, the Tunisian Union of Commerce and Industry (Union tunisienne du commerce et de l'industrie, UTICA), an owner's association very close to the government, the Tunisian National Women's Union (Union nationale de la femme tunisienne, UNFT), a women's organisation that has long served as an intermediary for the government's activities in the area of health and the transformation of the status of women, but which is finding it difficult to determine its role since the creation of a Ministry or State Secretariat (depending on the period) for the condition of women and the promotion of the family.

4 This refers to properties that were ceded or entrusted to the management of religious institutions, on the condition that the latter meet some need expressed by the donor. This resulted in a considerable area being earmarked: 750,000 hectares for public *habous*, and 1,500,000 for private *habous*. The lands owned by colonists then covered about 1,500,000 hectares, of which about 300,000 were to be bought back over the following years.

5 The monarch, originally the representative of the Ottoman occupying power.

6 The Tunisian dinar was then officially worth 18 French francs.

7 See G. Naccache, 1967, " Les coopératives agricoles de production du Nord," *La revue du CERES*, University of Tunis, no. 15, January.

8 In 1965 the government set up "cahiers de charge" for large landowners that obliged them to follow the government's agricultural choices. Some time before, most of the "absentee landlords" living in town, usually in Tunis, had agreed to stop dividing up their lands and renting them to small sharecroppers, and had moved to more modern modes of farming: renting or direct management of large areas.

9 It was estimated (in newspapers of the time) that peasants rented more than 150,000 hectares of land to large farmers. See GEAST, *Perspectives, stratégie et tactique, débat interne inédit 1970-72.* Tunis: Outrouhat, 1989.

10 M. Ben Romdhane, 1980. *L'accumulation du capital et les classes sociales en Tunisie depuis l'indépendance.* Doctoral dissertation, Faculté de Droit et des Sciences Economiques, Tunis.

11 The Destourian party, literally a liberal constitutional party ("Destour" means "constitution"), was created in the 1920s, and was the origin of the party that was the principal leader in the battle for independence, the "Neo-Destourian" party, which split off in 1934.

12 Clément Moore, 1965. *L'histoire politique de la Tunisie.* Paris: Mouton.

13 More precisely, 7 November 1987, the date of Bourguiba's fall and the official proclamation of democracy.

14 In 1972, the UGET, a students organisation, was dissolved de facto, after the left had clearly won a majority, despite the theoretical existence of a leadership named by the PSD.

15 *Le Petit Matin*, 21 October and 26 November 1961; *La Presse*, 31 October and 18 November 1961.

16 At the instigation of the authorities, a new organisation, the Union nationale de la coopération, had been launched in 1964 under the supervision of agents of the state, and it began at the local and regional level.

17 *La Presse*, 7 February 1971.

18 While it is especially difficult to obtain information about applicants for these lands, it seems possible that with the exception of a few privileged individuals, close to those in power, most of the firms that were granted these lands were made up, apart from agricultural technicians, of landowners whose property was often in the same region. The move to capitalist agriculture, based on finance from bank loans, mechanisation and marketing, developed.

19 The last decisions in this sense date from spring 1997, at a time when the party in power was increasing its strength relative to government institutions. It is not to be excluded that as the game of political balancing continues, these decisions might be amended.

20 The Tenth Congress was held between 14 and 15 March 1990 at Nabeul in Cap Bon.

21 We should also recall that the party in power, the PSD, had changed its name the previous year, and was now called the RCD. Perhaps the change of name also reflected an (ephemeral) will to assert the idea of change at any price.

22 The Fonds nationale de solidarité (National Solidarity Fund), controlled by the Presidency of the Republic, and recently provided with a Secretary of State, receives "voluntary" contributions from citizens, and finances a number of social projects. It is generally known by the number of its postal bank account: 26.26.

23 Their scarcity, indeed their total absence from public libraries, including the National Library and the National Documentation Centre, which specialise in periodicals, shows how little interest they attract.

24 These notions are developed in the appendix.

25 See GEAST, *Stratégie et tactique, débat inédit – 1970-1972*. Tunis: Outrouhat, 1986.

26 This repression, in the form of trial and imprisonment, ended only in 1980, proving that it was not due solely to Ben Salah and his "authoritarian socialist" orientations.

27 As in many trade union organisations, there was a certain rivalry between the officials who came from working class origins and those who came from "educated" professions, and who were civil servants (teachers, nurses, professors, et al.)

28 This meeting of the UGTT was held against the background of a civil war, in which the Bourguiba faction crushed its adversaries thanks to the armed support of the central trades union organisation. Banking on this role, the authors of the programme thought the time had come for a showdown with a single party in power that had no programme for constructing a new state.

29 Perhaps in order to explain the lukewarm character of the demands made at this time, we would have to explain that many of the UGTT's officers, even those from the working class, had joined the state apparatus, and that the UGTT itself, in order to ensure its financial autonomy, and also to set up organisations it could depend on, had started a number of enterprises under its management—a hotel, tour-operators, various co-operatives, insurance, and so on—thus moving toward an " American-style " unionism that also made it participate in economic life, and made it weigh more carefully its support for the demands of the (relatively) new working class. As for the former leaders, all, including their leader Ben Salah, were absorbed into government organisations, without setting any policy conditions, limiting themselves to applying Bourguiba's directives. Bourguiba himself said that his pragmatism forbade him to formulate any general programme until he adopted state planning !

30 Renamed Parti Socialiste Destourien, which became a de facto single party from January 1963, when all other political parties were banned.

31 In 1963, he had been summarily ousted as secretary-general by Habib Achour, who was at that time closer to the government.

32 This association has been duplicated by a Federation of the UTAP, which has taken away its membership and most of its leaders.

33 See André Métral, 1996, "De l' 'homo aeconomicus' au 'district industriel': destin de l'entrepreneur tunisois." *Correspondance*, no. 43, December 1996. Tunis: IRMC.

34 The Destourian Party (Destour means 'constitution', literally the Liberal Constitutional Party, set up during the 1920s, was at the origin of the party which led the struggle for independence, the "Neo-Destourian" Party, the result of a split in 1934.

35 The bureau consisted of Brahim Abdellah (president), who was subsequently to play an important political role, and other well-known militants of the time : Habib Mouehli, Ali Hamami, Amor El Mehdi, Ben Jemâa, Mahmoud Yassine, Mohamed Hedi Bel Haj, Abdelaziz El Bahri, Amor Ghannouchi, Amor Chouchane

36 See *Le Petit Matin*, Tunis, 13 September 1955, in the communique by the UGAT's BE was published.

37 The Higher Coucil's motion was published in the *Petit Matin* on 30 November 1955.

38 The farmers of Jendouba who were members of the UGAT and present in Kairouan, Tahar Azaiez (later president of the UNAT), Sadok Khalfallah, Abdelhamid Bergaoui, and Abderrahman Toukabri, were ordered to set up a section of the UNAT in Jendouba and to get as many members of the UGAT to join it as they could. This tactic, all the more effective because the UGAT did not resist it very strongly, was repeated everywhere.

39 As secretary general of the UGAT, Tahar Ben Ammar, the Bey's Prime Minister, signed the internal autonomy agreements with France in 1955. He was replaced as Prime Minister by Habib Bourguiba. A modern large landowner in the Tunis region, he was put on trial in 1958, in the context of the trial of "collaborators" with France, and found guilty. His "collaboration" with Bourguiba between 1954 and 1956, appeared to be more that of a hostage than that of an ally.

40 An agronomist, he had pursued a career in the Development Office of the lower valley of the Medjerdah, financed chiefly by American funds, and then became Secretary of State for Agriculture, M. Ghedira also acted for a year as Minister of Agricultural Production and Food Industries, while remaining president of UNA (1985-1986).

41 When he left the UTAP in 1992, T. Essid was named permanent secretary of the RCD (the new name of the Destourian party in power), in charge of militants, which shows the close ties between the state's supervisory structures and the "national organisations", even in these times of a multi-party system and the claimed autonomy of civil society.

42 A specialist in finance management, Abdelbaki Bacha was also president of the Fédération nationale des sociétés de caution des crédits agricoles, an arm of the UTAP. As a member of the central trade union organisation, from 1981 on he headed the UNA's research bureau, and was "lent" to the state's (the governorship's) supervisory structures in accord with the same principle of communication among the country's different structures.

References

Attia, Habib, 1978. *Les hautes steppes tunisiennes: de la société pastorale à la société paysanne.* Doctoral dissertation, University of Paris VIII.

Arraya, G., 1980. "Vers la disparition des activités rurales autonomes." *Le Monde Diplomatique*, August 1980.

Barat, Etienne, 1989. *L'association: son emploi rationnel.* Paris: Guillaumin et cie.

Barret-Kriegel, B., 1989. *Les droits de l'homme et le droit naturel.* Paris: PUF.

Bouayed, Anissa, and Chantal Chanson-Jabeur, eds., 1990. *Etat et mouvement syndical au Maghreb : table ronde des 14-15 mars 1989.* Paris: AUPA-GREMAMO, Cahiers du GREMAMO, 9,

Ben Romdhane, Mahmoud, 1983. *L'accumulation du capital et les classes sociales en Tunisie depuis l'indépendance.* Doctoral dissertation, University of Tunis.

Ben Romdhane, Mahmoud, 1984. "L'Etat et la paysannerie." *Le Mensuel*, no. 2, July 1984 (Tunis).

Bonnemère, Eugène, 1850. *Histoire de l'association agricole et solution pratique.* Paris: Dusaca.

Boudiaf, Messaoud, 1979. *Stratégie de développement agricole en Tunisie depuis l'indépendance.* Thesis, University of Lyon II.

Brichet, Robert, 1976. *Constitution et fonctionnement des associations et syndicats.* Paris.

Colonna, Fanny, 1987. *Savants paysans, éléments d'histoire sociale sur l'Algérie rurale.* Algiers: Office des publications universitaires.

Dimassi, Hassine, 1983. *Accumulation du capital et répartition des revenus, thèse de doctorat d'Etat ès-sciences économiques.* Mimeographed, Tunis.

Dumont, René, 1979. *Paysans écrasés, terres massacrées.* Paris: Laffont.

Ferguson, Adam, 1992. *Essai sur l'histoire de la société civile.* Trans. De Bergier (1783). Paris: Presses universitaires de France.

Gana, Alia, 1986. "Pluriactivité des agriculteurs et reproduction sociale dans les campagnes tunisiennes." *Annales de l'Institut de la recherche agronomique de Tunisie (INRAT)*, 60 (7).

Garricou, Lagrange, 1970. *Les rapports des associations avec les pouvoirs publics*, Paris: LGDJ.

Germion, Paul, 1976. *Le pouvoir périphérique.* Paris: Seuil.

Gervais, M. *et al.*, 1972. *L'univers politique des paysans dans la France contemporaine.* Paris: Armand Colin.

Gramsci, Antonio, 1978. *Cahiers de prison.* Paris: Gallimard.

Gramsci, Antonio, 1977. *Lettres de prison*. Paris: Gallimard.

Gramsci, Antonio, 1974-1980. *Ecrits politiques*, 2 vols. Paris: Gallimard.

Groupe d'études et d'action socialiste tunisien (GEAST-Perspective), 1970. *Les problèmes agraires dans la Tunisie actuelle*. Paris: GEAST.

GEAST, 1988. *"Perspectives": Stratégie et tactique, débat interne inédit 1971-72*. Tunis: Outrouhat.

El Amami S., J. P. Gachet, and T. Gallali, 1979. "Choix techniques et agriculture maghrébine : le cas de la Tunisie." *Peuples méditerranéens*, July-September 1979.

Humana, Charles, 1985. *Guide mondial des droits de l'homme*. Paris: Buchet.

Kassab, Ahmed, 1982. "L'agriculture tunisienne." *Revue tunisienne de géographie*, no. 9.

Kool, R. G. A., 1963. *L'agriculture tunisienne : analyse d'une économie*. Wgeningen: Veenman.

Lefebvre, H., 1970. *Du rural à l'urbain*. Paris: Anthropos.

Lesueur Emile, 1906. *Les associations agricoles en Tunisie. Rapport de mission adressé à M. le ministre de l'Agriculture*. Paris and Tunis: Office tunisien de colonisation.

Loghmani (Slim). *Le cadre juridique des organisations non gouvernementales en Tunisie*. Tunis: El Taller, s.d.

Meister, Albert, 1974. *La participation dans les associations*. Paris: Les éditions ouvrières.

Mendras, H., 1980. *Sociétés paysannes : éléments pour une théorie de la paysannerie*. Paris: Armand Colin.

Moore, C., 1965. *L'histoire politique de la Tunisie*. Paris: Mouton.

Muller, P., 1984. Le technocrate et le paysan. Paris: Les éditions ouvrières.

Naccache, Gilbert, 1994. *De l'idéologie du développement à l'idéologie des droits de l'homme*, unpublished.

Elloumi, Mohamed, ed. 1985. *Politiques agricoles, stratégies paysannes et développement rural*. Colloquium held in Tunis, April, 1985. Tunis: IRMC, 1995.

Poncet, Jean, 1961. *La colonisation et l'agriculture européenne en Tunisie depuis 1881*. Paris: Mouton.

Sethom, Hafedh, 1992. *Pouvoir urbain et paysannerie en Tunisie — Qui sème le vent récolte la tempête*. Tunis: CERES Production et Fondation nationale de la recherche scientifique.

Sicard, Hervé, 1965. "Les effets sociologiques et économiques de la suppression des habous en Tunisie." In *De l'impérialisme à la décolonisation*, dir. G. Ardant, K. Axelos, and J. Berque. Paris: Editions de Minuit.

Syndicat et société, 1989. Acts of the Seminar, Tunis, Centre d'études et de recherches économiques et sociales. Tunis: CERES.

Widmer, Charles, 1992. *Droits de l'homme et sciences de l'homme : pour une éthique anthropologique*. Geneva: Droz.

Zghal, Abdelkader, 1989. "Le concept de société civile et la transition vers le multipartisme." In *Changements politiques au Maghreb*, éditions du CNRS.

Reviews, Periodicals and Newspapers

Revue du CERES
Cahiers du CERES
Cahiers de Tunisie
Annales de l'Institut de la recherche agronomique de Tunisie
Revue tunisienne de géographie

Magazines de l'UNA-UTAP : *El Oumma, El Fellah*
Réalités (hebdomadaire bilingue d'informations)

In French
La Presse de Tunisie; Le Petit Matin (till 1964); *L'Action; Le Temps; Le Renouveau*

In Arabic
Es-Sabbah; El Amel;

Appendix

I - From the UGAT to the UNAT: Comments on an evolution determined from above

The UGAT, the Union générale de l'agriculture tunisienne (Tunisian General Union of Agriculture), was created in 1949 by farmers who were members of the Neo-Destourian party, but also of the "old" Destourian party,[34] who belonged to the regional farmers' associations encouraged by France. Its constituent congress, held on 1-3 April 1949, approved this Destourian orientation.[35] The Union also included large Tunisian farmers who were close to the French in their markets and their products. The UGAT, which had representatives in the various representative assemblies that succeeded each other during this period of the end of colonialism, supported Tunisian nationalism. In the Ben Youssef crisis of 1954-56, it supported Ben Youssef, which led the Bourguiba faction to create a new organization.

The break with Bourguiba's era was first manifested in a denunciation of the UGTT, Bourguiba's ally: in a letter to the then prime minister, Tahar Ben Ammar, the UGAT complained that the UGTT was encouraging the strikes by agricultural workers (June-August 1955). Later, the attack was more direct: first the Bureau Executif (BE),[36] then the UGAT's superior council,[37] fell in line with Ben Youssef and denounced the Franco-Tunisian conventions negotiated by Bourguiba that confirmed the country's internal autonomy.

The response by the Bourguiba faction was a call for the disaffiliation of the UGAT, issued on 7 November 1955 in the Sahel, and then the creation of the UNAT (Union nationale de agriculteurs de Tunisie) in December 1955 at Kairouan, in the presence of Bourguiba.[38] The UNAT, headed by Tahar Ben Ammar, and composed of militant nationalists who were not all farmers, steadily supported Bourguiba's faction until the Ben Youssef faction was defeated.

II - The Presidents of UNAT, UNA, and then UTAP: Urban actors and the State Apparatus

The list of the different presidents of UNAT, then of UNA, and finally the UTAP shows well enough to what point they were state personnel and not representatives of the farmers:

1955 : Tahar Ben Ammar, Prime Minister[39]
1956 : Tahar Azaiez, Member of the Political Bureau of the Neo-Destourian party
1970 : Mohammed Bellamine, Governor
1978 : Mohamed Ghedira, Secretary of State for Agriculture[40]
1986 : Taoufik Essid, Governor[41]
1992: Abdelbaki Bacha, Governor[42]

III - The UTAP's aims

The thirteen aims defined by the UTAP's statute cover activities of very different kinds, ranging from the defence and promotion of farmers to the organisation of the rural world, through the intermediary of structures of production and services, and by launching production projects or participating in the management of operations of general interest. In article 3 of the statutes, these aims are defined as follows:

1. Organising farmers and fisherman, representing them, defending and protecting their interests and rights, and acting with a view to developing their knowledge.
2. Taking care of everything that concerns them.
3. Acting with a view to improving the economic and social situation of its members.
4. Training and raising the consciousness of farmers, individually and collectively, orienting them and encouraging them to guarantee their production and to improve their productivity by making use of technical and scientific methods.
5. Creating and developing various mutual service structures, societies, professional groups, trading stations, and associations related the sector, such as water users associations, olive growers' associations, etc.
6. Intervening with the parties concerned in order to facilitate agricultural investments, and to procure the necessary financing in the form of credits, subventions, and incentives for farmers' needs in the area of investment and for the agricultural seasons.

7. Participating in studies and experiments seeking to develop and modernise the agricultural sector.
8. Participating in the conception and elaboration of national and other plans, participating in the monitoring of their execution, and acting to ensure their success.
9. Acting with a view to improving the standard of living as well as the environment and ecology in rural areas.
10. Acting to improve agricultural information through the various means available and to improve the management of the Union's means of information.
11. Pursuing efforts to establish agricultural centres at the regional and local levels.
12. Organising fairs and agricultural events, or participating in them at various levels.
13. Working to develop agricultural work in such a way as to adapt it to global economic changes.

After reviewing this list, it will not be surprising to learn that the launching of Chambers of Agriculture (Northern, Central, and Southern) aroused more than reservations on the part of the union organisation : a certain number of the objectives assigned to these Chambers overlapped with those the UTAP had set for itself, and there was thus a problem of determining responsibilities.

In fact, the text establishing the Chambers of Agriculture specifies that they have been created in order to strengthen the role of the profession in decisions relating to development, to officials, and to the future of agriculture and fishing. They are supposed to offer their opinions and offer suggestions regarding agricultural problems, in particular at the time when the national economic plans are being prepared, but at all levels as well. They are supposed to be directly involved in promoting the agricultural sector and in the councils and continuing education in this domain. They are authorised to manage certain development projects directly, and to supervise certain improvement projects, within the framework of decentralisation, whose aim is to relieve the ministry of agriculture of part of its responsibilities, and perhaps also in response to the hesitations foreign aid donors feel in giving funds to the state, not only in Tunisia. . They can also organise conferences and fairs, either by themselves or in partnership with other structures, and they are supposed to play an important role in training farmers.

These Chambers, structural organs of the profession, thus limit the UTAP's representativeness on this level, and take-over some of its prerogatives. This limitation of power is not merely theoretical:

given that most of the services provided by the UTAP to farmers have to be paid for, their decrease may also mean a deficit in the union's income.

The UTAP has ended up accepting the existence of the Chambers of Agriculture, by having its members participate in them, and thus "occupying" them to a certain extent. But it is still not clear that these Chambers now have the financial and technical means necessary to carry out their tasks. It seems they lack sufficient personnel to participate in all the deliberative bodies, or to supervise farmers on the level of training. A study made by the Ministry of Agriculture's "Direction générale des investissements et des encouragements" (General office of investments and incentives), made in February 1992 but available only in mimeographed form, "Réalités et perspectives des chambres d'agriculture" ("Reality and outlook for the Chambers of Agriculture"), emphasizes all the difficulties these Chambers encounter, particularly the insufficient number of technicians involved in these organs (3 engineers and 5 adjunct engineers, 10 workers in all), the lack of connections with other structures supervising agriculture and with each other, and the inadequacy of their financial resources. These difficulties considerably limit their ability to fulfil their missions, not only with regard to supervision and agricultural popularisation, but also with regard to negotiating with the administration (not enough representatives to be able to occupy seats in all the structures in which farmers should be represented) and all the other functions they are supposed to perform, in particular the management of development projects.

5

Factors Inhibiting the Participation of Peasant Organisations in the Democratisation Process in Senegal

Mouhamadou

Introduction

Over the past two decades, Africa has been experiencing major social movements: the struggles for democracy and the varied transitions process, including national conferences and military coups, labour protests against structural adjustment programmes, etc. It was in this socio-political and economic context that Senegal singled itself out by recognising a pluralist political system in 1981.

Moreover, in 1972, the year when the rural communes were created, Senegal had already entered into a phase of decentralisation, which was completed in January 1997 with the implementation of the regionalisation project. On the economic level, the state had already reduced its sphere of activity, with the devolution of power y to the economic actors, including the peasant organisations, and the introduction in 1994, of the New Agricultural Policy.

The objective of these various reforms is a more national democratic life that is more dynamic and participatory, with greater participation of the grassroots populations in decision-making within decentralised governmental structures such as the Rural Council.

Finally, the establishment of democracy raises a number of issues. As social categories representing the largest segment of the Senegalese population, the peasants organisations are provided here with a historic opportunity to protect the interests of their members, to participate in local decision-making bodies, to end their marginalisation and their non-development, and to renegotiate, on a more egalitarian basis, their relationships with the state. In order to do so, should they not gauge the political issues involved, and have the necessary resources to realise their ambitions, and finally, be able to preserve their autonomy vis-à-vis their own leaders, the state, and NGOs?

The history of the Senegalese peasants organisations can be summed up in a few significant facts. Thus, we can distinguish:

The Colonial Period

At the beginning of the century, the first peasants organisations appeared in the form of co-operatives known as "Native Provident Societies" ("Sociétés indigènes de prévoyance"), in the 1950s. These later became the "Rural Development Self-help Associations" ("Sociétés mutuelles de développement rural), and contributed to the strengthening of the cash crop economy, based on groundnuts. These organisations did not serve the producers, but the colonial power, which saw in them an effective means of controlling the rural masses.

Early Post-Independence Years

Throughout the 1960s, the nascent co-operative system made the co-operative an emancipatory and political awareness-raising tool for the peasant masses through training and organisation. The concern for eliminating the host of intermediaries involved in providing and marketing agricultural products led the state to set up government services such as the Agricultural Marketing Board and the Senegalese Development Bank, and to promote the establishment of peasants organisations with the help of the Rural Development Centres.

The Mid-1960s

The euphoria that up to that point had encouraged peasant emancipatory efforts began to die down and to yield to other political ambitions of the government. This was reflected in the use of co-operatives as vehicles for ideological propaganda and the instrumentalisation and "alienation" of their members.

This period witnessed, among other things, the establishment of Société d'Aménagement et d'Exploitation du Delta (SAED) in 1965 and Office National de Commercialisation et d'Assistance au Développement (ONCAD) in 1966, which were concerned, respectively, with irrigated rice growing and assistance to the peasants organisations in the Senegal River Valley, and with the supply and marketing of groundnuts, including the training of peasants in the Groundnut Basin. It must be acknowledged that ONCAD did not live up to the expectations of peasants organisations, because most of its resources were siphoned off in the management of its own organisation.

The 1970s

The 1970s witnessed the emergence of Producer Groups (PGs)set up by governmental supervisory bodies. These include the SAED associated PGs, and the Grassroots Producer Associations (GPAs) connected with SODEFITEX (Société de Développement des Fibres Textiles). The objective of these structures is to improve the training of producers and the distribution of the factors of production and the marketing of products.

Post - 1984

In 1984, the state created Economic Interest Groupings (EIGs) which may be defined as an agreement among two or more individuals or bodies wishing to undertake joint development activities of all kinds in the various economic sectors.

The emergence of the EIGs was facilitated by the New Agricultural Policy introduced in 1984. Such a policy promoted private initiatives, particularly in the agricultural and semi-agricultural sectors, as well as the roll-back of the state in favour of a greater devolution of power to peasants organisations. In addition, because of its flexibility, the EIG makes it easier to gain access to credits made available by the new banking institution: CNCAS (Caisse Nationale de Crédit Agricole du Sénégal).

Even though peasants organisations pursue similar goals, in particular defending their members' interests, the conditions under which they emerged vary. Currently, we can distinguish seven types of peasants organisations in Senegal:

1. Peasant Organisations initiated by Parastatals: The two most striking examples are the SAED-driven PGs and the SODEFITEX GPAs.

2. Restructured co-operatives : The co-operatives underwent very rapid development during the 1970s (there were 2,378 co-operatives in 1971), and later experienced numerous difficulties with the emergence of EIGs in 1984 and the removal of the Directorate for Co-operation in 1991. A few co-operatives or unions were able to survive, as is shown UNCAS, the National Union of Agricultural Co-operatives of Senegal.

3. Women's Promotion Groupings (WPGs): Such groupings can be compared with the peasant organisations. They emerged through the activities of the Ministry for Women, Children and the Family (MFEF), and some 3,614 groupings currently belong to an umbrella organisation called FNGPFS, the Senegalese National Federation for the Advancement of Women Groupings, created in 1987.

4. Economic Interest Groupings: The members of these groups, first created in 1984, are generally peasants, retired workers, laid-off workers, college graduates, and emigrants who have returned to Senegal. Such groupings are born of private initiatives.

5. NGO-driven peasant organisations: From 1973 (a drought year in Senegal) up to the present, NGOs have played a major role in rural areas, concentrating their efforts on young people and on women. Rather than work with the existing peasants organisations, some of these NGOs have opted to promote the creation of new ones. Thus, OXFAM-Great Britain has made an important contribution to the establishment of the Laaw Federation in the Department of Podor.

6. Development Project-driven peasants organisations: These organisations were born of projects in the area of credits, and seek to organise peasants in order to facilitate the recovery of their funds or the transfer of expertise. A case in point is the Federation of the Peasants of Amorphil Island, an offshoot of the Dutch project based in Cascas in the Department of Podor.

7. Village Development Associations (VDAs): Initially, these associations were concerned with sports and culture, but some of them have now become genuine peasants organisations. This development can be illustrated in the Senegal River Valley by ASESCAW, the Social-Educational and Cultural Association of Walo, an NGO that began in 1976 as a simple youth association known as the Ronkh Club.

Finally, peasants organisations exist at the local level as well as at the national level. At the local level, they are organised around the village, the district, the department, or the region. With the exception of CNCR (Conseil national de concertation des ruraux—the National Council for Rural Solidarity), which is national in scope, all the other peasants organisations referred to in this study are limited in their geographical scope.

The purpose of this study is to review the role played by peasants organisations in the democratisation process in Senegal. More precisely, it seeks to:

- analyse the contribution made by peasants organisations to the battle for democratisation, and to gauge its limits;
- assess the degree of internal democracy as a structure defending the interests of their members; and
- examine gender relations within these peasants organisations, in order to improve our understanding of the dynamics of gender relations.

In order to carry out this work, I have relied on the relevant theoretical literature and bibliography on the subject. I have also undertaken fieldwork consisting of interviews with leaders and members of peasants organisations. These interviews were conducted with the National Peasants Organisation and with leaders and members of local peasants organisations. In all, 195 persons belonging to thirteen such organisations were interviewed, including five leaders and ten members for each organisation.

Part I: Strengths and Weaknesses of Peasants Organisations in Senegal

1.1 Grassroots peasant organisations

Senegalese peasant organisations in general, and at the departmental and regional level in particular, have produced important results, which reflect their internal dynamics and their capacity for social organisation and mobilisation in defence of their interests. Today's peasants organisations differ in several respects from those of the 1960s and 1970s. Among their achievements, we can mention the emergence of a new peasant consciousness. This is particularly evident in the creation of peasants' organisations without any outside help or governmental influence.

In addition to the Association for the Development of Textile Fibres (SODEFITEX) and basic producers' associations, all the peasants organisations investigated are managed autonomously. With the gradual withdrawal of the state and the end of the welfarism, peasants are coming together to deal with their common problems, such as the supply of inputs, access to credit, and marketing of their products. Does this form of social mobilisation reflect the need to take things into their own hands? Does it express their wish to play the role of entrepreneur?

As an illustration of the proliferation of peasants organisations, we can cite the case of the Union of Economic Interest Groupings of the Delta, which was established in 1987, three years after the law on Economic Interest Groupings was passed. There are now 310 members of the group in the Delta alone.

Can we speak of the emergence of a genuine peasant consciousness, if this is incapable of producing the leaders it needs? For more than two decades after independence, peasants organisations were led either by people who were not involved in agriculture, or by people who served the interests of the state. Peasants organisations have now been able to produce their own leaders. It does not greatly matter if they are retired government officials or returned emigrés. The important point is that they live by working the earth, they share the peasants' aspirations, and they identify with them.

Since by definition, democracy means taking account of the interests of the majority, how can we explain the exclusion of women and youth from areas where decisions are taken and power exercised? These two categories, irrespective of their numbers, have always been marginalised in peasants organisations development programmes, and have been excluded from the centres of decision making. They have struggled to form a newbreed of leaders, defined by the criteria of ability, selflessness and willingness to serve, regardless of their age or social background. In some peasants organisations, women and young people have joined together in an alliance, in order to take over decision-making positions previously held by older men. Such alliances reveal an important psychological change among grassroots social actors.

The marginalisation of women for socio-cultural reasons has always been a reality. Even if they were given the chance to take part in meetings alongside men, it was difficult if not impossible for them to speak, and in the end, it was the men who took decisions on their behalf. Happily, the Women's Promotion Groupings gave them the chance to exercise control and to strengthen their position. The question then became one of applying this advance to mixed organisations, and it is only in this context that an examination of gender relationships is worth while.

Despite the paucity of women leaders, however, the few women who managed to join such organisations have a strong presence in them. For instance, some development activities are specifically designed for them, while many aid donors encourage men to give women greater responsibility. With regard to the internal workings of the POs, apart from those in the Casamance and Koungheul, one can note that they all function democratically. Leaders are chosen objectively, and even if those leaders seem to stay in office indefinitely, regular elections are at least being held. In addition, two of the peasants organisations studied have been able to impose sanctions against their leaders.

We should mention here that the unions or federations of peasants organisations have modern functional organs, such as general assemblies, boards of directors, executive committees, and administrative councils.

In addition, a desire to acquire local power has been shown by such unions as the Union of Groupings of Market Gardeners in Méouane, the Casamance Young Farmers Association, and ADENA, which all have a few rural councillors and regularly participate in the local and regional development committees. Village level associations still remain hesitant about participating. As we shall see later, there are several reasons for this hesitancy.

Although modest, these results were achieved by experienced leaders of the Senegalese Federation of NGOs (FONGS) and the National Council for Rural Co-operation, and with the methodological support of certain national and international NGOs which regard strengthening the association movement as a top priority. These various actors accept that counter-forces are necessary in order to stand up to the state and sustain democracy.

It would have been difficult to achieve these results without a favourable socio-economic, political, institutional, and media environment in Senegal. Since 1981, multiparty politics has prevailed and recent years have witnessed the birth of independent newspapers, private radio stations (including a rural radio station in Fissel-Mbadane in the department of Thiès), and the strengthening of civil society.

1.2 National peasants organisations

Until March 1993, when the National Council for Rural Solidarity was formed, the Senegalese Federation of NGOs, which now participates in this Council, was the only peasants organisation with a national scope. And that is why we should examine its democratic experience before we turn to that of the National Council for Rural Solidarity.

The Senegalese Federation of NGOs (FONGS) was established in 1976, and by 1991 had some 200,000 members in more than 850 villages in Senegal. Its contribution to the process of democratisation can be seen in activities such as strengthening the Senegalese associative movement through training and information, the development of vertical solidarity, the democratic nature of its governing structures, the introduction of greater transparency and managerial rigour, and its commitment to increasing the economic power of associations, in order to reduce their financial dependency on the state and on NGOs.

The desire to make these associations more autonomous is shown in the responsibilities vested on them; for example in the management of certain projects. It is also demonstrated by the defence of members of peasants organisations against the state, its official organs and against aid donors For example, FONGS is represented on the administrative council of the Senegalese National Bank for Agricultural Credit, and this gives it an opportunity to exert its influence on how credit policies planned for the rural world are defined.

During its four years of existence, the National Council for Rural Solidarity has managed to win the confidence of the federations of peasants organisations and to make a positive contribution to strengthening Senegalese democracy, notably by such remarkable achievements as the development of a broad framework, at the national level, for concertation and co-ordination among the nine federations representing the basic sectors of the rural world. By bringing together so many groups involved in development, the National Council for Rural Solidarity was sure to achieve worthwhile results. It is the only organisation of its kind in Senegal, and comprises:

- The Federation of Senegalese NGOs (FONGS)
- The National Union of Agricultural Co-operatives of Senegal (UNCAS)
- The National Federation of Economic Interest Groupings of Fishermen (FNAGIE/P)
- The National Federation of Groups for the Advancement of Women (FNGPF)
- The National Federation of Foresters (FENAFOR)
- The National Union of Livestock Breeders' Co-operatives UNCE)
- The National Federation of Horticulturists' Economic Interest Groupings (FNGIEH)
- The National Federation of Livestock Raisers' Economic Interests Groups (FNGIE/E).

Two other results worth mentioning are the defence of the interests of the peasant masses and participation in the definition of development and support for the rural world. In fact, in 1993, when Senegal's government adopted an emergency plan to restore financial stability by reducing salaries, the National Council for Rural Solidarity reacted in the same way as trades unions and civil society in general..

The National Council for Rural Solidarity took an active role in drawing up the declaration concerning agricultural policy and development in 1994, and in the negotiations on the Sectoral Adjustment Programme for Agriculture. The National Council for Rural Solidarity is a member of both the Higher Council on the Environment and the National Supervisory Committee for the study of the management of rural land and natural resources.

The National Council for Rural Solidarity also enjoys a certain degree of autonomy with regard to the state and donors, or at least it is not subject to their control. In 1996, during discussions with its partners on ways to re-launch agricultural activity, the National Council for Rural Solidarity suspended participation in all activities for three months, in order to bring pressure on the state to revise its positions, which it considered prejudicial to agricultural producers. This example should be emphasised, because it shows that counter-forces are beginning to emerge.

Referring to the National Council for Rural Solidarity's achievements, the authors of *Situation et évolution des organisations paysannes* (*The Situation and Development of Peasant Organisations)* argue correctly and pertinently that "Today, a peasants organisation is taking shape in Senegal which can claim to be representative of the majority of Senegalese peasants and thus position itself in all areas relating to the rural world' (Diagne 1995: 39).

In conclusion, let us point out that three of the leaders of the National Council for Rural Solidarity are women: the president of the FONGS, the president of the National Federation of Groups for the Advancement of Women, and the vice-president of the FNAGIE/P.

II - Typology of Factors Limiting the Participation of Peasants Organisations in the Democratisation Process

2.1 - Socio-cultural factors

The first part of this study on the history of peasants organisations drew attention to the role of the state and its implementing organisations, such as the National Office for Marketing and Development Assistance (ONCAD), the Association for the Improvement and Development of the Delta (SAED), and the Association for the Development of Textile Fibres (SODEFITEX), in the creation and functioning of peasants organisations.

The cultural heritage of some peasants organisations continues to influence how they participate in the life of their parent organisations. Peasants have always regarded rural co-operatives as intermediaries in the distribution of credit and equipment, and as a collection point, outside formal registration, which enables them to receive aid from the state. In other words, the notion of participation carries no conviction.

According to our investigations, many peasants organisations at grassroots level still have distant relationships with the state, which they consider, rightly or wrongly, as a power of which they have to be wary. The relations of these peasants organisations with the decentralised services of the state are limited to the resolution of purely administrative problems.

This negative perception of the state cannot but be reflected in the perception of political activity in general and democracy in particular.

Politics (and politicians) are regarded by grassroots members of peasants organisations as a means of exploitation and domination. Politics is not seen as having any civic mission or anything to do with raising consciousness and developing new citizens capable of making coherent political choices. There is no lack of examples to show that politicians never keep their election promises and that they are concerned only with their own interests. This is borne out by the low rate of participation in elections by rural populations.

Politics is still considered to be a matter for the intellectual elites, and the area of political activity confined to the towns. The fact that grassroots members of peasants organisations are unable to think of any benefits they might gain by taking part in local decision-making may show either their lack of awareness of the advantages democracy can offer or their total rejection of any political action.

For many grassroots peasants organisations, democracy is limited to freedom of expression, respect for the will of the majority, openness to others, and grassroots co-operation. Regionalisation is seen as a useless administrative reform. How can decentralisation succeed against the background of attitudes like these? What meaning can decentralisation have, without full and responsible participation by grassroots actors? In many ways, the state is regarded in the same way that peasants organisations regard women. Female members of mixed grassroots peasants organisations continue to be marginalised, despite all the political speeches about the need to promote the advancement of women. Cultural habits are tenacious and women seem resigned to their fate. Educational programmes within peasants organisations are intended primarily for men, and this deepens further the isolation of women, already handicapped by high rates of illiteracy.

The status of being a woman implies certain duties, such as managing the home, which make their participation in the life of peasants organisations more difficult. Thus, we can understand why there are few women leaders or rural councillors. In view of this situation, the low rate of participation by women shows that democracy within peasants organisations is still elitist, not to say sexist.

Finally, among socio-cultural constraints, we should note the way peasants view organisations which support development. Here, too, "peasant culture" appears as an impediment, since it regards these organisations as potential ways of getting money. The requests made to them by members rarely concern training in civics and much more often about the purchase of equipment or allocations for working capital. And yet NGOs, such as the African Network for Integrated Development (Réseau africain pour le développement intégré—RADI), have important programmes for training legal advisers in rural areas. These programmes may, indeed, be misunderstood by peasants organisations, but their priority remains the search for financial help and not for their own training. This preoccupation is perfectly comprehensible when one realises that for many peasants organisations, the main imperative is to survive and to satisfy their basic needs.

2.2 Cognitive factors

A few major factors need to be mentioned here. First of all, the illiteracy of members of peasants organisations. This is an important problem, which explains why peasants organisations participate so little in the democratisation process. Even though more and more of them are becoming relatively well educated, the great majority of peasants remain illiterate. All the peasants organisations investigated referred to illiteracy as one of the two main causes for their limited participation in the democratic debate. It could hardly be otherwise, since it is only during the rare election campaigns that political leaders deign to communicate in any real way with the 'voting masses' in their national languages. In other circumstances, French, the official language, is always used.

It should also be pointed out that few members of peasants organisations have learned to read and write their own languages. Any written message poses a problem, therefore, and necessarily limits participation in decision-making. It must be recognised, however, that political communication with the masses cannot be effectively achieved without the use of some written material.

For economic reasons, peasants organisations do not have sufficient means to conduct literacy programmes or other forms of popular education. Education and the raising of consciousness are not priorities for the peasants organisations we investigated. Their main imperative remains satisfying the economic needs of their members. There lies the dilemma. The peasants organisations have deliberately chosen to emphasise immediate income-producing activities, whereas education appears to be the surest way to increase their intellectual abilities.

The illiteracy of the members of peasants organisations also explains why they have to turn to educated people to run their activities, which makes them more dependent on such people and reduces their room for manoeuvre. Among other consequences of the illiteracy of their members, we may note the marginalisation of programmes of technical training and of democratic debate within peasants organisations. Training programmes for supporting organisations for peasants organisations are aimed mainly at the peasants with official positions, particularly the presidents and secretary-treasurers. Since this training seldom reaches further down, the gap between the leaders and the grassroots members grows wider. Even though peasants organisations reported that their administrative organs functioned relatively well, that does not mean that members participated in an effective and worthwhile way. The centres of decision-making in peasants organisations are not really the general assemblies that bring together most of the members, but rather the executive committees or administrative boards formed by the intellectual elites. The consideration of issues within peasants organisations remains generally poor, the centres of decision-making are distant from the members, and the leaders' relations with their members are vertical in nature. To show how this collaboration works, some peasants comment ironically that their new leaders are much the same as their old ones.

There are also problems relating to the leaders' aptitudes and attitudes. Like their members, the leaders of grassroots peasants organisations are often illiterate. With the exception of certain leaders at union level, the others have little schooling and weak managerial abilities. It is not easy to direct a peasants organisation without any preliminary training or professional experience. One of the first consequences of these limitations is the lack of any strategic vision at the level of the peasants organisation. Short-sighted leadership seems to be the general rule. Moreover, the low level of qualifications creates blockages because of the absence of initiatives and impetus.

Among the practices that do not help strengthen democracy in peasants organisations, are the centralisation of power, the blocking of normal decision-making functions and the tendency of leaders who have been sacked to set up competing peasants organisations. This tendency can also be observed among many political and trades union leaders in Senegal.

Finally, we should emphasise the peasants organisations' poor negotiating abilities, which, again, are partly due to members' illiteracy and their weak representation in local centres of decision-making. Not one of the peasants organisations we investigated mentioned its ability to influence qualitatively the decisions made within CRs, CLDs, and CRDs. It is as if their participations were purely symbolic.

Despite the democratic progress achieved by the CNCR, many problems remain. The state continues to fix almost by itself the rules of the political and economic game. Its associative partners do not seek, or at least they do not yet have the means to change these rules to their own advantage. In this area,

> It is as if everyone tried to adapt themselves within the defined limits of the agricultural policy, rather than actively seeking to change them. The peasants organisations are more concerned to set up sectoral projects on the edge of the agrarian system, which do nothing to solve the fundamental problems of agriculture' (Théba 19xx: 91).

Until some genuine counter-forces emerge, the state will continue to fix the price of agricultural products and will exploit its dominant position to the disadvantage of the peasants organisations.

2.3 - Financial and economic factors

All the peasants organisations we met at every level emphasised the financial and economic constraints they suffered from, which largely explains their limited participation in the democratisation process in Senegal. These constraints are reflected in the low level of agricultural investment, as well as in the obsolescence of equipment, and limited access to resources in general, notably to inputs, credit, and land. Indeed, many indebted peasants organisations are not eligible for credits from CNCAS. The land question remains closely connected with the persistence of customary law, with a misunderstanding of the law on the national domain, and with the increasingly significant presence of agro-industrial firms occupying most of the land to the detriment of small farmers. This is particularly worrying in the Senegal River Valley, where large land-holdings, as well as conflicts over land, are tending to increase. These financial constraints also account for the difficulties encountered in distributing, storing, and processing agricultural products. For some products, such as peanuts and cotton, the state still claims the right to fix prices for the producers, and the deterioration of the terms of exchange is a painful reality for the rural world. The financial and economic situation of the peasants organisations has been made worse by the state's abrupt disengagement for the past ten years and by the lack of any measures of compensation.

Finally, it should be emphasised that the NGOs' support programmes and development projects have mainly benefited peasants organisations at intermediate levels. Those at the grassroots encounter enormous financial difficulties that nearly cripple their normal activities. These difficulties are further increased by the insufficiency and irregularity of members' contributions, which also is very minimal. How can peasants organisations be concerned with democracy when they have a struggle to survive at all? Their state of poverty weakens their resistance to domination by the state and the NGOs, which hold the resources that the peasants organisations so desperately need. Trying to make peasants organisations into groups capable of constituting counter-forces to the state presupposes that they can at least satisfy their most urgent needs, without resorting to outside aid. For many grassroots peasants organisations in Senegal , the battle for democracy is subordinate to the battle against poverty.

2.4 Institutional and organisational factors

Institutional and organisational factors also have an effect on the low levels of participation of peasants' organisations. The Economic Interest Grouping (GIE), which is the most common legal form of peasants organisations, certainly offers many advantages, such as the simple way it can be set up, easier access to land (since rural councillors seem more responsive to demands for land that come from peasants organisations rather than from unorganised peasants), and easier access to Conseil nationale de concertation des ruraux (CNCR) credits, but it nonetheless raises some problems that contribute to weakening the associative peasant movement in Senegal. Because of the enormous opportunities they afford, many peasants, even though they are already members of peasants organisations, have rushed into setting up Economic Interest Groupings in an over-hasty manner, to try to lay hands on scarce resources.

As a result, a peasant might belong to several Economic Interest Groupings and the number of such groupings has increased considerably in a few years. The following comments on the effects of the law establishing Economic Interest Groupings bears eloquent testimony to this:

> This measure led straightaway to an exponential increase in the number of peasants organisations, from slightly less than 50 for the whole Valley at the end of 1987, to just over 1,300 Economic Interest Groupings in the Delta alone in 1993. The organisational dispersion of the rural world that has ensued, although it allows the producers freedom of action, also confirms their loss of economic power, because the thousand Economic Interest Groupings continue to function like the old co-operatives (SAED 1994: 61).

The proliferation of peasants organisations might have benefited their members if an organisational framework had been set up at the same time. Instead of increasing their economic influence and their negotiating power, this multiplicity has increasingly weakened them. One can thus find several unions of peasants organisations in a zone, without any framework to bring them together. The structure of lower and intermediate level peasants organisations still remains weak.

Finally, what is the nature of the relations of partnership between peasants organisations, the state, and the NGOs? In theory, the main partners of the peasants organisations should be the NGOs and the state and its organisations. These relationships are, however, somewhat distant and infrequent. Peasants organisations complain about being neglected in large-scale political and economic discussions, and about the lack of any helpful support programmes, and also about administrative delays. Some peasants organisations seem to have reduced democracy to a game for one player, where the state is the leading actor. For its part, the state seems little disposed to see the peasants organisations playing a genuine role in the process of democratisation. This is revealed by its lack of communication and by the marginalisation of peasants organisations in the process of considering regionalisation in Senegal, to quote one example. In many ways, this has been a social programme imposed on them.

Part II: The Issues in Perspective

The future of a democracy depends heavily on the active and conscientious participation of civil society. In order to avoid being an elitist democracy, it must involve the peasants organisations, if only because of their demographic and political weight. What opportunities are on offer to peasants organisations for increased participation in the process of democratisation in Senegal? What preconditions are necessary? What strategies should be put in hand? Three strategic axes for consideration and action seem to us to have priority, the strengthening of the capacity of peasants organisations, the preservation of their autonomy, and the improvement of their negotiating power.

Strengthening capacity

Assuming that Senegalese peasants organisations have the political will to participate effectively in the process of national democratisation, how could they succeed in doing this within a context of ignorance and poverty? If it is true that the Rural Council, for example, should not be a preserve of the elite, a foreign body whose role is limited to allocating and withdrawing land, to collecting rural taxes, and to issuing administrative documents, how could it be made a tool to promote the political, social, and economic well-being of the masses?

Since the low degree of participation by peasants organisations in the process of democratisation can partly be explained by the lack of a certain political culture, which in itself is laking due to illiteracy, it follows that efforts at education and consciousness raising should be made to remedy this. In this area, the state must renounce its hegemonic ambitions, the NGOs must be more politically active, and political leaders must play a role in educating a new citizenry that has nothing to do with obscurantism. Besides strengthening the political culture of peasants organisations, they must also be given a solid base of sound financial management, profitable investments, and improved productivity.

In brief, if the economic power of peasants organisations is enthanced, their sustainability and autonomy would be guaranteed. This strengthening of their capacity must be accompanied by making more effective use of existing human resources.

Those peasants organisations already involved in the battle for democracy within Rural Councils need to be involved at the highest level, since any serious participation requires proven competence among members of peasants organisations in such domains as land management, and the role and functioning of local government organisations, etc. Training should not be reserved for the leaders alone, but extended to all members, notably in such important fields as the training of legal advisers. In this area, the use of the law as an incentive for teaching people to read might significantly increase the intellectual level of members involved and give them means to defend themselves, control their leaders, and acquire new powers.

Finally, this education could be accompanied by exchange visits with peasants organisations that already enjoy a certain degree of political maturity and have made significant advances toward democracy. What good would any increase in capacity be, if peasants organisations remained soulless institutions that were still in thrall to political, economic, and religious powers?

Preserving Autonomy

Peasants organisations must at all costs preserve their autonomy and their identity, with regard to their leaders, the state, and aid donors. But to achieve this, they need a culture and not a cult of power.

The history of national liberation movements has taught us that the identity of the masses has always been scorned by their leaders, who have kept them passive and amorphous, though always mobilised for political causes in which they have no interest, and also that no state willingly gives up any of its prerogatives except under social pressure.

This distancing from the state, which has always made political hostages out of peasants organisations also applies to aid organisations, who may also harbour desires to dominate and make use of them. In other words, peasants organisations must learn to say no to certain forms of collaboration that are associated with 'aid with strings attached'.

The peasants organisation known as the National Association of Rural Family Households offers a good example of ability to preserve its autonomy. In a context of unprecedented financial crisis, it was able to keep aid donors from taking over wherever they wished to do so. Since it considers all its members to be on an equal footing, the National Association of Rural Family Households judged it preferable to follow its constitutional rules than to accept conditions set by donors who in the end, 'kept to the rules.'

The extent of autonomy that peasants organisations can enjoy also depends heavily on their own level of professionalism, and on the level of democracy in their workings.

Faced with globalisation, aid conditionalities, the withdrawal of the state and with the implementation of a policy of regionalisation, how can the great social demands made by members of peasants organisations be met with amateurism and short-term leadership?

Professionalism, far from being just a model or a state of development, should be a challenge, a way of making members of peasant organisations behave like entrepreneurs and managers of small agricultural enterprises, and being capable of negotiating with partners such as provider of services, local organisations, the state, credit institutions, and aid donors.

Professionalism requires a peasant—defined as someone who lives by working the land—to become a farmer, an agricultural producer. To achieve this without a cultural transformation of the peasants themselves, and indeed of society as a whole and of the state may prove difficult. As Michel Colin de la Verdière rightly emphasised, professionalism is rather complex and involves trades unions, politics, economics, and society:

> In the domain of trade unions, professionalism involves defending the interests of producers and representing them in institutional frameworks that often remain to be defined. On the political level, it requires greater participation in the formulation of agricultural policies and changing unsuitable legislative frameworks. From an economic point of view, it implies participation in the organisation and contractual management of channels for agricultural products, in order to ensure a more equitable distribution of added value. On the social level, it promotes the rehabilitation of agricultural activity and the construction of an agricultural identity as a recognised profession (de Verdière 1994: 2).

What about internal democracy? An organisation's ability to survive depends largely on its ability to meet the needs of its members and on the condition of its internal democracy, which should be an indicator of its ability to endure. Some peasants organisations are led by retired government officials, returned emigrants, and people new to the countryside, who see in the peasants organisations a way of making some money for themselves and getting access to some form of power. Rare are the leaders whose leadership styles are democratic or participatory. Paternalism and even authoritarianism are more the rule in the running of their affairs.

Some peasants organisations, moreover, have been made into political instruments, which has resulted in removing motivation from and demoralising some of their members. This is synonymous with non-participation and renunciation, and contributes to weakening peasants organisations.

Some leaders try to hold on to power by discouraging any changes in the office-bearers of their organisations. This leads to paralysis and laxity that enables them to monopolise power. To achieve genuine internal democracy, peasants organisations must acquire the ability to impose sanctions on their leaders. However, there is a level of ambiguity in the internal workings of organisations that do not provide for participation by young people and women? Even if young peoples' participation in the exercise of power is tolerated, many peasants organisations keep the oldest people as their decision-makers, however incompetent they may be. Women continue to play a symbolic role in mixed peasants organisations. Even if they fill the official posts of responsibility for women's affairs all the real power is exercised by men. Yet women have been more successful than men in areas such as savings and credit, even though most of these men still doubt their managerial abilities. Increasingly, women are becoming heads of households and directing agricultural operations (market gardening). It is high time that the question of gender was integrated into peasants organisations. This necessarily involves a revision of roles, powers, and the dynamics of relationships between the sexes. Since 1964, for example, the law on the national domain allows Senegalese women to own land, but it has to be acknowledged that women do not manage land ownership, because they do not have significant representation in the local decision-making authorities.

In the same way, the numerical weight of women should be reflected in the composition of administrative organs. It goes without saying that men are not prepared to see women take on new political roles, but women themselves have to set out to win power. Otherwise, they will take refuge in such organisations as the Groups for the Advancement of Women, from which men are absent. This does nothing to strengthen the peasant movement, which needs the influence and the strength of all its members, regardless of sex or age.

The main challenge facing peasants organisations, apart from strengthening their capacity and preserving their autonomy is improving their negotiating power. This is a function of the nature of the partnership among peasants organisations, and between the peasants organisations and their development partners, such as the state and developmental aid institutions.

While the acquisition of power locally is chiefly a task for grassroots peasants organisations, the strengthening of their negotiating power concerns primarily federations such as FONGS and CNCR.

The establishment of the CNCR reflects the political will of peasants organisations to acquire an effective tool for defending their own interests. Henceforth, the development of agricultural policies, setting prices for producers, and the formulation of national development plans ought to be carried out with the participation of peasants organisations through their national representation. Although it is a relatively new movement, the CNCR has achieved important results. And that is why it has been said that "Having a major social basis, the CNCR constitutes a political force that is far from negligible' (Diagne 1995: 46).

Its long - term objective might be the establishment of an ongoing dialogue with the public authorities and other development partners, in order to take into account the interests of the peasants organisations.

If one is aware that negotiation includes multiple stakes and interests that are often contradictory, one can grasp better the importance for peasants organisations of gaining local decision-making power and enlarging their network of relationships.

Even if peasants organisations are by definition non-political, they are nevertheless developing in a political environment from which they find it difficult to escape. In view of the numerous opportunities offered by regionalisation, for example, can one imagine peasants organisations being marginalised?

As we have already said, peasants organisations are popping up all over the rural world. Though there have been some regroupings at the local and national level, most peasants organisations are often unaware of each other, even those in the same geographical area, and have no common framework for advancing their interests. In view of their limited means and the nature of some of the problems they face, such as the degradation of the environment, peasants organisations that want to achieve durable development would do well to adopt common strategies. There is no point in having a strong national framework for carrying the battle forward if the peasants organisations at the grassroots remain weak, inadequately structured, and isolated from one another. The legitimacy and negotiating power of the national peasants organisations depend largely on how strong and how motivated are grassroots members. It cannot be otherwise, since the chief interlocutors of the national peasants organisations are the state and the donor community, while the grassroots peasants organisations have to deal with the state's decentralised structures and the NGOs on the ground.

The enlargement of the network of peasants organisations would, for example, allow the implementation of common educational programmes or common networks for dealing with certain questions that go beyond the competence of a single peasants organisation.

Efforts should also be made towards the NGOs and other organisations for aid to the rural world, not only in order to seek financial help, but also to consider strategic issues, such as the fight against poverty and the strengthening of democracy. Beyond their material preoccupations, the peasants organisations would benefit from being involved in the elaboration of strategic plans.

Finally, the enlargement of the network of relationships must not be limited to the peasants organisations alone. It should involve partners such as credit institutions, input providers, consumer groups, etc.

Conclusion

The straggles of peasants organisations are part of a global process of social movements. Before independence, the creation of peasants organisations responded to the need to satisfy foreign interests— those of the colonial power—and to use these organisations for political ends.

While it is true that at the beginning of the 1960s, the vocation of the new co-operatives was chiefly to emancipate the rural masses, who were still mainly illiterates, it must be acknowledged that the Senegalese state, in its turn, has not been able to resist the temptation to make these new organisations dependent on itself through its national supervisory structures, which are endowed with powerful human, material, and financial resources. This period coincided politically with the reign of a single party. The peasants organisations had no political ambitions other than taking advantage of this monopolistic situation.

During the 1980s, thanks to the Structural Adjustment Programmes and the withdrawal of the state from certain economic activities, the establishment of a multi-party system on the political level, and the passage of the law on Economic Interest Groupings in 1984, the Senegalese associative landscape underwent a historic change, marked by the proliferation of grassroots peasants organisations. It is now the strengthening of their organisational frameworks at the intermediate levels, the beginning of the battle for the conquest of local power, and the necessity of strengthening their negotiating power in dealing with the state and donors.

The results of the participation by peasants organisations in the democratisation process in Senegal vary widely, depending on the level of intervention and the actors involved. As far as internal democracy is concerned, we should recognise that there have been important achievements at the grassroots, with the emergence of a new peasant consciousness, greater autonomy with regard to the state, participation—weak, certainly, but still remarkable—in local decision-making authorities, and the birth of a partnership with development institutions. At the national level, Senegalese peasants organisations, first through the FONGS and then through the CNCR, have succeeded in strengthening the national associative movement, in taking a more active role in formulating policies of agricultural development, and in serving as interlocutors for the state and its development partners. The limiting factors are many and varied.

Peasants organisations suffer from several ills related to socio-cultural inertia, the illiteracy of their members, the absence of educational and consciousness-raising programmes, a misunderstanding of the legal environment, the aptitudes and attitudes of some of their leaders, the low level of profitability of agricultural operations, and the lack of professionalism in agriculture, etc.

The limiting organisational factors include the anarchic creation of peasants organisations with the aim of getting access to funds, the absence of organisational and co-operative frameworks, and the limits of partnership with the state and aid organisations.

Finally, the future development of Senegalese democracy is heavily dependent on the participation of all socio-economic actors, including above all the peasants organisations, which cover more than half of the population of Senegal. In order to achieve this, bold initiatives are required from the state, the peasants organisations, and the development partners.

References

De Verdière, Michel Colin, 1994, “La professionnalisation: du métier à l'organisation, intercoopérants,” l'agriculture africaine, no. 8, p. 2.

Denis Pesche, 1993, La lettre du réseau GAO, 2e trimestre.

Diagne, D., 1995, *Situation et évolution des organisations paysannes et rurales*, GAO Network (April 1995), p. 39.

Diouf M. and et Sène M., 1980, *Situation économique et perspectives des GIE dans le cadre des Organisations paysannes.*

IUED/SEREC, 1994, Guide d'approche des institutions locales, méthodologie d'étude des acteurs locaux dans le monde rural.

Jacob J.P and Lavigne XXX, 1994, *Les associations paysannes en Afrique.* Paris: Karthala.

Mondjanagui, A.C., 1994, *La participation populaire au développement en Afrique noire.* Paris: Karthala.

La lettre du réseau GAO (Groupement-Association villageoises-Organisations paysannes), 1993, 2e trimestre. 1995, Situation et évolution des organisations paysannes et rurales: Le Sénégal, réseau GAO.

SAED, 1994, 'Evaluation interne du programme irrigation IV', April, p. 61.

Sy C.T. Sy et al., 1988, 'Crise du développement rural et désengagement de l'Etat', ENEA.

Lachenmann, C., 1993, "Civil Society and Social Movements in Africa: The Case of the Peasant Movement in Senegal," *The European Journal of Development Research*, vol. 5, December.

Nindi, B.C., 1990, "Dilemmas of Development in Rural Institution in Sub-Saharan Africa," *Journal of Eastern African Development Research and Development.*

TNdiobo, M., *Réalités rurales et animation pour le développement. Les lieux/moments dans la CR de Coumbacara au Sénégal.*

Santos, A.D. Santos et alet Cie, 1998, *Petits exploitants ruraux: quels problèmes ? Quelles solutions?*

SAED, 1994, Evaluation interne du programme Irrigation IV, April.

Organisations referred to in this article:

National Association of Rural Family Households = Association Nationale des Maisons Familiales Rurales (ANMFR)

National Federation of Groups for the Advancement of Women = Fédération nationale des groupements de promotion féminine du Sénégal (FNGPFS)

Association for the Improvement and Development of the Delta = Société d'aménagement et d'exploitation du delta (SAED)

Association for the Development of Textile Fibres = Société de développement des fibres textiles (SODEFITEX)

Senegalese National Bank for Agricultural Credit = Caisse nationale de crédit agricole du Sénégal (CNCAS)

Senegalese National Federation of Groups for the Advancement of Women = Fédération nationale des GPF du Sénégal (FNGPFS)

National Council for Rural Solidarity = Conseil national de concertation des ruraux (CNCR)

Senegalese Federation of NGOs = Fédération des ONG du Sénégal (FONGS)

National Office for Marketing and Development Assistance = Office national de commercialisation et d'assistance au développement (ONCAD)

National Association of Rural Family Households = Association nationale des maisons familiales rurales (ANMFR)

Non-governmental organisation (NGO) = Organisation non-gouvernementale (ONG)

Part Two

Grassroots Participation and Empowerment

6

Peasant Organisations, Participation and the State: Two Case Studies from Rural Sudan

Khalid A. El Amin

Introduction

The multiplicity of Africa's complex economic, social and political crises is a major preoccupation of national and international development practitioners and thinkers. For the majority peasants the experience of the crisis is a struggle for improvement and survival. Forming the largest part of the world's poorest population, peasants not only experience the continent's persistent socio-economic crises, but their situation is worsened by the direct and indirect appropriation and centralisation of state power. Forming the majority of labour in agriculture and the main productive sector of almost all African countries' economies, peasants bear the brunt of Africa's mounting debt burden while attempting to produce enough to feed themselves and Africa's growing urban population. However, they generate surpluses, which are extracted for reinvestment and future development. In this context they can also contribute to the future curtailment of centralised state power and to building a more open and democratic civil society.

In no country on the continent have peasants, particularly the poorest among them, been so entangled by overarching state power as Sudan. Through direct tight control over production in the state-owned irrigated agricultural sub-sector and excessive direct and indirect taxation of the rain-fed traditional sub-sector, the Sudanese state maintains excessive power over the peasantry. In spite of these adverse conditions, Sudanese peasants have attempted to organise themselves through collective effort and using their own resources to improve their production and living conditions. Various forms of peasant organisations have evolved over the years. They differ in origin and form, in the tasks they undertake, their organisational structure, their membership basis and their relationship to the state and other outside authorities. Consequently the degree to which they have functioned as forums for peasants' participation and as bodies that serve and promote the interests of their members also differs.

This chapter examines the case of two Sudanese 'peasant' organisations — the Gezira and Managil General Tenants Union (GMGTU) and Kassala Vegetable and Fruit Farmers Union (KVFFU) — within the broader context of peasant responses to state encroachment and market penetration. The main argument is that the differential involvement of the state in the production affairs of peasant organisations determines the nature of farmers' participation within the organisations and the extent to which they serve the interests of members. State control over production processes, associated with its interest in revenue and foreign currency in the Gezira case, compromised the effectiveness of the GMGTU and constituted a barrier to tenants' participation, while the freedom enjoyed by Kassala farmers opened up their opportunities for participation.

The chapter first explores some conceptual issues related to the participation of peasants in development and their response to marginalisation through the formation of organisations to defend their interests. It assesses whether these organisations are dominated by rich peasant farmers and if so how, and whether this restricts broader peasant participation. The two peasant organisations were examined in terms of relevant historical background, production relations, objectives, organisational structure, participatory systems, relations with the state and national political power-holders, and the extent to which they served members and contributed to peasant empowerment.

Conceptual Framework and Scope of the Problem

Peasant subordination to the state in Africa has its roots in the colonial era and is inseparable from the continent's underdevelopment and persistent socio-economic and political crises. Whether manifested in tight administrative control over production conditions or heavy surplus extraction, peasant subordination has constrained the development of Africa's countryside. Authoritarian regimes have hampered increased agricultural and food production, the innovative capacity of peasants and peasant participation in local and national affairs. Thus the adoption of the participatory approach to development is regarded by many as one of the key solutions to mounting social, economic and political problems (ECA 1994).

While in economic terms peasant participation is considered to lead to improved agricultural production and productivity, in political terms peasant organisations could contribute to the process of democratisation and the curtailment of state power.

Participation as a concept has evolved over the years to describe widely differing situations. Mishra uses participation to refer to the role of members of the general public in 'influencing the activities of government in providing directly or indirectly for community needs' (Mishra 1984:15). For others, participation is a confusing and an 'ill-defined normative concept of social organization' (Dillon & Stiel 1987: 10) that can only have meaning within a defined social context. Oakley (1987: 4), while stressing the ambiguity of the concept, recognises that it is difficult to formulate a universal definition given the range of development paradigms, and that as a dynamic process it can have more meanings in practice than covered by most conceptualisations.

In 1970s and 1980s development discourse, participation was regarded as an essential factor for the success of rural development projects and community development programmes (Uphoff & Cohen 1979; Rahmato 1991). Accordingly participation meant peoples' voluntary collaborative contributions to development programmes implemented by the state and other outside agencies. This approach was regarded as a more efficient alternative than bureaucratic control, as achieving the desired results and as likely to lead to project sustainability and more equitable distribution of benefits.

Recent trends in development thinking assign greater importance to democratisation, and the concept of participation has correspondingly changed from the mere involvement and collaboration of people to self-reliance and the democratisation of the development process. Participation now denotes more than a contribution to project implementation, and encompasses people's active involvement in creating organisations, in taking decisions and formulating policies that shape their lives.

If peoples' participation in development is a desirable ideal it is also the antithesis of the reality of monopoly over state power. Democratisation and participation in the development process imply a radical redistribution of economic and social power. For the majority of poor peasants, it means the curtailing of state power and that of the rich rural classes to give more opportunity for their effective participation in decision-making and access to benefits. This can hardly be achieved without a long process of conflict.

The formation of an organisational framework within which peasant views are discussed and their concerns voiced is a prerequisite for effective peasant participation. However, the real and potential contribution of peasant organisations to effective peasant participation is that they enable peasants to take decisions, maintain control over development and attain self-reliance.

Participation can be handicapped by state interventions and by socio-economic and political divisions within the rural communities. Thus the analysis of peasant organisations' agency in promoting peasants' interests in both local and national politics must consider that peasants are neither a monolithic nor a structurally undifferentiated body. The evidence shows that African peasants have increasingly been differentiated through state and market penetration (Cowen 1981; Cliffe 1982), and rural class formation.

Even when peasants played a very significant role in the struggle for national liberation, united in opposition to the common colonial enemy, it was the rich peasants and those in control of the state who in the postcolonial period have controlled peasant organisations, dominated the leadership and the decision-making process, and directed such organisations to serve their personal interests.

In the Kilimanjaro area of northern Tanzania, for instance, unequal distribution of power and productive resources in coffee production within the peasant farming community enabled rich and powerful peasant farmers to dominate local organisations and tilted the decision-making process to their advantage.

In Mexico peasants resisted attempts by the state to control production and orient it to export crops through control over credit allocation. Peasant farmers who were organised in consultative committees were able to mobilise and take joint action against the state. Dissatisfaction with state policy was expressed by land occupation, rallies and pressure on agrarian authorities through occupying federal and municipal offices (Rello 1987).

In Colombia the expansion of coffee plantations was carried out at the expense of subsistence peasant production, and displaced peasant farmers organised protests in the 1970s through the National Association of Smallholder Peasants. This threatened to destabilise the Colombian state. State reaction was characteristically repressive and violent, but it served to cement the unity and coherence of the peasantry, who moved to the mountains and started to wage guerrilla warfare against the state (Cusicanqui 1987)

Gezira and Managil General Tenants Union Structure and Peasant Participation

The Gezira and Managil General Tenants Union started to function within a formal structure laid down by the 1954 Tenants Union Act, which remained unchanged until 1992 when it was amended. The GMGTU has a pyramidical hierarchy, with an executive committee (executive office) at the top, followed by a central committee, the executive committee of GMGTU group branches, the central committee of the group branches and similar offices for the block levels at the bottom of the hierarchy.[1] At the base of the GMGTU are grassroots tenants in the 'inspections' (each comprising hundreds of tenants running cotton and other plots in villages). Tenants in each village elect one or two delegates to make up the general congress at the block level. Village delegates who compose the tenants union block branch elect the central committee and the members of the executive office of the block branch.

The GMGTU block branches compose the group branches; with 18 group branches in all, standing immediately above the block branches. Each group branch comprises some seven blocks, and delegates from the block branches make up the central committee of each group branch.[2]

Members of the union bodies at each level of the hierarchy, other than the inspections' delegates to the block branch, are elected through a procedure of indirect representation. This starts from the executive offices of the block branches to the highest office of the GMGTU, the executive office. The delegates of the seven blocks composing the group (four delegates each)[3] of the SGB (Gezira scheme) administrative groups compose the central committees of the GMGTU group branches, which in turn (as in the case of block branches) elect a seven-member executive committee of group branches and the group's representative to the GMGTU top executive office in Wad Medani, the regional Gezira capital

The members of the executive committees of all the group branches comprise the all-GMGTU central committee, which elects the chairman of the GMGTU, the general secretary, the treasurer and their deputies from among the group representatives. The remaining branch representatives occupy a position in the GMGTU top executive office as members.

In addition to its function of electing the occupants of the top positions of the GMGTU executive office, the central committee of the GMGTU meets twice a year to discuss issues of concern to tenants, mainly related to agricultural problems. The GMGTU central committee puts forward recommendations to the executive office to implement and follow up with the relevant state authorities. A quorum of two-thirds of the central committee members can call for a central committee emergency meeting to discuss an urgent issue. The executive office can also summon an emergency central committee meeting, but only with the chairmen of the group branches.

Understanding the formal structure of the GMGTU is critical to understanding tenants' participation and their relation to the state and its organs at the Gezira scheme level and at the centre in Khartoum. The GMGTU's hierarchial structure corresponds locationally and organisationally to the formal administrative structure of the SGB. One of the major attributes of the GMGTU's formal hierarchial structure is that members in one layer of the hierarchy elect the occupants of office in the layer immediately above. Tenants' representatives to all GMGTU offices other than the central committee of the block branches are elected indirectly.

Female tenants are members of the GMGTU and thus eligible to vote for the selection of representatives to the block branch central committees through the GMGTU general congress. In 1961 the number of female tenants in both Gezira Main and Managil Extension was about 9 per cent; it had grown to 11 per cent in 1981.[4] However, females did not stand for election or selection as representatives to the block branches. Although legally females had as much chance of selection as males, social and cultural factors limit their active participation. Like the majority of tenants their effective participation was limited to the selection of representatives to the block branches. But these factors constraining female participation were surmounted by some educated females who participated in various social activities. However, female Gezira tenants were generally widowed women and mostly illiterate.

If we examine tenants' participation in the Gezira scheme we find that members who occupied GMGTU offices enjoyed a relatively higher degree of participation in decision-making, with relatively higher levels of effectiveness, than tenants occupying the lower levels of GMGTU bodies. While occupants of GMGTU executive offices participated in some of the decisions relating to the scheme as a whole and had political influence in national politics, tenants at the village level had a more local orientation and their participation through the union hierarchy was limited to selection of representatives to the GMGTU block branches every four years. Effective grassroots tenant participation in the election and accountability of GMGTU official bodies higher up the pyramid was constrained.

Since the union's electoral procedures were based upon a system of indirect representation, they persistently brought to the GMGTU offices, intermediate and top leaders who were unpopular among tenants at the grassroots level, because they did not directly participate in their selection. Members of GMGTU bodies other than the block branch central committee, from the block branch executive office and up to the GMGTU top executive office members, were not elected by tenants themselves but were all indirectly elected by tenants' representatives.

There were complex socio-economic and political factors whose interplay limited the effectiveness of tenants' participation in the selection of representatives to occupy the GMGTU formal structure at both intermediate and top leadership level. The main socio-economic and political developments, both at Gezira and at national levels, to have had significant impact on the Gezira economy and society and on the functioning of the GMGTU included the existence of numerous tribes, the important role assigned to the tribe and the loyalty it received from community members; and the spread of Sufi Islamic sects (tarigas) and their influence in the life and culture of the then subsistent peasant communities.

In addition, three important colonial and postcolonial socio-economic and political developments related to the above religious and tribal factors were critical. First was the growth of the market economy and the start of the process of social differentiation based on money and the acquisition of financial resources. Second was the policy of indirect rule adopted by the Brrket. The decline in cash returns from cotton and other crops and the increase in the price of consumer commodities led to the erosion of the purchasing power of the majority poor tenants. Hiring tractors for durrah, wheat and groundnuts was a burden that could only be cushioned by resorting to the better-off tenant groups for credit, and credit from rich tenants was also necessary to provide the necessities of life and badly needed cash.[5] These interactions between poor and rich tenants had strengthened a client–patron relationship that had an influence the selection of leaders to the GMGTU offices. This trend impacted negatively on effective tenant participation at all GMGTU levels.

The chain of support for intermediate and top GMGTU leadership, facilitated by the indirect electoral procedures, was enhanced by the GMGTU's economic activities and the allocation of tractors and combines.[6] Top GMGTU leaders, who decided on the allocation of the union's tractors and other agricultural equipment, used their position and influence to favour allies and followers among the intermediate tenants' leadership which elected occupants of the GMGTU top offices. Tractors and combines acquired by intermediate leaders (group and block leaders) were in turn hired to tenants on credit to enhance the leaders' political support base.

Despite divisions and factionalism among tenants, which were a source of weakness that made the GMGTU leadership and Gezira tenants susceptible to state manipulation and control over the scheme, tenants were unified in their discontent at their economic position within the scheme. They demanded an increase in tenants' sales proceeds, participation in the administration of the scheme and the diversification of crops to stabilise tenants' income in response to fluctuations in cotton prices.

It was the militancy shown by a radical faction in struggles for these demands, which were met by the state, that gained the GMGTU the support of rank and file tenants, among whom traditional tribal and religious loyalties dominated.

Tenants enhanced their participation through collective action in resisting state demands for surplus extraction, gaining an increase in tenants' cash income and in putting tenants' representatives on the board.

Tenants demanded better representation in the top SGB administrative body, because the tenants' representatives on the board were a minority (three out of ten) and consequently had an insignificant impact on the decisions of the board. The other seven members of the board were all technical experts and high-ranking officials. Discussion within the board preceding decision-making involved technical issues in which tenants' representatives were not well versed. The seven board members were also members of other state bodies and had a high loyalty to the state. This marginalised the three tenants' representatives. State and SGB interests in cotton had dictated centralised control mechanisms that left little room for tenants' representatives to influence major decisions on crop mix, crop rotation and other inputs.

Although they had been officially recognised by the state as partners and not wage earners, in reality tenants had no effective decision-making role either at the top levels or the lower levels of the SGB administrative hierarchy. Factionalism and divisions based on party loyalties and party affiliations frequently surfaced among tenants, leading to destabilisation of the union, limiting the effective participation of tenants in general.

From the mid 1970s to the early 1980s, however, a number of laws were enacted by the state to facilitate and materially support involvement by GMGTU executive officeholders in economic activities on tenants' behalf.[7] This served to undermine the tenants' collective resistance as they were co-opted by elected officials. For instance, Badr's union executive office in 1982 accepted changes to the land and water charges and individual account (LWC&IA), which had been resisted by tenants for some time. This significantly altered the tenants' relations with the state and the manner of their participation in the Gezira scheme, which had been established by the British in the 1920s. The sharing formula for production, which had defined the relationship between the state, tenants and the scheme's administration, was abolished. According to the new LWC&IA rules tenants would bear the costs of production for all crops, and pay land and water charges plus the SGB administrative costs through deductions from each tenant's individual account. However, despite this change in the production cost structure, the state still made decisions on the crop mix and rotations.

The effects of this change motivated both rank and file tenants and their leaders (particularly intermediate leaders) to demand greater participation in the decision-making process from the field level up to the top SGB hierarchy.

In the first place, as the cost of production incurred by tenants increased and their cash returns declined, the tenants' indebtedness increased. Rapid increases in production costs resulted from input price rises and inappropriate mixes of inputs[8] imposed by the SGB. Tenants had little knowledge of inputs costs, and they suffered from delays in delivery and untimely application of inputs and the performance of some mechanical operations. All this led to pressure by tenants' intermediate leadership for tenants to have close supervision of production costs so as to improve net returns.

Secondly, the tenants associated the increase in their costs of production partly with SGB administrative charges. Tenants argued that there was no need for administrative control because they had long experience with crop husbandry and they knew their interests better.

A third problem arose because the payment of water charges did not tally with the amount of irrigation water required and received, and the water was not timeously provided. Tenants participated neither in the distribution of water nor the maintenance of irrigation infrastructure, and water charges were levied irrespective of the frequency of watering. Although tenants paid annual water levies, the irrigation system was not properly maintained and consequently some tenants suffered from damage to crops and financial loss.

But despite increasing grassroots tenant pressure for compensation in cases of loss and expenditure on the maintenance of irrigation network, both problems still remain unsolved.

The control of decisions on crop mixes became more problematic after the implementation of the LWC&IA. While charges on land and water plus other taxes (see below) ensured revenue for the state, the state monopolised the decision-making, especially in cotton because of its imprtance in earning foreign currency.

Gezira cotton represents between 45 per cent and 60 per cent of Sudan's agricultural export earnings by value, making it the major source of foreign currency. Yet while tenants bore the production costs and faced the risks, they had no right to decide on what to grow to maximise their returns. Accordingly, they increasingly deviated from the set crop mix and crop rotation, despite the risk of fines.[9]

In addition to the above charges, taxes and dues levied for water and land by state bodies contributed to the tenants' low cash returns and stirred up resentment. Annual export tax on cotton, for example, amounted to 10 per cent of the gross proceeds of cotton, while the local crop tax levied on each crop produced by the Gezira tenants was also 10 per cent. There were also indirect taxes in the form of custom duties on production inputs.[10] And deductions were also made by the union executive office from tenants' individual returns for government organisations and in support of jihad. This increased tenant indebtedness.

In most cases deductions were made by the top leadership without the tenants being informed,[11] which created scepticism among tenants and reinforced mistrust of an already unpopular leadership. This generated tension especially when the SGB officials were given the authority to use state security forces to appropriate durrah — food store by tenant families—in repayment of debt.[12] Persistently indebted tenants were threatened with eviction.

State control over Gezira tenants that limited access of tenant farmers to what they produced even after the deduction of all costs, is best illustrated by forced wheat delivery. Tenants had to deliver their wheat output to the state at a pre-set price.[13] Wheat was harvested in the presence of the security forces, and taken away by SGB staff. This practice was started during the

Nemeri regime, and continued by the transitional government that assumed power after its overthrow in the popular uprising of April 1985. Forced wheat delivery to the state without immediate payment of tenants' wheat cash returns led to the spontaneous June 1989 strike, which ended only when the present military regime took over, paid tenants' wheat returns and modified the practice by enforcing only the delivery of a quantity of wheat equivalent to its production costs plus charges.

These problems exerted great pressure on the GMGTU leadership at all levels. The executive office representing the top leadership took on a dual role by attempting to lessen state control while at the same time mediating and facilitating acceptance of state policy among tenants. Members of the executive office persistently urged tenants to repay their debt to the state but were seen to have done nothing for member tenants whose food had been taken by the security forces and while forced wheat deliveries were still maintained.[14]

The position of the GMGTU executive office on the eviction issue was in line with state policy of dismissing tenants with low yields and who continuously ran into debt.[15]

GMGTU intermediate leadership at the group and block branch levels, which was closer to grassroots tenants, consistently articulated tenants' problems, concerns and demands, but without response from the top leadership. The failure of irrigation, compensation in cases of crop loss, rescheduling of tenants' debt, reduction of state taxes on crops, cutting of production costs and the low level of participation in the decision-making process were some of the problems facing tenants that were frequently brought up by the intermediate leadership in its joint meetings with the executive office.

Although many of the intermediate leaders were mostly better-off tenants the reduction of production costs and taxes, and the improvement of irrigation networks were problems which affected all tenants. Thus this level of leadership and tenants united in resisting their marginalisation. The GMGTU top leadership, on the other hand, was more occupied with enhancing its power vis-à-vis the SGB, by having more control over the administration of the scheme, than with the problems of rank and file tenants. Given the present alliance of the state and executive office members, state interests continued to prevail over the rights of grassroots tenants. In this way, state involvement both in production and in manipulating the GMGTU remained an important factor in constraining tenants' participation.

Kassala Vegetable and Fruit Farmers Union

Kassala town, in far eastern Sudan close to the Ethiopian and Eritrean border, around which extensive horticultural farming has developed, was established in 1840 by the Turkish governor Ahmed Pasha as a frontier post. The town later emerged as a regional centre and an important point on the trade route linking Ethiopia and the Sudan (Steenwinkell 1981:14).

El Gash is a seasonal river which flows from the Ethiopian highlands, runs through Kassala from south to north and divides the town into eastern and western parts. Although the river flows only during July–August, it recharges shallow and deep underground aquifers, which encouraged well construction and the development of small-scale horticultural farming. Horticultural production is practised in the immediate vicinity of Kassala; north and south of the town, along the eastern and western banks of the river.

Horticultural farming around Kassala started during early 1920s with the encouragement of the British administration, after the first sagia (irrigation area using oxen-driven water wheels) was licensed in 1921. Between 1935 and 1944 citrus fruits including oranges, grapefruits and limes were introduced, and the planting of fruits was further encouraged by the establishment in 1944 of the government fruit tree nursery that distributed seedlings to farmers (Thrower 1954; Fisher 1954). Interest in farming particularly among migrants from northern Sudan led in 1948 to the formation of the Land Distribution Commission, which regulated and supervised land distribution to sagia applicants (Steenwinkell 1981: 35).[16]

Land was leased to farmers for a period of 45 years for a nominal rent of two shillings per feddan per year. However, once leaseholder farmers had constructed their wells and installed the sagia certain customary rights were established. In practice these customary rights amounted to absolute individual land ownership and were recognised by the government, although in legal terms land remained under state ownership (Ibid.: 36). In effect the leaseholder was considered to be the landowner since the landowner had in practice an unrestricted right of land use. The leased land could be sold and inherited and leased to others.

The horticultural production areas, which have been developed over the years around Kassala town, may be divided into five geographical zones. The earliest and largest irrigated areas are El Sawagi El Ganoubia (Southern Sawagi) and El Sawagi El Shamalia (Northern Sawagi), which have been developed since the 1940s; both are west of El Gash seasonal river. In addition, the relatively smaller areas of Khur Shaygyya, Wad Sharafie and El Sawagi El Shargia (Eastern Sawagi) were developed during the 1950s and 1960s.

Horticultural production in Kassala was primarily geared to the production of high-value vegetables and fruits which were marketed in Kassala, Port Sudan, Khartoum; sometimes in El Obied in western Sudan, and across the border in Ethiopia and Eritrea. Onions, tomato, leafy vegetables, raddish, cucumber, okra, carrots, beet, water melon and melons were the main vegetable crops. A variety of fruits was produced, including oranges, limes, grapefruit, guava, mangos, bananas and dates.

Kassala fruit and vegetable production has become relatively highly capitalised. In addition to modern irrigation techniques, farmers made use of the best current practice. Herbicides, insecticides, pesticides, fertilisers and animal manure were used to improve vegetable yields. Improved seedlings and cross-bred varieties of fruit are planted. Better crop production and crop husbandry, which increases output, plus the need for profitable marketing channels, led to attempts by farmers to seek ways to export fruits and vegetables to Saudi Arabia and other Gulf countries.

These developments in horticulture were accompanied by the gradual evolution of certain production relations, as explained below.

The forms of production that evolved were based on private land control and independent farming, which unlike the case of Gezira involved little or no state control. This factor impacted on the structure of Kassala Vegetable and Fruits Farmers Union (KVFFU) and the degree of participation within it and its relation with the state. The dominant forms of agricultural practice were family farming for smallholders, share cropping for some farmers and hired agricultural labour for relatively large farmer owners on fruit farms.

One major characteristic of horticultural farming in Kassala was a wide disparity in the size of land holdings. Although the size of farms varied between four and 180 feddans, the majority of farmers owned between four and eight feddans.[17] In general ownership of irrigation facilities differed between individual farmers, according to economic ability and the size of plot(s). While some small farmers used the traditional methods of sagia, others owned modern water pumps and electric motors. Among the better-off farmers ownership of water wells, pumps and water-lifting electric motors varied from one to six facilities.[18]

In terms of share cropping, owners provided land and irrigation water (with pump, fuel and maintenance), while the cost of inputs other than sprays, fertilisers and seeds was shared between owner and share cropper according to their shares in net cash returns.[19] The share cropper supplied labour (his own and his family's) for all agricultural operations and kept water channels clean from weed. The shares of the partners in the profit of all vegetables other than onions were 50 per cent each between share cropper and owner, the same share as in the cost of inputs. The distribution of onions and bananas was 40 per cent share in net cash returns for the share cropper and 60 per cent for the owner. Hired agricultural labour was used on fruit plots owned by larger farmers.

Although there was some overlap between the different categories of farmers, there were clear distinctions between rich and poor farmers. The few rich farmers represented around 5 per cent of all farmers in the Kassala area. They owned large-sized farms (40-180 feddans),[20] had bored deep wells, owned fruit tree farms in addition to vegetable crop plots and used both hired labour and share cropping. For rich farmers who were also large landowners, agriculture was one among many other cash-earning economic activities, including trade and the export of vegetables and onions (purchased from small farmers) to Eritrea and Ethiopia. On the other hand, small and middle farmers used sagias and diesel-powered pumps on open, shallow wells to irrigate small and medium farms that varied in size between five and eight feddans. These were worked mainly by family labour and supplementary wage labour during peak seasons. The crops cultivated by poor and middle-level farmers included some fruit but mostly vegetables that had a quick cash turnaround.

Seed capital invested in this costly form of irrigated agricultural production had different origins. Some early owners of sagias were able to make profits from the cultivation of vegetables, which they invested in pumps and purchased land to increase their holdings and turned to fruit crop production. This happened in Southern and Northern Sawagi.[21] In other cases farmers' sons who migrated to the Gulf and Saudi Arabia sent remittances to buy land and establish new farms and/or develop family farms. In some instances capital accumulated from involvement in other businesses by Kassala inhabitants, for example, commerce, was invested in horticultural production once the sizeable profits made by pioneer farmers were observed.

The emergence of the Kassala Vegetable and Fruit Farmers Union in 1968[22] was a response to problems that accompanied developments in horticultural production. The KVFFU has made a significant contribution both to agricultural production and socio economic development in the Kassala area. Both the union and the farmers it represents have enjoyed organisational freedom. While individual farmers had a choice in decisions on all matters related to agricultural production (e.g. the crop mix, crop rotation and the mix of inputs), the KVFFU had freedom of choice and action to undertake whatever measures it considered necessary to solve farming and farmers' problems.

The expansion of the cultivated area, the adoption of water pumps and the increased use of modern methods, which certainly contributed to increasing agricultural production and farmers' income, were accompanied by the emergence of new problems and difficulties perceived to be beyond the ability of the individual farmer to solve. Moreover, the solution of some of these problems often involved contact with state agencies. The KVFFU was formed both to contribute to solving agricultural problems and to represent farmers as a collective body, voice their concerns and articulate their interests and demands when dealing with government authorities.

From its inception KVFFU membership was open only to male farm owners. Although females and share croppers and agricultural labourers might benefit from the KVFFU's activities and facilities, they were not entitled to KVFFU membership. Share croppers and farmers considered themselves to have different interests, and share croppers had their own union. However, once share croppers owned farms, they were automatically accepted as KVFFU members. But females were not entitled to KVFFU membership even if they owned land.

Female exclusion derived from a long tradition of segregation between the sexes and the non-acceptability of females participating in what was considered to be a male job. Moreover, membership was not only restricted to the farmer member (the male head of the household), but extended to other male sons who were working with him on the farm. The male heirs of a deceased farmer who continued to work on the farm were also entitled to membership.

During the period 1968-92, the selection of KVFFU officials did not follow a formal election procedure, with the choice of representatives based on a tradition of agreeing on selections before meetings were held. Farmers in Kassala's five horticultural areas held a general meeting in which a choice of representatives to the central committee was made, in accordance with a list of chosen individuals agreed upon before the meeting. The representatives (members of the central committee) then chose by consensus the chairman of the union, the secretary, the treasurer and their deputies.

Since 1992 the KVFFU had functioned according to the 1992 Nomads and Cultivators Organisations Act, which gave it some formality.[23] But there remained a number of differences between KVFFU actions and the provisions of the Act, given specific conditions in the area and the prevailing political climate in the Sudan since the late 1980s.

The 1992 Act stipulated that the union should be constituted in a pyramidal structure, starting from the bottom with branch unions, each of which would send delegates to the general congress, a central committee and an executive office. The union branches had their own central committees and would send representatives to the central committee of the general union and also one representative to the executive office of the general union.

The adoption of this kind of pyramidical organisation might be relevant to the Gezira scheme, where the large number of grassroots members (164,000 of them) and the centralisation of the GMGTU accompanied by a system of indirect representation led to the adoption of a five-layer hierarchy. But in Kassala's union, where an estimate gave only some 1,300 members,[24] some layers of the hierarchy prescribed in the 1992 Act were in practice skipped altogether. Thus unlike the GMGTU's general congress, which was composed of representatives, the general congress of the KVFFU was made up of grassroots member farmers who directly elected delegates to branches who then formed the central committee of the union and its executive office.

Thus the selection of the KVFFU central committee members was made directly by grassroots farmers rather than the four or five steps of indirect selection adopted by the GMGTU.

The KVFFU's top leadership (the executive office) was composed of 11 members 'elected' by the 40 - member central committee from among its ranks. The 11 members of the executive office elected from among themselves a chairman, secretary, treasurer and their deputies. Each one of the five remaining executive office members became a chairman of one of the specialised committees attached to the executive office.

The agricultural services committee handled all issues related to agricultural production problems and the promotion of agricultural development including inputs and equipment as and when necessary. The livestock committee, introduced since 1990, was mainly concerned with the improvement of livestock production (chiefly cattle and poultry). The dispute settlement committee settled disputes without resorting to courts. The members' affairs committee handled contacts with government officials, state departments, companies and other relevant organisations and agencies.[25]

The structure and composition of the dispute settlement committee differed from that of other committees. It had no permanent structure and its membership changed according to the location of the conflict under its consideration. For the Kassala horticultural area there were three chiefs (sheikhs) who mediated in the settlement of disputes. The committee was composed of the chairman and members, one of the three chiefs (according to the location) and two members from the share croppers' union. The two representatives of the share croppers and the farmers were chosen from the area in which the conflict between a share cropper and a farmer had occurred. So neither was the number of the committee specified nor was its membership permanently constituted, and subsequently the composition of the committee changed as the area of conflict changed.

KVFFU bodies from grassroots farmers to the top executive leadership functioned through holding meetings at three levels — the general assembly (individual grassroots KVFFU members at branch levels),[26] the central committee (the 40 - member delegates of the five union branches) and the executive office and its specialised committees. The general assembly, which was composed of rank and file farmers at each branch level, convened every four years to elect its assigned number of delegate members to the central committee. Elections were held on a one-member-one-vote principle.

The KVFFU leadership had greater freedom of movement than in the case of the GMGTU, partly owing to the absence of both direct state involvement in production and parastatal organisations in Kassala horticulture. In addition, state interests were not as heavily involved as in the Gezira cotton production. Farmers could decide on their own crop mix, crop rotation, inputs, the marketing of their crops and moreover keep in full cash returns for their crop.

Nonetheless, in its efforts to deal with farmers' problems the KVFFU did come in contact with the different state institutions responsible for agriculture and service provision, regardless of the government in power. In many instances the KVFFU leadership, relying on member farmers' own resources, made contributions to the cost of the services rendered and provided facilities to the concerned government department. For example, the KVFFU contributed financially to the research programme of the Ministry of Agriculture for the innovation of new fruit varieties. Another example was the support KVFFU rendered to the state-owned nursery to enable it to provide farmers with improved seedlings of fruit varieties.

Historically, the preconditions for the formation of the KVFFU were created by the development of risky privately owned irrigated horticultural farming. Here farmers took the decisions and risks, and reaped the benefits. The KVFFU emerged as a result of internal pressures and social needs, and while some external agricultural factors influenced its development it was not created by an external national political organisation nor was its formation connected to national politics in the same way as the GMGTU. Rather it was set up to solve agricultural problems and provide services necessary for the promotion of agricultural production.

With the relatively small size of the area under cultivation and the relatively small number of farmers KVFFU served, national political forces had not been as heavily involved as in the Gezira sheme. Farmers' pressing problems and needs at the local level assumed primacy over factional national politics, party and other political affiliations that have subverted other Sudanese 'peasant organistions'. Relations with the state in this case were manipulated and used to meet farmers' demands and solve farming problems rather than the KVFFU becoming susceptible to state manipulation and serving its interests. Under conditions of totalitarianism, the KVFFU's limited political presence was a source of strength rather than weakness. Unlike the GMGTU, the KVFFU did not present a potential political threat, and the agricultural activities of its members were not as detrimental to the economy and the state as those of the Gezira tenant farmers'. Therefore the KVFFU was not as closely watched nor its activities as keenly supervised. Looser central direction was facilitated by the KVFFU's overriding concern with direct farming issues.

KVFFU Contribution to Promoting Farmers' Interests and Their Participation

The KVFFU had made some substantial contributions to the promotion of agricultural, social and economic development, and remained actively involved with state bodies in handling problems encountered by Kassala's vegetable and fruit farmers. A major problem faced by the union both in the past and present was the provision of a guaranteed supply of fuel for diesel-powered water pumps. The economic situation in Sudan had led to the scarcity of commodities such as fuel, but the KVFFU succeeded in using its relation with the state from the Nimeri period to the present to secure a quota of diesel fuel. This was allocated in its name and received by KVFFU directly from Port Sudan. The KVFFU was also actively involved with state bodies in attempting to reduce regional state taxes levied on fuel, which were around 14 per cent per barrel.[27]

In its efforts to provide a radical solution to the fuel problem the union made a serious attempt to extend electricity to the irrigated areas and to encourage the adoption of electric motors, all through farmers' own financial resources. The extension of the electricity network to some of the farming areas started in 1987, when each group of farmers received a transformer that supplied them with electricity at a subsidised rate of Ls 50/kW. This successful response to the problem of scarcity and high prices of both fuel and spare parts for water pumps had however been limited by the fact that only few farmers actually used the electric power.

Moreover, KVFFU helped farmers to access spare parts and obtain production inputs such as insecticides, pesticides, spraying equipment, improved seeds and fertiliser.

The union co-ordinated its activities in the provision of inputs to farmers with local multi-purpose co-operatives at a subsidised rate. The co-operatives also had technical experts to advise farmers on appropriate input use. The KVFFU obtained some of the inputs on credit from the Farmers Bank, and the Agricultural Bank of Sudan or on cash payment from El Rahad Agricultural Corporation, Wadi Halfa Agricultural Corporation or from private companies in cases of emergency.

Horticultural production in Kassala area was tied to the market and faced costly production inputs. This created the need for capital, and capital availability was thus one of the major constraints facing particularly the poor small farmers, since access to bank credit depended on the farmer's ability to provide collateral. KVFFU's contribution to solving this problem was by facilitating farmers' access to bank credit through the issuing of 'guarantee warrants'[28] to individual farmers as a substitute for collateral.

The seasonality of vegetable and fruit production resulted in the flooding of Kassala's and other Sudanese markets with perishable fruit and vegetables. With the continuing expansion of the cultivated area, output had increased, with a consequent drop in prices and decline in farmers' cash returns. Other marketing channels in Saudi Arabia and the Gulf were now sought in co-operation with the Arab Authority for Agricultural Investment and Development (AAAID). Earlier border trade with Eritrea and Ethiopia, organised by the KVFFU in co-operation with government authorities was aborted following tensions between Sudan and its two eastern neighbours.

Heavy and dual taxation had been one of the major issues of concern to farmers. Accompanying the policy of decentralisation, which had been put into effect since the early 1990s, numerous state, regional and local administrative bodies were created which, in the absence of central financial support, sought local sources of revenue to meet their administrative costs. Numerous examples of duplicated local and federal taxes were levied on Kassala horticultural production; including a land tax, production tax, crossing tax, zakat and income tax. Apart from income tax, other taxes levied on agricultural produce amounted to around 33 per cent of the crops' cash value by 1994.[29]

KVFFU persistently lobbied to reduce and unify tax channels, which resulted in the Council of Ministers passing a resolution in 1994 that limited taxes on irrigated agriculture to 15 per cent of the cash value of crops (10 per cent production charges plus 5 per cent zakat). It took two years of lobbying with the local authorities, during which KVFFU's relations with the farmers were strained, for KVFFU to be able to have the resolution put into effect in 1996. In addition, the KVFFU also helped reschedule farmers' income tax and debt to banks by contacting the relevant authorities on farmers' behalf.

One of the activities of the KVFFU executive office was the resolution of conflicts between farmers and share croppers. The dispute settlement committee would receive a complaint from either the share cropper or the farmer. As an example, a farmer could allege that a share cropper had failed to keep irrigation channels and sub-channels clear of weed, while a share cropper could complain to the union's executive office that the owner had failed to provide the necessary inputs timeously. The committee provided its good offices to mediate and finally ratify a settlement. But should mediation fail, the dispute was taken to court. In practice, most of conflicts between share croppers and farmers were settled through the committee.

For all its wide-ranging contributions to solving production problems, KVFFU as a forum for real participation of those involved in farming had been limited in terms of membership, debate, elections and other KVFFU activities. Making land ownership a condition of membership limted numbers and excluded share croppers. And in strict legal terms even the so-called land or farm owners were not land owners. The distinction then becomes a differentiation between owners of capital and providers of labour.

Though female membership was legally allowed as some females do 'own', or more accurately control, land, yet female participation was limited by tradition and cultural factors. In legal terms females who 'own' land were considered KVFFU members, but in reality their membership was redundant because females participated neither in elections, in meetings nor other KVFFU activities. So far this seemed to be accepted by women themselves and there were no attempts on women's part to break traditions and cultural barriers.

Farmers' participation within the overall organisational structure of the KVFFU was through the representatives of grassroots farmers having direct contact with KVFFU's central committee members and the latters' direct accountability to farmers. The central committee, which was directly elected by grassroots farmers, met every three months to discuss issues and problems raised by members and articulate concerns to the executive office to take action. This enabled the central committee to follow up the activities of the executive office and assess actions being taken on pending issues. Moreover, members of the KVFFU general congress, the grassroots farmers, met annually to discuss issues, voice concerns and put forward recommendations to the central committee and the executive office as guidelines for action. However, because meetings were infrequent, the number of members relatively large (around 1,300) and because meetings were held for only one day, participation in discussions by grassroots farmers was limited. Nonetheless, the direct contacts between farmers and the leadership which these meetings offered was a forum for participation by the rank and file farmers to openly voice their concerns and views to the leadership.

Participation in the selection of representatives by individual grassroots farmer members of the KVFFU was partly determined by the position of the member within the KVFFU organisation and structure. Occupation of a position within the KVFFU provided more opportunities for participation in the selection of members of the KVFFU bodies. The higher the farmer's position within the union, the more frequently he participated in the selection of occupants of KVFFU offices. Similarly, the degree or the extent of participation in the decision-making process and in other KVFFU activities corresponded to the position the member occupied within the KVFFU structure. Members of the executive office participated more in decision-making and in other KVFFU's activities than ordinary members who met once a year and took part only in the general discussion of issues and the formulation of general guidelines.

Although participation in the selection of KVFFU officials was practised according to vote, prevailing national political conditions did have some influence on the degree and form of participation in the selection of these officials. During parliamentary governments, party loyalties and KVFFU leadership overlapped, but the union's work mainly focused on farming and local problems, not national political issues. During military governments (the present one included) state intervention was more direct in influencing the choice of those who occupied top KVFFU's offices. Although meetings were called and elections might be held, the outcome was influenced by forms of covert political manipulation that limited effective participation by grassroots farmers in the selection of the leadership. However, in Kassala and for reasons described above, the choice of leaders followed tradition rather than government imposition, and this influenced the list prepared by top state officials.

The economic position of Kassala farmers influenced the extent to which they were able to participate in and benefit from KVFFU activities. The rich farmers participated more in decision-making and other KVFFU activities than poor and middle-level farmers. Since KVFFU top offices were mostly occupied by rich and better-off farmers, the correlation between economic ability, prospects and scope of participation is clear. However, most problems of irrigated horticultural farming in Kassala were shared by all categories of farmers (e.g. the supply of diesel oil, reduction of state taxation, provision of improved seeds and agricultural equipment). Thus actions taken by the executive office members on these issues reflected common concerns.

Share croppers' participation in KVFFU activities and selection of KVFFU officials was limited by their exclusion from membership, which was conditional on control over land, but the share croppers did share benefits brought about by KVFFU. Improved seeds, agricultural equipment, diesel oil, fertiliser, insecticides, herbicides, durrah and other consumer goods acquired by the KVFFU were distributed through co-operatives and benefit share croppers as well as farmers. Women, who were excluded from participation in KVFFU activities through tradition and culture, also shared such benefits. Services and facilities such as vocational training provided for women in the field of food processing, small-scale industry and income-generating activities were another benefit from KVFFU support.

Conclusions

The establishment of the Gezira scheme in the 1920s marked a radical change in the socio-economic and political structures of the region. Former subsistence producers were incorporated into the market through the production of cotton for the British textile industry. State activity was instrumental in bringing about that change. Land control was shifted from peasant owners to the state through a complex arrangement of rent and lease that turned formerly independent peasants into cash crop producer tenants on their own land. The state controlled the means of production and the production process. The administrative structure created to control production was hierarchical and centralised to ensure efficient cotton production in accordance with the quantity and quality required. Former independent peasants, now turned into tenants, were told what to grow, when to grow and how to grow it within a framework of tightly administered and controlled time schedules. The dominance of state and capital interests in cotton led to the marginalisation of tenants and tenants' interests.

The KVFFU differed from the GMGTU in origin, organisation of production and production relations, relation to the state, membership, electoral system, the degree and level of participation provided for its grassroots membership and the extent to which it served its farmer members. KVFFU—state relations were limited to the union's obtaining state and parastatal help in achieving these objectives and at a later stage in reducing taxes. Although there were state attempts at politicising the KVFFU, it was the union that managed to manipulate or use its relation with the state to achieve its objectives of serving farmers' interests.

The form, degree and effectiveness of participation within the KVFFU had more to do with factors internal to it than with state involvement.

The state was not directly involved in Kassala's horticultural production, and this gave room for relatively greater participation within the KVFFU. Unlike the GMGTU, minimal state involvement in the KVFFU was facilitated by the more localised nature of KVFFU activities and influence plus its relatively slight weight in national politics and the economy. Thus even under totalitarian regimes Kassala vegetable and fruits farmers managed to keep their leaders on the state's selected list of executive office members. The ability of KVFFU leaders to strike a balance between national political affiliation on the one hand and their own farming demands on the other helped the KVFFU leadership remain non-politicised and to focus on KVFFU objectives. For instance, it had managed to acquire a guaranteed diesel oil quota, extend electricity, reduce state taxes and get technical experts to assist farmers.

In contrast to the GMGTU, the KVFFU's top leadership was more directly elected by grassroots farmers. In addition to meetings in which farmers' concerns were voiced, the KVFFU leadership was also in more direct contact with its grassroots. The direct contact between grassroots and leaders had contributed to the responsiveness of the KVFFU leadership to ordinary farmers' demands. Although better-off farmers participated more than poor and middle-level ones, KVFFU was a far more participatory organisation than the GMGTU. However, some forms of participation were limited by the KVFFU's own rules and traditions. In terms of representation participation of others involved in farming like share croppers was limited by the restriction imposed on membership through land ownership. Women who controlled land were excluded not because there was a member ship restriction policy based on just gender but more for social and cultural reasons. In terms of sharing benefits, farm owners benefited more from KVFFU activities than others. However, share croppers and females shared access to resources and opportunities, the provision of which had been facilitated by the KVFFU. Still better - off farmers benefited most from KVFFU activities, but not to the exclusion of others.

The two case studies show the extent to which state involvement in peasant organisations can affect members' participation and socio-economic development. The more the state is involved, as in the GMGTU, the less room is allowed for the involvement of the 'cultivators', and the less the state is involved, as in the KVFFU, the more room is allowed for participation.

Notes

1. The Gezira Tenants' Union Act, 1954; also interviews with Gadalla Hassan Mohamed, member of the GMGTU executive office, and Musa Mohamed Ali, chairman of Wadi Shaair group branch, *Wad Medani*, August 1997.
2. Ibid.
3. Each group branch comprises seven blocks. Each block branch sends four representatives, making 28 members of the central committee of the group branch. Personal Interview with Musa Mohamed Ali, chairman of Wadi Shaair group branch, *Wad Medani*, August 1997.
4. Women tenants were 6,475 out of a total of 73,796 tenants in 1961 and 13,057 out of 102,247 in 1981. This increase was due both to the area expanding by adding the Managil Extension and to subdivision of tenancies (Shaw 1965: 15).
5. Interview with a Managil tenant, Khartoum, August 1997.
6. GMGTU internal files, *Wad Medani*.
7. In 1974 Agricultural Bank of Sudan lending regulations made 'guarantee warrants' issued by farmers' and tenants' unions an acceptable form of collateral. In 1976 the Peasant Farmers' Organisations Act legalised the pursuit of economic activities by the unions' executive offices, including the appropriation and disposition of tenants' resources without it being accountable to the state or to tenants. The economic activities of the GMGTU executive office included wheat milling plants and a textile plant (El Badi 1985; Agricultural Bank of Sudan 1974; GMGTU 1976).
8. For example, insecticide air spraying, which was arranged by the SGB and its cost deducted from the individual tenant's returns, was regarded as very costly. If tenants had the choice they would have used the cheaper labour-intensive hand sprayers. Personal interview with Rahmt El Mula Mohamed Ahmed Abu Zeid, chairman of El Tahameid GMGTU group branch; Ahmed El Zein, chairman of the GMGTU South group branch; and Busati Saad Mohamed, chairman of the GMGTU Maatug group branch, Wad Medani, August 1997. Also see *El Gezira*, 4.7.1996.
9. Personal interview with Ahmed Abdel Wahab El Zein, chairman of a GMGTU group branch; and Musa Mohamed Ali, chairman of GMGTU group branch, August 1997.
10. *El Gezira*, 19.8.1976; personal interview with El Shaikh El Tom Osman, deputy chairman of GMGTU executive office, Wad Medani, August 1997.
11. *El Gezira*, 25.7.1996.
12. *El Gezira*, 26.12.1996; interview with an El Managil tenant, Khartoum, August 1997.
13. *El Gezira*, 4.7.1996.
14. Personal interview with a Managil tenant, Khartoum, 20.8.1997.
15. *El Gezira*, 27.2.1997.
16. Most likely land was not registered in freehold ownership according to the 1899 Title to Land Ordinance (see above).
17. Horticultural Department, Statistics on Land Holding, internal files, Ministry of Agriculture, Kassala.

18. Ibid.
19. Weeding is a task of manual labour, which is considered by both sharecropper and farm owner to be the responsibility of the sharecropper. If the sharecropper chooses to use herbicides instead of carrying out the operation manually, this is at his own expense.
20. Horticultural Department, internal files.
21. The reinvestment of capital realised or accumulated from agriculture back into agriculture is clear from the case of El Sawagi South and El Sawagi North. Both encompass some of the earliest vegetable and fruit farms and now are the richest. They mainly specialise in fruit production (Steenwinkell 1981).
22. The KVFFU was given different names at different times. Immediately before 1994 it was called the Sawagi and Project Owners' Union.
23. Since the takeover of the El Bashir's military government to the time of writing elections had still not been held. Although a much more formal structure of the union was laid down, representatives were in effect selected by the government according to the formal structure. Thus, a list of would-be farmers' representatives was passed to the regional authorities, meetings held and the list announced to farmers. Elections according to the new structure were due in March 1998.
24. The estimate is based on a total of 15,000 feddans divided by 5 feddans (the average holding) to arrive at a number of 300 farmers. The actual number of farmers is in fact far less than 300 if one takes into consideration the disparity in land ownership between rich and poor farmers. If each farmer has three sons working or involved with him in farming then the total number of KVFFU members was in the region of 1,200.
25. Informal interview with Ahmed Faki Adam, the KVFFU secretary, Kassala, July1997; informal interview with Mohamed Ahmed Atta, the KVFFU chairman, Khartoum, August1997.
26. The general assembly holds annual meetings in which issues related to farmers' production problems discussed, such as diesel oil shortages, high taxes and dues, the provision of inputs, improved seeds and seedlings, low prices for produce and the possibilities of new marketing venues.
27. The Kassala price was Ls70,000 per barrel in addition to Ls10,000 state tax. The farmer accordingly pays Ls 80,000 per barrel after state tax.
28. Agricultural Bank of the Sudan, Agricultural Credit Regulations, ABS internal files, ABS Headquarters, Khartoum, 1974.
29. Ibid.

References

Abdel Hamid, A. (1966) Agricultural Labour in the Gezira Scheme, a report (in Arabic), SGB, Barakat.

Abdel Rahman Mohamed, A. (1982) 'The coexistence of Kassala town and the horticultural area', DSRC diploma dissertation, from the first (1956) and second Sudan (1973) censuses, University of Khartoum.

Agricultural Bank of Sudan (1974a) Credit Regulations (in Arabic), ABS Internal Files, ABS Headquarters, Khartoum.

—— (1974b) 'Agricultural Credit Regulations', ABS Internal Files, ABS Headquarters, Khartoum.

Cliffe, L. (1982) 'Rural class formation in Africa', in H. Alavi and T. Shanin (eds.) *Introduction to the Sociology of Developing Societies,* London, Macmillan Press.

Cowen, M. (1981) 'The agrarian problem: Note on the Nairobi discussion', *ROAPE*, no. 20.

Cusicanqui, Silvia Revera (1987) *The Politics of Colombian Peasant Movements: The Case of the National Association of Peasant Smallholders,* Geneva, UN Institute for Social Development.

Deloitte Haskins and Sells (1990), Gezira Rehabilitation Project: Debt, Credit and Marketing Study, draft final report, Annex 5, Table 2, Khartoum.

Dillon, B. and Stiel, M. (1987) 'Making the concept concrete', *Ideas RRDC Bulletin*, UNRISD Participation Programme, March.

ECA (1994) *African Charter for Participation,* Addis Ababa, ECA.

El Badi, M. (1985) The Tenants' Movement in El Gezira Scheme and Managil (in Arabic), Khartoum.

El Buni, A. (1984) 'The British administration and the tenants' movement in Gezira 1925-1956' (in Arabic), unpublished M.Sc. thesis, Institute of African and Asian Studies, University of Khartoum.

nd. Rehabilitation and Modernization Project, vol. III, Table D.1.4.

Farah, H. A.(1973) The Gezira Pattern of Development within the Context of the History of Sudanese Agrarian Relations, Dakar, UN Institute for Economic Development and Planning.

Fernando, Rello (1987) *State and Peasantry in Mexico, A Case Study of Rural Credit in La Languna,* Geneva, UN Institute for Social Development.

Fisher, F.W. (1954) *Citrus Culture,* Khartoum, Ministry of Agriculture.

Gaitskell, A. (1959) *The Gezira Scheme: A Story of Development in the Sudan,* London, Faber.

Galal, Mahmoud Yousif (1997) 'The Gezira Scheme: The Greatest on Earth Under One Management'.

Ghazi, Sabbagh (1981) 'Institutional constraints facing co-operatives of horticultural farmers in Kassala', DSRC diploma dissertation, University of Khartoum..

Haileab, Ephrem (1982) 'The effects of the Port Sudan—Khartoum road on the Kassala region', DSRC diploma dissertation, University of Khartoum.

IBRD (1966) Gezira Study Mission, main report, IBRD, Washington, D. C.

Khalid, A. El Amin (1990) 'The state and forms of capitalist development in the Sudan:The case of Gezira', unpublished Ph.D thesis, Department of Politics, University of Leeds.

Mishra, S. N. et al. (1984) *Participation and Development,* New Delhi, NBO Publishers.

Oakley, P. (1987) 'State or process, means or ends: The concept of participation in rural development', *Ideas RRDC Bulletin,* March.

Rahmato, D. (1991) 'Peasant organisations in Africa: Constraints and potentials', Dakar, CODESRIA Working Paper 1/91.

Rello, Fernando (1987) *State and Peasantry in Mexico, A Case Study of Rural Credit in La Languna,* Geneva, UN Institute for Social Development.

SGB (1964) Annual Agricultural Report (in Arabic), SGB, Barakat.

—— (1980) Annual Agricultural Report 1979/1980 (in Arabic), SGB, Barakat.

Shaw, D. J. (1965) 'The Managil south west extension: An extension of the Gezira Scheme', *The Netherlands Bulletin*, no. 9.

Simpson, xx (1957) 'Land tenure aspects of the Gezira Scheme in the Sudan', *Journal of African Administration,* VIII, 1.

Smith, C. (1989) Peasant resistance to the 'grow more coffee' compaign in Kagera region Tanzania, *Canadian Journal of Development Studies*, X, 2.
Steenwinkell, J. (1981) Land Distribution and Land Tenure in Kassala Eastern Region, DSRC ISS Joint Survey Report, DSRC, Khartoum.
Syahuka-Muhindo, A. (1995) 'The Rwenzururu movement and the democratic struggle', in M. Mamdani and E. Wamba dia Wamba (eds.), *African Studies in Social Movements and Democracy*, Dakar, CODESRIA.
Thrower, I. B. (1954) *Commercial Horticulture in the Sudan*, Khartoum, Ministry of Agriculture.
Uphoff, N. T and Cohen, J. (1979) *Feasibility and Application of Rural Development Participation*, a state of the art paper, Cornell University.
Warburg, G. (1978) *Islam, Nationalism and Communism in a Traditional Society: The case of Sudan*. London, Cass.

Primary data

Reports and internal files

El Gezira newspaper, various issues
GMGTU (1954) The Gezira Tenants' Union Act, Internal Files, GMGTU, Wad Medani.
—— (1976) The Peasant Farmer Organisations Act, Internal Files (in Arabic).
—— (1979) The Presidential Decree on Production Relations in the Gezira Scheme, Internal Files (in Arabic), GMGTU, Wad Medani.
—— (1992) Nomads and Cultivators Organisations Act, Internal Files (in Arabic), GMGTU, Wad Medani.
—— (n.d.) Managil Tenants' Union's Principal Order, Internal Files, Wad Medani.
—— (n.d.) A memorandum submitted to Awad's Committee for the revision of production relations in the Gezira Scheme, Internal Files (in Arabic), GMGTU, Wad Medani.
Kassala Horticultural Department (n.d.) Internal Files, Ministry of Agriculture, Kassala.
—— (n.d.) Statistics on Land Holding, Internal Files, Ministry of Agriculture, Kassala.

Informal interviews

Ahmed A. El Zein, 1997, Chairman of the GMGTU South Group Branch, *Wad Medani*, August.
Ahmed Faki Adam, 1997, KVFFU secretary, Kassala, July
Busati, Saad Moh, 1997, chairman of the GMGTU Maatug Group Branch, *Wad Medani*, August.
El Shiekh, El Tom, 1997, deputy chairman of the GMGTU Executive Office, *Wad Medani*, August.
Gadalla, Hassan Mohamed, 1997, member of the GMGTU Executive Office, *Wad Medani*, August.
A Managil tenant, 1997, Khartoum, 20 August.
Mohamed, Ahmed Atta, 1997, chairman of KVFFU, Khartoum, August.
Musa, Mohamed Ali, 1997, chairman of Wadi Shaair GMGTU Group Branch, *Wad Medani*, August.
Rahmt, El Mula Moh Ahmed Abu Zeid, 1997, chairman of El Tahameid GMGTU Group Branch, *Wad Medani*, August.

7

Peasant Organisations and Grassroots Empowerment in Nigeria: Limits and Possibilities*

Musa I. M. Abutudu

Introduction

Most authoritarian regimes in Africa have been enmeshed in a profound crisis of legitimacy since the 1980s. This crisis has heightened agitation for fundamental human rights, democracy, good governance and accountability across the continent. Some commentators believe is in the 'transition to democracy'. However, there is cause for concern that the form of negotiation of political power that this response portends and the organisation of political authority it promises may shut out the peasantry and continue to confirm it, in spite of its numerical preponderance, as the most exploited and marginalised social category in contemporary Africa.

For democracy to be meaningful in Africa, its form and scope has to be made relevant to the living conditions of the peasants and other populous groups (Mamdani 1986: 38). This in turn might suggest that the peasants' own structures may constitute the appropriate channels for enhancing their productivity and empowerment. As peasants increasingly devise or strengthen various organisational forms to cope with economic decline, can these structures also be seen as possible building blocks in their own empowerment? Are peasant organisations the form of rural civil society? What is the nature of their leadership? What processes and mechanisms of accountability are put in place within these organisations? What are the forms of articulation between peasant organisations, local power structures and the state? Do the peasant organisations relate to any larger (national) democratic project? It must not simply be assumed that peasant organisations are inevitably democratic in their operations. They may in fact exhibit as many authoritarian tendencies as democratic ones.

An attempt to come to grips with the above questions necessitates an examination of peasant organisations in terms of their types, forms of internal differentiation, the principles of membership and participation, the character of local power structures and the state, and the relationship between these structures and the functioning of the organisations. This study identifies and examines a variety of peasant organisations in the Etsako area of Edo state of Nigeria, with a view to determining their potentials for the empowerment of the peasantry, and the assertion of their identity within a national democratic project. Data for the study was sourced from the various Etsako clans, particularly the Ekperi clan, because of the central location of the 14 villages that make up this clan. The study approach was an attempt to highlight the issues and expose trends, tendencies and possibilities rather than to provide a thorough geographical and social representation of the Nigerian picture.

The Peasantry and Peasant Organisations in Africa

Peasants are usually seen as being essentially concerned with the need to produce and earn enough from agricultural production, wage labour and other occupations in order to meet subsistence needs (Allen & Williams 1982: 67). Thus, while peasants are predominantly involved in smallholder agriculture, they may also supplement their livelihood with other occupational endeavours in the rural areas. The peasants may produce essentially for subsistence but 'they are also involved in a wider system in which they are economically exploited, and politically and ideologically dominated' (Momba 1989: 331). It is indeed their incorporation into a wider economy that compels African rural cultivators into relations of exploitation and domination (with merchant capital and the state) and transforms them into a peasantry (Bernstein 1977; Friedmann 1980; Saul & Woods 1987). The 'suspended' African state, erected on the thesis of the 'uncaptured' peasantry (Hyden 1980), therefore becomes increasingly tenuous.

State power, as Catherine Boone argues, actually penetrated the rural areas and annexed them for state objectives. While this was the general trend, it was the peasants' own social structures that determined the form assumed by state penetration. In fact, she argues that 'the fact that rural social structures differed markedly across cases' means that peasant societies will offer 'different possibilities for alliance between state and rural elites' (Boone 1995a: 27).

Thus, while the exploitation of the peasantry remains the hallmark of the colonial and postcolonial state, the political mechanism deployed for its control was in fact substantially generated from materials tapped from its very social structure. In tapping such materials, there is no assurance of the state's fidelity to the originals and the checks inherent in them. The colonial state, for example, deliberately erected a system of political domination on a magnification of some of the most objectionable features of 'traditional' African political systems, or embarked upon a deliberate mischievous reading of those features that laid emphasis on tapping their authoritarian content or imbuing them with such content where none existed. Indirect rule, the name by which this political edifice was known, 'came to be the mode of domination over a "free" peasantry' (Mamdani 1996: 17).

These developments are usually not uncontested by the peasantry. However, the nature of the structures from which the actual form of state penetration is constructed also ensures that peasant societies 'differed in their capacity to resist the aggrandizing designs of the post-colonial regimes' (Boone 1995a: 27; 1994).

The nature and degree of inequalities and differentiations within the peasantry will in general be expected to diversify political possibilities within a peasant community and its relations to the wider society. Typically, the differentiation of the African peasantry 'entails the stratification of peasants into rich, middle and poor peasants' (Momba 1989: 332). This is further operationalised in terms of the concentration of wealth, differential access to land, the proletarianisation of sections of the peasantry, the capacity to diversify into more lucrative cash crops, and so on (Ibid., 331). For some scholars (Cliffe 1977; Shenton & Lennihan 1981), these inequalities are so strong that they stratify rural Africa into social classes. However, for Williams (1976) they may not be sharp enough to amount to distinct social classes.

Differences expressed economically are very significant in the analysis of the peasantry and political possibilities within it. However, these differences may be complemented, reinforced or even find expression in other forms such as generation or gender. In fact, it may also be the case that economic differences not necessarily apparent have to be traced through cultural forms to capture their impact on peasant organisations. It is precisely at this level of cultural processes that peasant groups may attempt to counteract the possible negative effects of accumulated economic privileges on their internal cohesion. Peasant organisations may explicitly put in place specific measures to direct attention to the necessity of equalising economic advantages among members while the expressions of other forms of difference remain largely ignored and unchallenged. Peasant organisations tend to reflect these differences in their structures, just as they may constitute the principles or the foundation of membership.

A multidimensional view of differentiation among the peasantry is significant in another sense. Different foundation or membership principles offer different possibilities for political engagement with local authority structures or the state. This points to the necessity of analysing peasants' relations with the state not only in terms of the differentiation within the peasantry but also in terms of the organisational categories in which it usually expresses itself.

The current economic crisis in Africa has drawn attention to the various organisational forms through which the peasants have attempted to ensure their continued viability (Rahmato 1992).

Peasant organisations, according to Rahmato, are 'formal or informal bodies, involving peasant men and women in which the major objective is the pursuit of mutual benefits in return for mutual obligations' (Rahmato 1993: 3). The formal kind is usually registered, structured and hierarchical. On the other hand, the informal type does not have a structured leadership and is not accorded official recognition. The informal type suggests another, possibly different type usually left out of consideration. This is the large, unorganised mass of an informal and spontaneous character that is usually formed in response to particular dynamics and demand-sustained political action (Zghal 1995: 127; Moyo 1995: 11). While these may assume the form of confrontation with the state in some cases, they are likely to be more common as protest movements directed at political abuses at the local level.

In general, the distinction between the formal and informal largely hinges on state recognition. The 'presence' of the state seems to inform another common, but largely implicit basis of differentiating between peasant organisations. This is based on the dichotomy between those whose origins is traceable to the peasants themselves and those whose impetus is largely external (Nindi 1990; Kerr 1978). In the latter case, the peasant organisation is seen as a means of enhancing agricultural productivity and preparing the rural population 'for political autonomy and imbibing in it the principles of concerted democratic action' (Buchler 1975: 16). Nindi, however, points out that such intervention by the state actually tends to turn rural co-operatives into 'lower elements of the state apparatus' (Nindi 1990: 50). State intervention may in effect alienate the peasants from their own organisations while imposing some external agenda on them. In contrast, organisations founded on the peasants' own initiatives will tend to attract higher levels of participation.

As has been suggested, however, both autonomous and dependent peasant organisations 'can adopt varying strategies which temporarily shift from independence and resistance of state agencies to collaboration with the state' (Moyo 1995:12). Furthermore, an uncritical view of the autonomous peasant organisation may be unmindful of the possibility that various forms of inequalities and differentiations are reflected in these organisations. Whether these differences are of a class, gender or generational nature, differentiation can diminish the possibility of horizontal linkages among the peasantry, foster factionalism and weaken or even preclude the possibility of effective group action (Sudarsky 1978: 127). A capacity for self-reliance and autonomy, therefore, need not be taken as self-evident in peasant organisations. Various forms of inequalities may act as constraints to the democratic impulse in the countryside. However, attempts by peasants to assert themselves as actors in their own history, sometimes hidden, and at other times open (Isaacman 1990; Friedmann 1990), may be reactions to precisely such forms of inequalities, exploitation and domination. These factors may produce two contradictory reactions in the rural areas. One may be to act as a constraint to the democratic impulse. The other is to provide materials for organised resistance to arbitrary power.

Democracy, Empowerment and Peasant Organisations

It is generally held that 'participation' is a key element in democracy and empowerment. However, the precise form it must assume to meet the test of empowerment appears to be more ambiguous. Liberal democracy is content to see the act of enfranchisement, multiparty politics and periodic elections as necessarily definitive of participation and empowerment. However, democracy so conceived is not only very limited, it generally only leads to the empowerment of rival factions of the bourgeoisie (Mamdani 1986; Nindi 1990). Furthermore, the reduction of democracy to multiparty elections in post-cold war Africa has led to its annexation to the service of the 'world's most notorious autocrats ... who are now able to parade democratic credentials without reforming their repressive regimes'(Ake 1996: 130). Liberal democracy, then, simply advances a set of static conditions which hardly inform us of the goals of democracy.

Here Barrington Moore's conception of democracy proves to be instructive. Moore conceives of democracy as a process of continuous struggle by the people 'to do three related things: 1) to check arbitrary rulers, 2) to replace arbitrary rules with just and rational ones, and 3) to obtain a share for the underlying population in the making of rules' (Moore 1966: 414).

This conception of democracy does not necessarily emphasise formal structures. Rather its accent is on a process with the specific objectives of opening and liberalising political space. Governance in effect becomes meaningful to the extent that it is directed by the interests and aspirations of popular constituents.

This chapter identifies four crucial elements of grassroots empowerment, all of which embrace means and ends in highly interwoven and reciprocal relationships. First is participation in decision-making and implementation processes. Generally, a degree of self-awareness and self-organisation is critical to participation (Ghai 1988). For participation to empower, however, it must mean 'a real opportunity to make decisions instead of participating only to legitimize the preconceptions of officials' (Ake 1996: 143). It also implies a capacity to constitute a leadership and hold that leadership accountable.

Secondly, there is the element of autonomy, which refers to the correspondence between the preferences of social entities and their actions (Nordlinger 1981: 8). Autonomy enables social entities to set goals for their organisations (Bratton 1994: 236) under the condition of a relative absence of external compulsion, but not necessarily external (or internal) constraints.

The third element expresses itself at the material level. Most of the activities of peasant organisations are given expression at this level. It entails the capacity to acquire working tools, processes of improving living conditions and the bonding process in which solidarity among the peasantry is at once expressed in material terms and cultural production.

Finally, there has to be an awareness and a willingness to assert an identity within a larger democratic project. This means that the peasantry should be able to articulate its peculiar concerns on a national basis, and in doing this construct horizontal linkages with other popular forces in the wider society

Etsakoland: A Background

The Etsako area in the north of Edo state, one of the 36 states in Nigeria, is located in the south-central part of the country. The state is bounded to the south and south-east by Delta state, to the south-west by Ondo state, the north-east by Benue state, to the north by Kogi state, to the east by the River Niger and to the north by Okene in Kogi state. Within Edo state, it shares boundaries to the south-west with Owan East local government area and to the north-west with Akoko-Edo local government area. To its south-east is Esan North-East local government area.

Etsako is made up of 13 clans, divided into three local government areas. Each of the clans is made up of widely scattered villages, except for Auchi whose component villages exhibit a compact settlement pattern. The land area is approximately 2,590 square kilometres (Haruna 1991: 14). The 1991 national census put the size of the population at 274,829.

The people of Etsako generally trace their origins to ancient Benin from where migration began in the 14th century and continued until the 19th century (Bradbury 1957). These migrations are recurrently suggested to reflect attempts to escape from the difficult and demanding tributes exacted by the Benin monarch. The people's oral tradition has it that post-migration political organisation was influenced by the need to avoid the kind of centralised authority that tended to exact punitive tributes. This explains the general absence of centralised political authorities in the area until the second half of the 19th century following the Nupe invasion.

Christianity and Islam are the dominant religions in Etsako, the former making inroads in the wake of colonial conquest, while the latter made its appearance largely by way of the Nupe invasion. Although the pure and exclusive practice of African traditional religion is hardly professed by anyone, it is quite possible that a substantial number of those who claim adherence to Islam or Christianity have not given up beliefs and practices associated with indigenous religions

The major economic activity of the people of Etsako is smallholder peasant agriculture. Fishing is also pronounced among inhabitants of communities situated on the banks of Osimile River (River Niger). These are the riverine parts of the Weppa Wanno and Ekperi clans. The advent of colonialism led to the diversification of agriculture into cash crop (export) production, with cocoa, rubber and palm produce as the major crops.

Export production and colonial agricultural policy had a significant impact on land tenure. Prior to colonial rule, land was communally owned in Etsako. It still remains substantially so. However, a strong tendency towards individualisation was emerging (Abutudu 1991, 1994), which had become even more pronounced since the Land Use Decree promulgated by the Federal Government in 1978. Pressure on land generated by the economic decline since the early 1980s also contributed to this trend.

A major change in land tenure brought about by colonialism is traceable to the colonial policy of creating forest reserves. This policy, systematically prosecuted in Etsako division from 1938 (Shokpeka 1990: 229-30), had the effect of removing large tracts of land from the control of the various communities. The reserve policy was highly unpopular and was resisted by peasant farmers. However, it was still possible to accommodate the resulting pressure on land by moving into the outer farm lands, which were usually far removed from the population settlement. The reserves, now opened up, actually made the state into a direct 'landlord' of the peasants in most communities in Etsako. Thus land allocation committees, a feature of practically all communities in contemporary Etsako, were put in place by peasants (through state directives in most cases) to relate the peasants to the new landlord.

State policy and the Nigerian economic crisis since the early 1980s also impacted on land tenure in another way. The Land Use Decree of 1978 through which the military sought to create a strong agrarian bourgeoisie, and the foreign exchange crunch of the 1980s that seriously affected the capacity of many large industrial concerns to source their raw materials externally, combined to push large business concerns into agriculture in Etsako. Thus, areas suited for rice and maize production (such as Weppa Wanno, Aviele and Ekperi) came under pressure, as large-scale land alienation resulted, leading to the wide-ranging displacement of peasant farmers or threats to them, as in Ekperi (Abutudu 1991). Peasant farmers' resistance, especially in Ekperi, was equally dramatic.

Etsako, then, had its peculiarities, but it was also shaped by Nigeria's political economy. Cash and food crop production are well developed in the area. Its rich soil for the production of rice and maize made it a strong attraction for migrant farmers from within and outside Edo state. Its social organisation displays a rich mix of cultural forms, a pointer to its multiple cultural heritage.

Peasant Organisations in Etsako

A variety of peasant organisations exist in Etsako. Some are formal in structure, others informal. The organisational forms reflect the range of issues and interests the peasants seek to protect, project or contest. This study focuses on four major forms in which peasant farmers express themselves organisationally. These are mutual support and welfare organisations, age grades, community development organisations and protest movements.

Mutual Support and Welfare Organisations

The most common organisational form among peasant farmers in Etsako was the mutual support and welfare organisation, typified by resource-pooling, mainly towards finance and labour needs. Membership could be based on friendship, same generation, lineage, gender or some special interest such as hunting. This chapter focuses on three specific cases of this type to examine its dynamics, and variations within it. These are the Aluaye group, the Omoayekhe Social Club and the Erhaewvhipko group.[1]

The membership of the Aluaye group in Ekperi was exclusively female.[2] This group was formed probably in the early 1960s. The impetus for its formation is traceable to the necessity of acquiring a certain large metallic basin (locally called *agbada*). This basin was crucial to the day-to-day activities of the women—for storing grains or for fetching water from the stream. Priced then at 15 shillings and given extremely low incomes, this vital item required a considerable period of disciplined saving to acquire. It was the need to save and acquire the agbada that provided the immediate occasion for the formation of the Aluaye group.[3]

The reasoning behind this initial impetus has since shaped the activities and evolution of the group. Contributions were specifically targeted at particular household needs. Although practically all members were farmers, most items acquired by the members through their collective savings efforts were not directly tied to the procurement of farm tools. Neither were the savings distributed to members in the form of cash. Rather, such efforts generally took the form of putting something aside from farm income to acquire basic 'modern' household items.[4]

The significance of this is that it fostered a high sense of community between members, provided a democratic way of determining their felt needs, and then a creative way of meeting these needs. Members collectively decided on the item to be purchased at a particular time. The fact that all had whatever item was jointly decided upon is a pointer to the underlying egalitarian spirit of the organisation. It is also a pointer to the determination of the members to improve the quality of their lives.

In terms of leadership structure, there were only two officers— the chairlady and the secretary. Actually, there was only one officer and that was the chairlady. The secretary was her personal appointee. The chairlady acted as treasurer and took personal responsibility for the purchase of any item decided upon and distributed them to the members. This may sound like a concentration of authority. Indeed, her own insistence that she could not be removed from that position by anyone, including the clan head, may suggest this.[5] However, the chair could not impose any decision on the group. While she was the founder, and therefore, chairlady, her real function was to implement collective decisions. Her position on any issue hardly counted more than any other member's.

This leadership form may exhibit another pattern elsewhere in Etsako. In Auchi, for example, mutual support groups among women were very often lineage-based. This meant that the position of chairlady was automatically the preserve of whoever was designated as the head of the women in a particular lineage. The right to make that designation was the preserve of the male head or daodu of the lineage. The latter in turn was chosen by the clan head. In other words, the headship of these groups might be derived from the general authority of the clan or community, and not from the membership.

In contrast, most other mutual support groups had more formal and elaborate structures of authority.

The Omoayekhe Social Club (OSC) had a formal constitution, which made provision for an eight-man executive committee, headed by a chairman. Formed in 1991, largely in response to a deteriorating economic situation, the OSC was an all-male group with a membership (in 1997) of 35. It had an executive committee elected for a term of three years. That was the constitutional position. However, the more common practice had been to roll over on the grounds of good performance. Equally, the members had shown no hesitation in removing from office any individual who was not performing well in that office, or who committed a serious offence against the community. For example, the inaugural chairman of the OSC was impeached while less than two years in office and then expelled from the group for 'violating the rights of a member and committing an offence against the community'. In general, an individual was not expected to put himself forward for a position. It was the members who advanced an individual as a candidate based on how well they thought he could fit into a particular role.

Three levels of resource pooling were identifiable in the OSC. The first was co-operative farming, locally referred to as *eko*. The co-operative farm was a collective farm belonging to the members. All the members must go to work on this farm every Saturday of the week. Any member who was unable to turn up must either send in a non-member as replacement for that day or pay the cash equivalent of one day's labour to the group's coffers. Exemption was only granted in cases of serious illness. The harvest from this co-operative farm went to the common purse. But there was another form in which the harvesting was done. In this case, a day was usually appointed on which every member was expected to turn up in the farm. Whatever he was able to harvest on that day was strictly his to dispose of as he saw fit.

This blend of the co-operative spirit and individualism ensured a high level of participation at the workplace—the farm. Indeed, turnout in formal meetings at home tended to average 50 to 60 percent, and this was mostly made up of executive members. In peak farming season, it fell drastically. However, the weekly work in the collective farm or the annual joint harvest was generally guaranteed to bring out almost the entire membership. One effect was that the farm became the real forum for group meetings.

The second level of resource pooling was in the form of contributing labour for work on an individual member's farm plot. Here again, all members were expected to be present. In lieu of this, a member might send a substitute or pay one day's labour equivalent in cash to the host member. This labour-pooling mechanism cut labour costs and enabled an individual member to put a relatively large area under cultivation.

The third level of resource pooling was financial. Contributions were weekly and monthly (N5 per week and N20 in the last week of every month). With this and the revenue from the collective farm, the group had been able to develop a strong welfare scheme. Interest-free loans were given to members with readily identifiable problems.

The OSC was also involved in the maintenance of community infrastructures and the clearing of farm routes. In addition, the group prided itself on 'keeping a close tab on community affairs', a responsibility its members claimed as members of the Ekperi community and backed up financially.[6]

The Erhaewvhikpo group is another example of the mutual support and welfare organisations in Ekperi. Unlike the Aluaye and Omoayekhe groups, which are gender-specific, the Erhaewvhikpo group was made up of both men and women. With a membership of about 150, it was headed by a joint chair (chairman and chairlady). They were assisted by a joint vice-chair, that is a vice-chairman and a vice-chairlady. The other positions of leadership were secretary and treasurer, both of which were open to all members.

The question of periodic change of the officers of the organisation did not feature as a significant issue in the operation of the organisation. On the other hand, it had been traditionally sensitive to gender in its leadership structure. Thus, while members were simply elected into the executive committee through acclamation by the generality of the membership, there was also a convention that the top two positions must be jointly held by one male and one female in each case. Furthermore, members were not expected to put themselves forward for office. Rather, they would be drafted on the basis of good character and proven ability, and except in cases of gross misconduct and abuse of office, they could expect to serve indefinitely.

The Erhaewvhikpo group was an old organisation whose co-operative efforts were reinvigorated in the context of the Nigerian economic crisis. The co-operative activities of the group focused on labour-pooling for work in members' farms and a welfare scheme for members. The labour-pooling mechanism was well developed. In it, the members divided themselves into six groups each consisting of 25 persons, both men and women. Members of each of these sub-groups took turns in working in each other's farm plot. On the basis of this arrangement it was possible for the members to attend to the farm work of six members simultaneously on any given day. As in the Omoayekhe group, a member who was indisposed to participate must pay the equivalent of one day's wage labour in cash to the host farmer, or send an adult non-member as replacement.

This labour-pooling mechanism was blind to gender. Men and women had their individual farm plots, and the labour pooling and rotation mechanism did not discriminate on the basis of the sex of a member.

The welfare scheme was based on two basic sources of finance. The first was the weekly contribution of N5 per member. Financial resources were also generated from the group's cultural dance troupe. This cultural troupe was to some extent commercialised, although like many other organisations of its kind, the cultural arm also served the important function of preserving and developing traditional musical and dance forms.[7]

The resources generated from membership dues and the activities of the cultural troupe were used for granting soft loans (with interest rates of 2 to 5 per cent per annum) to members. Non-members might also benefit from such loans, but a member must in this case stand as guarantor. When members had difficulties paying the school fees of their children (one of the most pressing problems in an area where modern education is highly valued) the loans became interest-free.

Quite apart from its co-operative thrust, two other significant points about the Erhaewvhikpo group should be noted. The first is gender consciousness in the internal administrative structure of the group. This is not necessarily unique in Etsako. In fact, mutual support and welfare schemes with mixed membership generally emphasise internal arrangements in which gender equality is guaranteed. This is not simply a matter of representation in the executive committee irrespective of the position actually occupied in it. It is more of a guarantee effected by means of a joint chair at the very top of the organisation. This also spilled over into the labour-pooling mechanism, although in this case gender awareness was given concrete expression by means of a policy of gender-neutrality. The gender-neutrality of the labour-pooling system would however appear to be a phenomenon whose evolution had been speeded up by more recent developments associated with economic and social changes that saw women increasingly engaged in the cultivation of crops (such as rice and yams) traditionally seen as male preserves.

The other significant point concerns the cultural troupe. The cultural troupe may or may not be a regular part of a peasant organisation. However, its real import was in giving vent to the artistic talents of peasant farmers, bringing out their self-representation, and their construction of their economic and social environment through music and dance. Peasant organisations, then, should not be seen alone in terms of activities or activism directly tied to agriculture. Cultural and artistic creativity as well as their reproduction are important matters in such organisations.

Age Grades/Groups

It is not apparent that age grades can automatically qualify as peasant organisations. For one thing, some or even many of those who belong to particular age groups, especially the younger ones, do not necessarily make a living through smallholder agriculture. In fact, many of such members are employed in professional and high-tech jobs in the modern sector of the economy or the state. In this sense, they may be more appropriately seen as communal organisations. However, a sizeable proportion of members of most age grades are usually in employment or part-employment as smallholder cultivators. It is these same individuals usually still resident in the rural areas who give content, continuity and meaning to the age grade system. Quite often, these organisations turn out as instruments for articulating and resolving issues of significance to the peasant farmers. In fact, their organisational structures, their issue-focus and the patterns of relations subsisting between them and local power structures may be strong pointers to certain aspects of exploitation of the peasantry and their own modes of contesting these.

The age grade system was a prominent feature of Etsako communities. It was predominantly a male institution in which those born within a specified period were initiated into manhood. Membership of a particular age grade was for life. In Ekperi, the age grades ranged from one to twenty-two. Initiation into the first age group means admission into adulthood. Under colonial rule, this became taxable adulthood. It also meant that one was now obliged to participate fully in communal works, such as clearing of farm routes, building bridges across rivers and streams, construction of market places, and so on.

The age grades were well adapted to mutual support and welfare functions. In Ekperi, they pooled labour and rotated it in members' farm plots. They met periodically, and thrift contributions were made and rotated among members. Target savings were also very popular. The bicycle, the major means of transportation in rural areas, was a popular focus in target savings among age grades over the years. However, the motorcycle, which had become very popular in rural transportation (also for its direct commercial transportation value), was now the focus of target saving schemes, especially among the younger age groups.

The leadership of each age group was usually chosen by its membership. However, there were specific criteria guiding this. For example, the various villages that made up Ekperi clan had to be represented in the executive committee. Except for the titular position of 'leader' of the age group, which traditionally went to the oldest member of an age set from the oldest village in the clan, all other positions(chairman, vice-chairman, secretary, financial secretary, chief whip and judge) were elected by the members themselves. There was a slight but significant variation in Auchi, where the most powerful position of chairman was chosen from among the ruling houses by the clan head. In other words, the head of any particular age grade must be a member of the 'aristocracy'.

The execution of communal works formed an essential part of the functions of age grades. These functions often involved the provision of the basic infrastructure for peasant agricultural production such as the construction and maintenance of rural roads and bridges connecting the population settlement with farm areas. The participation of members of age sets in these public works was compulsory, and absenteeism attracted a stiff fine, usually collected by participating peers. This was rigidly enforced. In a different context, however, the concept of communal works by age grades had come under challenge by members. In many Etsako communities, communal public works programmes included periodic labour in the 'royal' farms, that is, the farms of the clan head. Some of the clan head enthronement ceremonies specifically enjoined this duty.

For example, one of the rituals performed on the eve of the coronation of a new clan head in a particular case involved his being given a hoe and a cutlass (machete) to do some symbolic manual work. After this, these instruments of smallholder agriculture were taken from him and painted white. Henceforth it became an abomination for him to ever engage in farm work, or any manual labour for that matter. His farm work had become the affair of the community (Ikharo 1981).

The age grade system was the community's arm for carrying out tilling and cultivation.

This practice had hardly been controversial until the mid-1980s when it was challenged by some members of the younger age grades. The system of free labour embedded in the assimilation of the 'royal' farms into the community work system actually appeared to mirror the basic form of local hierarchies, whose economic nature was buried in a thick layer of customs and traditions. A development that largely came into prominence in colonial times, the challenge mounted against it was also very much a product of its time.

The economic crisis since the 1980s might have oriented people to greater collectivism as a coping mechanism, but it also seemed to have induced some opposition to the tradition of collective performance of a public service whose product was bound for private appropriation. It is critical to note that the age grade system that acted as the vehicle of this phenomenon also provided a relatively easy way to challenge it. Matters of policy making, implementation and discipline were internal to every age set. Once the enforcement of the rules of absenteeism was contested and successfully argued in regard to non-participation in the farm of a clan head, the latter was effectively on its way to decoupling from the concept of communal public works.[8] The fact that the enforcement mechanism relied on sanctions that could only be enforced from within an age set meant that its successful challenge would result in a discontinuance. Generally, there had been no problem among the younger age sets in embracing this position, although there are some among the older ones who were wont to hallow the older tradition. This apparent generational friction was largely immaterial to the outcome since those who really bore the burden of communal work had adopted the path of disengagement from its contested dimension.[9]

Community Development Organisations

The community development organisations were all-embracing in their issue focus, and took the progress, development and 'modernisation' of the entire community as their cardinal objectives. These objectives embodied strong peasant agendas. These organisations also tended to form the bedrock of political engagement within the community and often they lay strong claims to representing the interests of the community to the state. The Ekperi Youth Movement (EYM) and the Ekperi Community Development Association (ECDA) are typical examples.

The EYM straddled different age grades with a membership that was exclusively male. These members were generally farmers. Thus, like the age sets and the mutual support organisations, the EYM had a well-developed labour pooling system. Again, like the age sets, it also performed the functions of clearing community access roads, constructing bridges across streams and carrying out other communal works. Contributions from members and fines imposed for disciplinary cases constituted its main financial base. This in turn was channelled towards a welfare fund for members. However, the real significance of the organisation lay in other areas. These were its political arrangements, its high-profile role in the community and its relationship to state structures.

The EYM had a 10-member executive committee that was responsible for running its affairs. Election for all positions were held every three years for all positions. Apart from the fact that these elections were usually contested, there was also a provision that no one might serve in any position for more than two consecutive terms. This provision had been rigidly enforced over the years. In this respect the EYM seems to have broken new ground in the manner in which it constituted its leadership and stuck to the letter of its constitution in its efforts at democracy. However, the nature of its high-profile intervention in the affairs of the community had sometimes led to charges that some in the community would obviously see as being contradictory to that conclusion.

The youth organisation claimed responsibility for maintaining peace and order in the Ekperi community. To this end it operated a 42-member vigilante group that its executive members claimed had state recognition, and worked closely with the Nigerian Police Force.

The organisation also initiated 'by-laws' for the administration of the entire community. Such 'by-laws' were presented to the clan head for approval and, once this was given, they were vigorously enforced by the group. In one case, the EYM initiated impeachment proceedings against the chairman of the ECDA, the umbrella clan organisation, and also a prominent member of the community in his own right for allegedly violating one of such by-laws (ECDA 1995: 8). When this failed to unseat him, the fact that the alleged offender paid the fine did not prevent the EYM from utilising the earliest opportunity of the ECDA annual elections to mobilise and vote him out of office.[10]

In addition, the EYM had taken on the role of settling disputes within the community. In fact, it claimed to have successfully intervened at one point or the other in disputes involving the clan head and his village heads and members of their communities, and disputes between elders and youths (EYM 1994). In certain cases, it had even assumed the powers of compulsory arbitration within the community.

This high-profile activism had not gone down well with certain social categories, especially the elders, who seemed to see the whole development as being at the expense of their own authority. The view was also strong within the rank and file of the membership of the EYM that the powers of the organisation had been hijacked by its leadership. But as those who took the latter position within the EYM also maintained, there was a remedy for them in the organisation's constitutional provisions on elections, and they intended to take full advantage of these at the earliest available opportunity. The same went for those who were not happy with the EYM's decisive intervention in the ECDA elections of early 1997 that led to the defeat of a prominent community leader and its chairman.[11] The promise to return the latter to office at the earliest possible opportunity suggests a struggle among communal factions, constituted and allied in various ways. It is also a pointer to their belief in the efficacy of the electoral machinery in leadership succession and matters of political accountability.

The EYM's high profile in communal affairs may have been promoted by two basic factors. The first was the power vacuum created by the death of the clan head in 1991. With that position yet to be filled, and the elders seemingly diverted by the factional politics associated with the clan head selection process, the EYM had simply stepped in to fill the vacuum. Secondly, there was the organisation's association with certain state structures. The EYM is formally affiliated to the National Youth Council of Nigeria (NYCN), an organisation created and basically run by the military government and closely associated with pro-military politics in Nigeria. The EYM is also closely associated with the Edo State Ministry of Information, Youth and Culture, and the National Orientation Agency (NOA). The ministry indeed plays a crucial role in co-ordinating the activities of the EYM with other youth organisations in the local government area and the state.[12]

During the military-rulled political dispensation in Nigeria, the ministry, the NOA and the NYCN were the critical elements in the mobilisation of 'grassroots support' for the military regime in Nigeria. This 'support' was most often dramatised in the phenomenon of the 'rented crowd'. These were well publicised (by state information organs), state-sponsored 'public' demonstrations in which participants were generally believed to be paid. Repeated at various levels of government, the 'rented crowd' phenomenon enabled the military regime to credit itself with popular support for whatever political line it was pursuing.[13]

The close connection with the state did not seem to have rubbed off positively on the agricultural activities of the members of the EYM. While the organisation had often called for state assistance in making farm tractors available to its members, the members equally conceded that the government at any level had never actually made these available, although promises were often made to that effect. Indeed, under the current political dispensation, the benefit that could possibly accrue from the kind of relationship forged between the group and the state agencies would necessarily be at the level of individual 'settlement', and at a comparatively minor scale at that.[14]

However, the contact with governmental agencies and the close association with the police through its vigilante activities had provided, or been read as, some sort of official backing for the EYM's increasing role in community governance. The main responsibility for community governance in Ekperi, however, lay with the development associations. Each of the constituent villages of Ekperi had a community development association (CDA). Most of these had been in existence under one name or the other since the 1970s or earlier. However, these, along with the umbrella organisation (the Ekperi Community Development Association, ECDA), formally took on the nomenclature of CDA when in 1987, the Directorate of Food, Roads and Rural Infrastructures under the regime of General Ibrahim Babangida directed rural communities nationwide to form development associations.

The annual conference was the highest body of the CDAs. The ECDA held its annual conference in December of each year. All resident Ekperi men and women were expected to attend, although the women actually tended to stay away. There were branches 'abroad', each of which was represented by two delegates at the conference. Issues such as education, electricity, water and health schemes were tabled, discussed and decisions taken on them. Land matters and the presentation of financial reports by various committees also featured prominently

The ECDA was run by an executive committee, elected for a term of three years. There was no specific term limit to any office, although incompetence and financial misconduct (rumours of which tended to be rife) could easily lead to impeachment or form the basis of intense electoral rivalry. In general, the executive positions of the ECDA were usually the subject of intense political rivalry. One major significance of these rivalries, however, was that the various factions involved in them usually rationalised their positions in terms of the need to ensure accountability and probity in the organisation.[15]

The struggle for power was often quite pronounced in the relationship of the development association to the traditional ruler or clan head. For example, the immediate past clan head claimed the right as the traditional ruler of the community to dissolve the popularly elected executive committees of the CDAs. The associations, on the other hand, claimed the right unhindered to constitute their committees as they saw fit. Between 1986 and 1987 this resulted in a dissolution—reconstitution struggle between the largest village CDA and the clan head. This contestation between the traditional authority and the CDA was actually an issue of whether the latter should function at the pleasure of the former, or whether they were autonomous instruments of community governance in themselves. This in turn was actually a dispute over the control of community resources associated with land.

Two major issues seemed to have dominated the agenda of the ECDA in recent times. The first was that of adequate political representation for the Ekperi clan in the local government council of Etsako-Central and Edo state in general. This had meant intense lobbying efforts and petitions to various agencies of government, including the National Electoral Commission of Nigeria. The issues here were actually presented as equitable sharing of resources between different communities. The second issue related to the selection of a new clan head, which had remained vacant since the demise of the last occupant in 1991. Here the position pushed by the ECDA with the

Edo state government reflected that of the dominant faction in the politics of clan head selection.

These two cases are pointers to the close interaction between the ECDA and the state agencies as well as the character of these interactions. However, the real forms of relations that defined the relationship with the state and then with the traditional authority structures lay elsewhere. This can be approached by way of examining how the resources of the ECDA were generated.

There were three principal sources of income. First, the constituent villages were taxed according to their population. The branches 'abroad' were similarly assessed and taxed according to their numerical strength.

A second major source of income was the imposition of levies on various occupational groups and on community resources patronised by non-members of the Ekperi community. For example, a levy was paid by non-members who came to purchase (bulk) agricultural products such as rice or cassava. Equally, Hausa-Fulani herdsmen in the area paid an annual rent for pasture land. Palm wine tappers were similarly levied. Non-members of the community who came to hunt game in the forests of the Ekperi paid a fee for the yearly permit. Timber contractors were charged a community fee over and above what they paid to the state coffers. Each of these activities had a co-ordinator who was appointed by the annual conference and charged with the responsibility of monitoring and collecting the prescribed fees or dues from the respective occupational groups.

The third source of revenue was the rent paid on farmland leased to non-members of the community. For the purpose of this, each constituent village had a land allocation committee (LAC), which was constituted by the conference of the CDA. The LAC was usually made up of seven members: a chairman, a secretary and five others.

The rent the LAC collected might come in any of three forms. An indigenous member of the Ekperi community might claim ownership of a particular piece of agricultural land. He or she might allow non-members of the community to cultivate the land without charging any rent on it. However, if a rent was charged, it must be paid to the community through the village LAC. In fact, in Ekperi, an individual who leased land to non-members of the community and failed to turn the proceeds to the community was deemed to have stolen from the community and would face sanctions.

The LAC also directly managed communal land. The allocation of the latter to those who were not natives of the community attracted a rent that was payable to the community coffers.

However, the real activity of the LAC centred on the management of the forest reserves constituted under colonial rule. With increasing pressure on agricultural land, these forest reserves were gradually being made available for cultivation to peasant farmers.16 Access to this land was made possible through a formal application to the LAC, which in turn applied to the Forestry Department of the Ministry of Agriculture and Natural Resources. Applications made well before the commencement of the planting season had to be accompanied by a fee, currently put at N100 per acre.

The LAC applied for land from the Forestry Department based on the number of applicants and the area of land each required. The Forestry Department demarcated the size of land required and handed it to the care of the LAC, which in turn parcelled it out among the farmers. The LAC paid the N100 per acre fee to the Forestry Department. This was actually what the members of the community paid. The LAC's own revenue was derived from charging the non-indigenous members at twice the rate of what it paid to the state. The LAC granted credit facilities to peasant farmers who were unable to pay the prescribed rent at the time of application. Such credit facilities were non-discriminatory, interest-free and repayable after harvest.

The LAC's position as an intermediary between the state and the farmers as well as its role in making credit available to facilitate access to land were important elements in agricultural production in Ekperi. For example, in Ugbekpe (one of the 14 villages in Ekperi), about 50 farmers took advantage of the facilities of the LAC in the 1997 planting season alone.[17] However, the intermediary role of the LAC obscures one important fact: that the state had effectively translated itself into a landlord by charging a rent on land originally confiscated from the same peasant communities by the colonial state. This practice, found in many other Etsako communities, adds a qualitatively different dimension to the idea of the peasantry in this area, in the sense that it becomes increasingly difficult to hold up free access to land as the basis of the assertion that the category is not an exploited one.

The LAC (or the ECDA), however, did ameliorate the harsh undertones of this and therefore helped to contain it politically.

The traditional authorities were also very much involved in the appropriation of the resources generated by LACs. In some Etsako communities, the mode of appropriation was fairly stable and generally not acrimonious. Here the practice was usually to divide the amount generated in any year into three equal parts. One part belonged to the community and was managed on its behalf by its development association. The second part belonged to the village head, while the third went to the clan head, who would have this facility replicated from all the villages under the clan.

In certain cases, the traditional rulers constituted the LAC singlehandedly and also appropriated all of its proceeds. However, in many cases the issue of how to dispose of the resources generated through the activities of the LACs remained contentious between the development associations and the traditional rulers.

This might take the form of a power struggle between the two sides expressed in terms of whether the right to constitute the executive organs (and subcommittees) of the development associations was a popular one or a royal prerogative.

A Protest Movement: The Clan Head Boycott Movement

There was a rash of large-scale land alienation in Etsako from the early 1980s. This process was set in motion by two major forces. The first was underscored by the efforts of the military regime of General Olusegun Obasanjo to create an agrarian bourgeoisie in Nigeria, efforts most vividly epitomised in the promulgation of the Land Use Decree in 1978. The second was the economic crisis which began in the country in the early 1980s.

These twin forces facilitated the movement of corporate organisations, top military officers and bureaucrats (serving and retired) to the rich, arable lands of Aviele, Agenegbode and Ekperi which were reputed to be well suited for the production of rice and maize. The acquisition of land for the kind of large-scale agriculture envisaged by the corporate organisations and bureaucrats required the active consent and collaboration of the traditional ruler of the community. This had already been demonstrated in the cases of Agenegbode and Aviele clans, where peasants' protests against this development had generally taken the form of recourse to the judicial machinery of the state.[18]

However, when faced with a similar situation, the people of Ekperi adopted a different approach. In the early 1980s, the clan head had to contend with his people over an allegation that he was in the process of giving out large parcels of agricultural land to some wealthy individuals and corporate bodies. This, for the peasant farmers, was unacceptable. For one thing, the state forest reserve policy had already effectively alienated over 50 per cent of the best land, and succeeded in turning the peasant farmers into tenants in their own land. Secondly, they had seen what a similar land alienation in neighbouring Agenegbode had done to the peasant farmers in that community, who were pouring in increasing number into Ekperi and beyond in search of arable land for cultivation. Thirdly, the attempt by the clan head to unilaterally give out the land suggested an arrogation of power beyond his traditional or customary competence.

It seemed as if the clan head was trying to give a practical bent to the largely symbolic position that all community land was stool land. The fact of the matter was that he held the land in trust for the community, which does not in any way imply or suggest an authority to alienate.

This was precisely what the peasant farmers of Ekperi sought to impress on their clan head, who was equally bent on exercising what he saw as the prerogative of his office. Spearheaded by two well-known critics of the clan head,19 the peasant farmers mobilised and made two demands on their clan head: that he halt all actions pertaining to the land deal, and apologise to the community for the insensitive attempt not only to alienate the land but also his arrogant and overbearing disposition towards his subjects on the issue. For the clan head, the first demand was unacceptable, since he saw himself as only exercising a prerogative of his royal office. The second was an abomination, since royalty does not go on bended knee before his subjects. From his perspective, it was always the other way round.

This response pushed the struggle into a new phase. This took the form of a general resolution by the people to boycott all activities in respect of the office of their traditional ruler, avoid his person, and assume that they no longer had a clan head. All these were done without actually pronouncing him deposed.

The boycott movement attracted general support, and had to be underwritten by a high sense of popular anger in the form of the 'community curse'. This was indeed a potent weapon of ensuring the solidarity and cohesion of the peasant farmers against their own traditional ruler. The fact that it could be invoked simultaneously in the two forms in which it is considered most potent was a strong demonstration of the popular will.

In Ekperi as in much of Etsakoland, there were two major forms the community curse could take. One was by the oldest women in the community, the other by the youngest male age group. Their simultaneous invocation was rare, and to have achieved this in this particular case underscored the sense of anger on the issue that prompted the ban on any form of fraternisation with the clan head. The boycott was in effect for about three years (1983-6), during which period the clan head was virtually ostracised by his subjects.

To prevent an authority vacuum from being created, the people selected a relative of the clan head to oversee matters relating to disputes settlement and some other pressing functions of the office of traditional ruler while waiting patiently for the clan head to back down. Through the boycott, the peasant farmers achieved the immediate objective of preventing the alienation of the land. However, this did not end the matter as the apology (which was really the guarantee that the objectionable action would not recur) was yet to be made. Traditional rulers from other parts of Etsako attempted to effect a settlement which did not incorporate the apology. This failed. However, it was another interested 'third party', the Ekperi 'people abroad' (that is, resident outside Ekperi, the diaspora) who eventually prevailed upon the clan head to tender the required apology, and have the boycott movement brought to an end with the revocation of the community curse.

Peasant protest movements against the authoritarian tendencies embedded in their traditional political structures are not uncommon in Etsakoland. Such protests come in various forms, including in some cases violent confrontation.20 An interesting aspect of these revolts was the role by the community's diaspora in effecting settlement. While these people were part of the community, the fact that they were not in the vicinity of the engagement area enabled them to claim some distance from the issues at stake, and therefore play a mediating role that might also be credible in view of their stake in the community. This role might itself represent an expression of the popular position, as the Ekperi example illustrates.

However, a wrong reading of tradition, or giving it an authoritarian, 'king vs subject' bent can easily discredit such intervention. Thus, in a similar case in another Etsako community (which was actually violent), an apology demanded by the members of the community from their clan head could not be contemplated by their 'diaspora', because a king, according to them, cannot apologise to his subjects. In this case the stalemate had, not surprisingly, lingered for over six years.

Empowerment and Democracy at the Grassroots

The struggle for empowerment is one for the redistribution of power in a given society. It therefore addresses itself to the issue of exploitation, marginalisation and exclusion. An awareness of this power asymmetry on the part of the marginalised and excluded, or a recognition of those forces that may produce marginalisation and exclusion, is a precondition for such a struggle. However, this awareness must be given practical content at two levels—the material or economic, and the political.

At the material level, the peasant farmers in Ekperi displayed imaginative organisational abilities. The acquisition of working tools, the provision of basic, even if rudimentary infrastructure for the requirements of peasant production, and then household needs were some of the most significant outcomes of self-organisation. The mechanisms of labour-pooling, savings and welfare schemes provided some basis for the construction of the autonomy of the peasant farmer. These activities seemed to supply not only the bulk of the funding required for small-scale agriculture, but also provided the peasant farmer with some insurance against usurious money lenders.

The labour pooling mechanism was especially significant in another sense. It seemed to have equalised opportunities to own and cultivate plots without reference to gender. Indeed, it was as much a reflection of this development as it was its facilitator. The gains from labour-pooling were immense, especially in terms of increasing the capacity to cultivate a larger area at a comparatively reduced cost. But the particular advantage to women should be noted. It facilitated access to labour on the same terms as men. This should be expected to have a positive effect on women's empowerment within the household. In fact, most of those interviewed stressed that the fact that women now engaged in the cultivation of rice—the main commercial food crop—had practically transformed many of them into the economic heads of their households. At this latter level, it would appear that it was cultural beliefs rather than economic power that had continued to underwrite the asymmetry in gender roles.

The autonomy of the peasant farmer is usually seen as largely a product of free access to land. However, the emerging picture in Etsakoland suggests that land was increasingly under pressure from business and bureaucratic interests allied with the local or traditional aristocracy. The state forest reserve policy exerted pressure in the same direction. While the former had indeed become quite visible in various parts of Etsako, the people of Ekperi were able to confront it decisively and nip it in the bud through a spontaneous protest movement. In the other case, however, access to land was guaranteed to peasant farmers on an annual tenancy basis, a situation that effectively turned the state into a 'landlord'. Here the organisational forms of the peasant farmers had essentially taken the form of managing the process, not questioning or confronting it.

Most of the peasant organisations pursued purely economic objectives. This, as pointed out earlier, tended to enhance self-reliance in the membership. While this could conceivably generate more self-confidence at interpersonal levels, it might not quite translate into a higher level of political participation at the community level, especially in the more politicised organisations such as the EYM and the CDAs. Men dominated these institutions to the point of near-total exclusion of women. Thus, even as the material disparity between men and women was narrowed, women's participation in community political institutions remained very low. However, the evidence also pointed to the fact that in organisations of a mutual support character and with male and female membership, there were also explicit provisions for gender-based 'power sharing'.

The leadership of peasant organisations can emerge in various ways. Customs and traditions are often critical. In some cases, the initiator became the leader, usually on a permanent basis. This might give room for authoritarian displays within such organisations, but hardly financial corruption. The latter was invariably met with the requirement of prompt repayment of the misappropriated sum, and then prompt removal from office. The possibility of authoritarian displays was not very pronounced in peasant groups that were mostly of an economic character. This might relate to the largely routine character of their activities and their face-to-face nature. However, it sometimes happened that a leader unilaterally decided to by-pass the order of rotation of savings. While these were usually vigorously protested by those affected, such leaders (whose position is usually derived from an authority outside the organisation) tended to have their way.

The age grades and the community development organisations had more formal ways of constituting their leadership. However, while periodic elections were stipulated for the choice of officers, there were hardly fixed terms, except for the EYM, whose rules were clear on the two-term limit. Once elected, officers might expect to continue in office indefinitely. However, as the case of the ECDA shows, perpetual stay in office was being strongly challenged. In fact, elections had become fierce, and bore the firm imprint of factional politics in the community. Unlike most other organisations, elections to the community development organisations were becoming instruments of ensuring domestic accountability.

Accountability within peasant groups may be finance-related or a matter of fidelity to social norms and political objectives. While financial misappropriation was hardly an issue in the mutual support and welfare organisations, records of the community development organisations show that it was a factor that constantly agitated the minds of the membership. The financial resources of these organisations were comparatively large, and they were also engaged in activities, as several of those interviewed maintained, through which funds could be misappropriated. It is a measure of the importance of the issue of accountability that community factions built their electoral fortunes in the ECDA around it.

For example, while one faction mobilised and had the chairman (and the secretary) voted out on the grounds of violating community by-laws, the other faction maintained that the opposing faction was piqued because the former chairman and his secretary did not give any room for financial misappropriation through the tapping of 'hidden' sources. For the members of the current opposing factions, therefore, the next election would be fought on the question of financial accountability.

One common observation was the visible degree of interlocking leadership among various groups. For example, the chairman of the Erhaewvhikpo group was also the chairman of the strategic land allocation committee and the chairman of the local chapter of the state-sponsored Multi-Purpose Co-operative Society while that was still functional in Ekperi. This might create the impression of some kind of oligarchy running across the various organisations. While this may not be discounted, the routine, 'administrative character' of the activities of most peasant organisations and the limitations on their resource distribution capability acted as a rein on the possible accumulation of power that might arise from the practice.

One interesting aspect of the interlocking leadership was that it involved individuals who had some formal education, and probably had some stints in the urban areas, but were back in the countryside and fully involved in agriculture. What this basically adds up to is that some degree of formal education is essential for leadership in peasant organisations.

The peasant farmers displayed a remarkable facility for political organisation at the local level. This was quite evident in the way they organised to confront their traditional authorities when the latter attempted to alienate land on a large scale, contrary to customary rules governing land matters. This, along with many of the routine activities of many peasant organisations, was a strong check on the growth of arbitrary power at the local level. This suggests a vibrant rural civil society. However, this vibrancy does not necessarily translate to a successful check on arbitrary powers across the board in Etsakoland. In other words, while peasant protests against land alienation and misuse of power were not uncommon in Etsako, the level of success achieved in Ekperi might not be a true reflection of the larger picture.

Many communal traditional authorities had succeeded in the alienation of vast tracts of land. The fact of continuing contestation of these in various ways (mostly in the law court) was an indication of elements of a civil society in the affected areas.

Many state-owned farms were privatised in recent times in various parts of Nigeria, just as large tracts of land under state control were handed over to private individuals and organisations. Free access to land as a source of peasant autonomy is therefore increasingly under stress in the midst of the structural adjustment-fuelled development.

Quite often, the state got away with it precisely because it only had to contend with atomised rural communities whose traditional rulers were also allied with various state interests. While civil society in Ekperi, for example, had experience in checking the ambitions of its traditional ruler, similar confrontation with the state, to be successful, would require a project of rural civil society on a wider basis. It is precisely at this point that we find 'transcommunal' peasant organisations hard to come by. Where they are available, they are essentially state-driven, unstable and usually viewed with suspicion.

An example was the Multi-Purpose Co-operative Society in Edo state. A consumer co-operative society with official backing by the state government, it floated a scheme to form a co-operative bank in 1987. The Ekperi branch of the society enthusiastically bought the shares of the proposed bank allocated to it, and equally encouraged members to purchase individual shares, which they did. That was the last they heard about the proposed bank, which never got off the ground, and the share money was never refunded.[21] The local chapter invariably collapsed following this experience.

Experiences of this nature had tended to deter peasant farmers from pursuing projects beyond the confines of the immediate community even when there was an awareness of the potential benefits of such projects to the larger community as well. On the other hand, it was evident that some vigorous organisations formed and run by the peasant farmers themselves could break out of their locality, as the EYM indicated. But here again, a different mode of state penetration became visible. Thus the military government found some peasant organisations very useful in its gigantic project of creating a fictitious legitimacy. The interesting point here is that the peasant organisations involved were connected to the state propaganda agencies and not to those agencies actually related to agricultural production. The leaders of the groups always pressed for state assistance in terms of access to certain agricultural inputs, and promises were made to that effect by the political propaganda agencies in matters in which they obviously lacked competence and expertise. But as should be expected, none of such promises had ever been fulfilled.[22] They had nevertheless been effective in annexing the peasants to the political objectives of the regime.

The 'connections' to the regime hardly conferred any advantage to the collectivity, but they did give some visibility to the leadership of the organisations in community affairs.

Finally, relations to the state were usually mediated by political figures from the community operating at the national or state level. There were hardly 'peasant interests' from this perspective. Peasant interests were rather submerged in 'constituency interests' (defined vaguely in geographical terms and propagated in communal rhetoric). These figures were quite often closely identified with the political agenda of the military and were hardly sensitive to the possibilities of an autonomous peasant project. The peasant groups were not unaware of this. This is precisely why the more critical among them called for caution in relying on such political figures.[23]

Conclusion

A variety of peasant organisations existed in the area of study. This variety was also reflected in the way leadership in these organisations was constituted and the nature of tenure in them. In some, perpetuity in office was promoted by a diverse set of factors. However, this did not in any way allow for corruption or a personalisation of power. On the other hand, some organisations, especially those categorised as community development associations, increasingly relied on periodic elections as a means of leadership selection and holding the leadership accountable.

Generally, participation was high. This was even more so in those organisations that were of a routine character and provided direct economic and welfare benefits to their members. In such organisations, which also had a mixed membership, there was an obvious consciousness of 'power-sharing' between the male and female members. This was strictly enforced in the constitution of the leadership. However, this was not the case in the more politicised groups such as the community development associations. In spite of the fact that women were not barred from participation in these groups, they hardly attended meetings and were hardly to be found within the leadership structures.

Many peasant organisations were quite autonomous from their traditional rulers. But quite often too, this was an autonomy that had to be constantly defended. Points of confrontation, as should be expected, were usually associated with the alienation of agricultural land by traditional rulers. The modes of dealing with this development differed among peasant groups in different communities. The results were also mixed, but it is clear that this issue formed the most important point of contestation between the peasant organisations and traditional rulers. The widespread efforts by the latter to assume greater power in land matters were of course reflections of developments in the larger Nigerian political economy and the state.

The peasant organisations demonstrated a remarkable ability to construct and shape their autonomy and struggles in traditional and cultural forms. In their fundamental forms, these struggles also had a direct bearing on checking the arbitrary exercise of power at the local level. However, the emerging pattern of relations with the state as epitomised in the EYM suggested a different tendency. This relationship, obviously replicated with various groups nationwide, provided fuel for the phenomenon of the rented crowd, under which the military regimes in the past few years sought to annex various organisations to their authoritarian agendas, including very crude attempts at self-succession. In this form, the state did not only mediate but also appropriated an opportunity for transcommunal interaction among peasant organisations.

The emergence of a rural civil society was discernible in Etsako. This had displayed critical self-awareness and remarkable resourcefulness in confronting arbitrary rule at the local level. However, it also accommodated forms that tended to promote the authoritarian project of the state.

Notes

1. These names actually indicated the underlying philosophy or mission statement of the particular group. *Aluaye*, for example, means 'progress' while *Erhaewvhikpo* means 'our clothes will not catch fire'. *Omoayekhe* means 'we toil so that our children may have a future'.
2. The exception was the secretary who was male, he was accepted because of his literacy.
3. Interview with the founder and chairlady of the Aluaye group, July 1997. The meetings of the group were held every seven days, with the founder's residence as meeting place.
4. Such items included large cooking pots, basins, family-size beds and mattresses, and chinaware.
5. For example, against the strong opposition of the chairlady, most members had at a point insisted on taking loans from their contributions. Every member except the chairlady took the loan. When repayment became difficult for members, they also decided that repayment was not necessary in any case since the money was theirs in the first place, again against the objections of the founder. The result was a serious financial disruption of the group as it was unable to meet its target for some time. That was the only time loans had ever been given out by the organisation. The lesson from that near-collapse led to a decision to keep the group strictly to its original focus.
6. In a recent launching of an appeal fund to construct a community hall in Ekperi, the OSC pledged to contribute 10 bags of cement and two bundles of zinc to the effort.
7. The cultural troupe was usually invited to perform on special occasions by individuals or by the local government. The group had even made a musical record.
8. In the Etsako clan in question, the whole issue resulted from the protest of a youth who refused the payment of a fine imposed by his peers on the grounds that working in the royal farm did not qualify as community work. He argued further that the royal farm did not enhance his capacity to maintain his family since he had no access to the product. In any case, the royal family could easily afford to hire wage labour. Coming in the context of the economic crisis of the mid-1980s, this argument against a form of labour that was not reciprocated struck a positive note in his peers.
9. The fact that traditional rulers no longer depended much on direct agricultural production as a source of income may have contributed to the largely 'no-contest' basis of the resolution of the issue.
10. The by-law in question prohibited the use of poison or dynamite to kill fish in ponds or streams. The fine for violation was N400 and a goat. The alleged violation took place in 1994. The elections were held in March 1997.
11. The former chairman's supporters insisted that its defeat would not have been possible if not for the fact that key members had been diverted by their involvement in another election—the nationwide local government elections which took place at about the same time.
12. Such co-ordination included the conduct of elections into higher levels of the youth organisations, that is, at local government and state levels.

13. The main strategy of mobilisation was for the NOA to invite about 50 members each from various groups for 'seminars' at the state capital or local government headquarters. These might end up as pro-military rallies. According to those interviewed, two of such 'seminars' were scheduled for the month of July 1997.
14. The members interviewed claimed they had not received any gratification from the state. They insisted that the fact they were responsible for all expenses when they attended 'seminars' outside Ekperi made it difficult for them to mobilise a significant proportion of the number usually invited.
15. The EYM mobilised and got both the chairman and secretary of the ECDA voted out in March 1997. The EYM was evidently motivated by the accusation that the ECDA chairman had violated a 'by-law' of the community initiated by the EYM. But supporters of the former chairman claimed he was voted out because he did not allow those who wanted to misappropriate the community's financial resources to do so.
16. In Ugbekpe, the headquarter village of Ekperi, it is estimated that about 70 per cent of the arable land was held by the state as forest reserve. Interview with the chairman, Ugbekpe LAC, July 1997.
17. Interview with the chairman, Ugbekpe LAC, July 1997.
18. In Aviele clan, for example, at least two land cases were pending in court as at 1991. One was instituted against the clan head by the village of Ughioli for the alienation of its land to a corporate body; the other was instituted by the village of Ubiane against a corporate body for the illegal acquisition of its land.
19. These were Mr Aliu (now deceased) of Ugbekpe village and chief Inu Legogie, a major contestant for the currently (1997) vacant position of clan headship.
20. Iviuku, a village in Uzairue clan, Etsako West, is a case in point. The people went to 'war' (firearms were freely used) in 1991 against their village head and his kindred over the mode of disposition of compensation paid by firms engaged in quarrying in the community. The matter was yet to be resolved as at the end of 1997. In Auchi, a boycott of the clan head in the mid-1980s was initiated by the lineage chiefs over matters of resource distribution and consultation. This lasted for over six months, with settlement eventually effected by people of Auchi 'abroad'.
21. Interview with chairman of the Ekperi chapter, July 1997.
22. Interviews, Ekperi, July 1997.
23. See, for example, ECDA 1995: 3.

References

Abiri, U. I. (1986) *An Outline History of Etsako*, Auchi.

Abutudu, M. (1991) *Crisis of Accumulation and Tenurial Transformation in Etsako, Bendel State*, research report submitted to the Social Science Research Council of Nigeria (Ford Foundation Project).

Abuttudu, M. (1994) 'Crisis, adjustment and transformation in land tenure: Recent trends in the commercialization of agricultural land in South-Central Nigeria', *Journal of Rural Development and Administration* 27, 4, Autumn.

Ake, C. (1996) *Democracy and Development in Africa, Washington*, D.C., The Brookings Institution.

Allen, C. and Williams, G. (1982) (eds.) *Sociology of 'Developing Societies: Sub-Saharan Africa*, London, Macmillan.

Beer, C. E. F. (1976) *The Politics of Peasant Groups in Western Nigeria*, Ibadan, Ibadan University Press.

Bernstein, H. (1977) 'Notes on capital and peasantry', *Review of African Political Economy*, no. 10.

Berry, S. (1984) 'Oil and the disappearing peasantry: Accumulation, differentiation and underdevelopment in Western Nigeria', *African Economic History*, no. 13.

Berry, S. (1985) *Fathers Work for their Sons*, Berkeley, University of California Press.

Boone, C. (1994) 'States and ruling classes in post-colonial Africa: The enduring contradictions of power', in J. S. Migdal, A. Kohli and V. Shue (eds.) *State Power and Social Forces: Domination and Transformation in the Third World*, Cambridge, Cambridge University Press.

Boone, C. (1995a) 'Rural interests and the making of modern African states', *African Economic History*, no.23.

Boone, C. (1995b) 'The social origins of Ivorian exceptionalism: Rural society and state formation', *Comparative Politics* 27, 4.

Bradbury, R. E. (1957) *The Benin Kingdom and the Edo-Speaking Peoples of South-Western Nigeria*, London, International African Institute.

Bratton, M. (1994) 'Peasant—state relations in post-colonial Africa: Patterns of engagement and dis-engagement', in J. S. Migdal et al.(eds.), *State Power and Social Forces*, Cambridge, Cambridge University Press.

Buchler, P. (1975) *Agrarian Cooperatives in Peru*, Berne, Paul Haupt.

Cliffe, L. (1977) ,Rural class formation in East Africa', *Journal of Peasant Studies* 4, 3.

ECDA (1995) Minutes of Ekperi Annual National Conference held at Ogochi, Ekperi, 28-29 December.

ECDA (HQ) (1996a) EC/ECH/1/96: Report on Ekperi Clan Head by the Interim Steering Committee, 14 July.

Ekperi Leaders of Thought (ELOT) (1996), ELT/DW/EC/96: Request for More Wards in Ekperi Clan (Letter to the Chairman, NECON, the National Electoral Commission, Abuja, Nigeria), 15 December.

EYM (1994), An Address Presented by Ekperi Youth Movement on the Occasion of Ekperi 1994 Annual Conference Holding at Ugbekpe, Ekperi, on 28-29 December.

Friedmann, H, (1980) "Household Production and the National Economy: Concepts for the Analysis of Agrarian Formations", *Journal of Peasants Studies*, vol. 7, No. 2, January.

Friedmann, S. (1990) *Peasant Intellectuals: Anthropology and History in Tanzania, Madison*, University of Wisconsin Press

Ghai, D. (1988) *Participatory Development: Some Perspectives from Grassroots Experience*, Discussion Paper no. 5, UNRISD, Geneva.

Haruna, H. (1991) *Early History of Etsako*, Lagos, Etsako Club 81.

Hyden, G (1980) *Beyond Ujamaa in Tanzania: Underdevelopment and the Uncaptured Peasantry*, Berkeley, University of California Press.

Ikharo, A. O. (1981) 'The founding and origins of Auchi clan', mimeo.

Isaacman, A. (1990) 'Peasants and rural social protest in Africa', *African Studies Review* 33, 2.

Kerr, G. B. (1978) 'Voluntary associations in West Africa: "Hidden" agents of social change', *African Studies Review* 21, 3.

Mafeje, A. (1993) 'Peasant organisations in Africa: A potential dialogue between economists and sociologists — some theoretical and methodological observations', *CODESRIA Bulletin* 1: 14-17.

Mamdani, M. (1986) 'Peasants and democracy in Africa', New Left Review, no. 156.

Mamdani, M. (1996) *Citizen and Subject: Contemporary Africa and the Legacy of Late Colonialism*, Princeton, Princeton University Press.

Momba, J. C. (1989) 'The state, rural class formation and peasant political participation in Zambia: The case of Southern Province', *African Affairs* 88, 352.

Moore, B. Jr (1966) *Social Origins of Democracy and Dictatorship: Lord and Peasant in the Making of the Modern World*, Boston, Beacon Press

Moyo, S. (1995)' Peasant organisations and research issues on democratisation in Africa', paper presented to the CODESRIA Multinational Working Group Methodology Workshop on Peasant Organisations in Africa, 21-23 November, Dakar.

Nindi, B. C. (1992) 'Dilemmas of development in rural institutions in Sub-Saharan Africa', *Journal of Eastern African Research and Development.*, vol. 22 p. 45-59.

Nordlinger, E. A. (1981) *On the Autonomy of the Democratic State*, Cambridge, Mass., Harvard University Press.

Post, K. (1972) 'Peasantization and rural political movements in West Africa', *European Journal of Sociology* 13, 2.

Rahmato, D. (1991) *Peasant Organisations in Africa: Constraints and Potentials*, Dakar, CODESRIA, Working Paper 1/91.

Rahmato, D. (1993) 'On peasant studies: A reply to Archie Mafeje', *CODESRIA Bulletin*, no. 2.

Saul, J. and Woods, R. (1987) 'African peasants', in T. Shanin (ed.) *Peasants and Peasant Societies*, New York, Oxford University Press.

Shenton, R. W. and Lennihan, L. (1981) 'Capital and class: Peasant differentiation in Northern Nigeria', *Journal of Peasant Studies* 9, 1.

Shokpeka, S. A. (1990) 'Local government and development in Benin Province, 1938-1960', unpublished Ph.D thesis, University of Benin.

Sudarsky, J. (1978) 'Rural cooperatives, neo-patrimonial regimes and interventions, Part 1', *Journal of Rural Cooperatives* 6, 2.

Stahl, M. (1989) 'Capturing the peasants through cooperatives: The case of Ethiopia', *Review of African Political Economy*, no. 44.

Ugbekpe Community Development Association (UCDA) (1993), Conference Minutes, 26 December.

Williams, G. (1976) 'Taking the part of peasants: Rural development in Nigeria and Tanzania', in P. C. W. Gutkind and I. Wallerstein (eds.) *The Political Economy of Contemporary Africa*, London, Sage.

Williams, G. (1980) *State and Society in Nigeria*, Idanre, Afrografika.

Zghal, A (1995) 'The "bread riot" and the crisis of the one-party system in Tunisia', in M. Mamdani and E. Wamba dia Wamba (eds.) *African Studies in Social Movements and Democracy*, Dakar, CODESRIA.

8

Peasants Organisations in the Cape Verde Islands: From Individualism to Associative Forms of Organisation in the Rural World

Claudio Alves Furtado

Introduction

During the 1990s, most African countries underwent a connected process of political and economic reform. African societies were democratised by setting up multi-party systems and by carrying out pluralist elections. With the creation of new regimes and new political systems, a variety of new organisations arose, including political parties and a rush of newly developed national civil associations (Lachenmann 1993). On another level, the state disengagement from certain sectors of the economy, and the opening up to economic liberalism and private enterprise in order to ensure the neccesary macro-economic balances prescribed by international financial institutions, brought with it deregulation and a destructuring of the whole economic and social fabric. In fact, the precipitous fall of macro-economic indicators, linked to the burden of servicing foreign debt, caused many countries to feel obliged to carry out policies of structural adjustment that had serious social, political, and even economic consequences.

While the dynamic of associations is seen, at least in African countries, as an element to restructure these societies, the economic situation to which they are subjected—growing unemployment and under-employment, the liberalisation of prices of products, and an end to import restrictions, which has led to sharpening competitiveness and competition—has forced associativism to acquire a new impetus and a new dimension within the framework of a plural society, which is neo-liberal from a political and economic point of view. In this context, social movements acquire an increasingly important role in civil society's processes of dialogue with the state or within social groups.

The rural world has had a great impact on economic reforms, both at the level of agricultural production and further up and down the economic chain. The difficulty of obtaining credit and access to the factors of production, as well as constraints on marketing, with the opening of domestic markets to agricultural and livestock products from other countries has, among other problems, brought many agricultural operations to collapse or even disappear. However, it has also allowed the emergence of new strategies for mobilising and organising the rural world, in particular peasants organisations.

In Cape Verde, the same economic and political problems exist, although aggravated by the country's geographical situation as islands in the Sahel zone. The fact that the country is composed of islands has consequences for the configuration of its space, especially rural space. The small amount of arable land results in the pressure on land taking on specific significance, notably with regard to the price of land, particularly of irrigated land. We also have to keep in mind that the archipelago's Sahelian climate determines the level of rainfall, the season for agricultural production, droughts, etc.

The present study focuses on the social organisation of the Cape Verdean rural world, and in particular on the dynamics of peasants organisations in the archipelago, in a society whose economic fabric, politics, demography, and culture are undergoing rapid and profound change (Furtado 1993).

Specifically, this study seeks to understand the process by which peasants organisations become autonomous with respect to the various social actors, notably governmental and supervisory structures, as well as political parties. Our analytical perspective is based on the assumption that although a theoretical-epistemological supporting construct is needed, we have to begin from concrete reality in order to theorise about this reality. That is, we have to learn about the social dynamics of the Cape Verdean agrarian world, in order to have a basis on which to improve our understanding of peasants organisations as social movements.

Thus in the framework of this study, after a brief historical review intended to draw up an inventory of traditional rural peasants organisations and movements, we will then outline a statistical inventory of the peasants organisations which currently exist in the country, and attempt to analyse the dynamics and the context of their emergence. This methodological approach will allow us to understand and capture the dynamics and strategies of peasants organisations, and at the same time, explain why they remain invisible socially, politically, and economically. The approach will also enable us to describe the challenges which they face from the demographic and economic changes rocking Cape Verdean society in general and its rural world in particular.

A Theoretical Framework of Peasants Organisations

In the historical and sociological literature, the associative movement in the countryside goes by several names. These divergent nomenclatures sometimes conceal not only differences of a semantic nature but also of content and meaning: peasants organisations, rural organisations, local organisations, organisations for basic development, etc. (Rahmato 1991). For some authors, the associative movement in the rural world implies chiefly rural producers seeking community development, in particular the improvement of otheir agricultural operations, at a time when the state is not able to respond effectively to their needs. For other authors, it includes everything, although they are mainly concerned with the ongoing endogenous development of a community, very much like the traditional attributes of non-governmental organisations. For still other authors, who are devoted to mutualism, mutual aid must also be involved.

While it is true that the concepts of a peasant and of peasants organisations have a history, and even an archeology, which, as Mafeje has so well shown, may not coincide with African agrarian reality, we must restore the operational nature of the two concepts, to the extent to which socio-agarian reality in Africa, and in particular in Cape Verde, demonstrates that the appearance of the peasantry as a social and historical category occurred in a context in which agrarian production and the agrarian economy were being strongly commercialised for export, or, as Correia and Silva add, an agricultural economy was accompanied by commercial stock-raising (Albuquerque and Madeira Santos 1992). As a result, peasant production, embedded in the social relations of production, was always situated in a perspective that was not characterised by simple subsistence, especially when we are talking about irrigated farming. Thus it seems that the theoretical and operational construction of the categories "peasants" and "peasants organisations" would preferably consist of the very process of producing knowledge about the reality being analysed. However, an initial definition must be made, despite the fact that the operational aspects of the context are always better explained in the framework of the study and analysis of the subject of research.

The analysis of peasants organisations as social movements must not be limited to the view based on a dichotomy between so-called traditional organisations and so-called formal organisations, for example, co-operatives.

First, we must try to see how they resist the various forms of domination, by rejecting some outside proposals while at the same time seeking to influence other spheres and domains in the society concerned.

In Cape Verde, few studies have examined social movements in general and peasants organisations in particular. The existing rewiew tend to analyse the Cape Verdean rural world in the framework of public agrarian policies. References to peasants organisations focus more on the fact that they are an instrument for carrying out policies rather than on their nature as autonomous institutions within the framework of broader social relationships. Often, these references are more concerned with defending the thesis that the preconditions for the emergence and intervention of peasants organisations do not exist, given that stubborn Cape Verdean individualism will always tend to gain the upper hand over associative initiatives. This Cape Verdean individualism is thus supposed to lead to the inability of peasants to organise themselves in order to resolve problems at the level of production, at the level of the community, or in the defence of their own interests against third parties (Stockinger 1990 40-4; Pires *et al.* 1986:14). Here, we should emphasise that according to Amilcar Cabral, the chief theoretician and strategist for the liberation movement of Guinea-Bissau and Cape Verde, the whole process of fighting for liberation should centre on two essential actors: the lower middle class and the peasantry. Cabral points out that 'so far as the struggle is concerned, the latter [i.e., the peasantry] constitute the primary class in the Cape Verde Islands. This class of peasants, in the true sense of the word, had a powerful reason to fight: they own no land. And once the objective of the battle is defined as giving land to workers, they have a reason to join in it' (Cabral 1978: 113). Moreover, the peasantry was considered a strategic partner in the struggle for independence. From 25 April to 5 July 1975, the political mobilisation carried out by the PAIGC, which sought to obtain a majority in the legislative elections that preceded independence, was centered on the slogan "the land belongs to those who work it" (Furtado 1993; Teofilo 1993).

In addition to an analytical position that seeks, on the basis of a closed theoretical construct, which may also be an ideological one, to shape reality within this same construct, most studies on peasants organisations also tend to conduct a biased analysis of the social and agrarian reality in Cape Verde. The present study will demonstrate the ideological bias of the previously mentioned theoretical perspectives, by revealing the associative dynamics which cover several stages in the agrarian history of Cape Verde, and will also determine the political and socio-economic contexts to which peasants organisations belong. Nonetheless, our analysis does not conceal the fact that the political, social, and economic conditions have sometimes constituted serious obstacles to the expansion of peasants organisations and to their playing a leading role.

Historical reconstruction will provide the basis for understanding peasant associativism and then for refocusing our discussion on the most recent period of Cape Verde's history. Thus we will examine in particular the post-independence era, distinguishing two main periods: the first stretching from national independence to 1990, when the change in the regime and the political system occurred, and the second continuing to the present time.

Within the framework of this study we will define peasants organisations as rural social movements that may take the form of co-operatives for agricultural production and/or processing, or formal or informal associations for production or for providing services.

Methodology

Our study of peasants organisations in Cape Verde was carried out by drawing on the complementary methods and techniques of research and analysis. We have made an inventory of all the peasants organisations, relying on the data in the census of the co-operative movement carried out in 1995, which we have complemented by conducting an inquiry into peasant associations.[1] This inquiry was carried out on the national level and implemented by rural organisers.[2] Thus we were able, not only to gauge the size of peasants organisations, their development through time, and their distribution in space, but also to identify those peasants organisations which, in a second stage of our research, were to be studied in greater detail. This methodological perspective, connected with a historical review of the associative movement, made it possible to have a better understanding of the phenomenon in the country's recent history, and also to determine the changes the movement has gone through, which are not linear but have characteristic oscillations, in a dialectical relationship with the overall social dynamics.

In addition, an open interview was conducted with the leaders of ten peasants organisations chosen at random on the basis of the data of the inquiry, taking into account the legal form of the peasants organisation (whether a co-operative or an association) and the domains in which it was active (agriculture, stock raising, herding in fields and woods). Finally, a group dynamics involving members of three peasants organisations was developed, with the aim of better grasping the internal dynamics of their functioning, gender relationships, in particular the level of women's participation, and relationships with partners outside the peasants organisation being analysed. These interviews were conducted on the peasants organisations' own terrain, by taking advantage of rest periods or weekends when joint recreational and cultural activities were being pursued. Ndiobo described this way of work as follows:

> Talking with the peasant at his home, in his field, with the flock near the mosque after prayer, is first of all a way of respecting him, and also allowing those who have the "right"

> to speak to say what they know. The audience is not the interviewer with his criteria determined in advance, but the "usual village audience (Ndiobo 1997: 3).

There is no inventory of peasants organisations in Cape Verde. Only those that have been legally formalised and recognised can be easily classified, insofar as the process of legal recognition is being published in the official gazzete. However, we should emphasise here that in 1995 the National Institute of Co-operatives (INC), in collaboration with the International Co-operative Alliance, carried out an inventory of all the cooperatives existing in Cape Verde, whether formal or informal (INC, March 1995). Thus, since the data provided by the inquiry are classified by areas of activity, we can identify the co-operative organisations that are concerned with agriculture, stock raising, and herding.

As peasants associations were not listed, we had, as we have already noted, to undertake an inventory of all those existing in Cape Verde. This was accomplished during the second half of 1996 and published in 1997.

Peasants Organisations in Cape Verde: General Presentation

The Associative Movement: Historical Roots

Many specialists regard Cape Verde society as the product of a confrontation of diverse cultures and peoples, although the latter may be divided into two groups: the Portuguese and the Africans (Mariano 1958; Duarte 1951; Almada 1989; Hopffer Almada 1991; Semedo and Turano 1997). Since the islands were previously uninhabited, populating them with slaves coming from the African continent and with European colonists was a syncretic process, simultaneously ethnic, racial, and cultural.

In the framework of this syncretism, associative traditions, both those coming from the African continent and those coming from Portugal, had fertile ground to work in. During the golden period of the slave trade and commerce, the large landowners had no difficulty finding enough workers, but with the ending of the prerogatives proffered on Cape Verde by the Portuguese Crown, a crisis ensued in the country. The great landowners were forced to divide up the land and work it by share-croppping. Freed slaves began to cultivate these lands, whose produce they shared with the landowners, in addition to the work they were obliged to do on the lands directly exploited by the owners.

The geography of the islands, which prevents any kind of mechanisation, along with the decapitalisation of new farmers of arable land, meant that a system providing mutual agricultural and other services must be established.

The result was Djunta Mon. This is a mutual service institution that is valuable for agricultural work. This way of organising work extends to the construction of houses and other community projects. Djunta Mon still remains the most widely used form of organising work in the countryside, despite a growing labour shortage caused by the rural exodus and by emigration, which have increasingly emptied the countryside. In fact, insofar as Djunta Mon involves an exchange of non-remunerated services, the attractive salaries offered by civilian builders are helping draw young people away from agricultural work.[3]

It has also to be observed that associative movements and movements for the defence of class interests were systematically prohibited and repressed. For instance, Tabanca has been banned several times. Sena Barcelos emphasises that in 1895 this kind of public demonstration would have been forbidden. New legal measures are said to have been taken with a view to prohibiting activities connected with Tabanca. Thus Ordinance no. 52 (26 April 1923) prohibited public demonstrations. Although this prohibition was suspended in 1927, the effort to coopt Tabanca and its organisation continued until independence.

Agrarian history shows us that in the rural world, particularly on the island of Santiago, there were strong peasant social movements that firmly opposed both the authoritarianism of the great landowners and the colonial power. Peasant revolts in Ribiera dos Engenhos, Achada Falcäo, and Praia in the nineteenth century and at the beginning of the twentieth century provide convincing proof of this. Moreover, it was in the tradition of peasant resistance to power and to the colonial situation that Amilcar Cabral viewed peasants as the basis for political and armed struggle for liberation, both in Guinea-Bissau and in Cape Verde (Cabral 1978).

National Independence and the Promotion of Associative Movements: From Voluntary Service to Compulsion

The period immediately preceding Cape Verde's independence (April 1974 to July 1975) constitutes a period of agitation and important political movements. The main political parties fought with each other in order to make themselves interlocutors with the Portuguese government in the process of negotiating independence. In addition, they sought to establish themselves at the national level and to win voter support for the election of the parliament that was to proclaim independence. Three political parties competed for voters: the Parti africain pour l'indépendance de la Guinée et du Cap-Vert (PAIGC), which had engaged in armed struggle in Bissau, the Union du peuple des îles du Cap-Vert (UPICV), and the Union démocratique capverdienne (UDC). In this political battle, all kinds of weapons and all kinds of arguments were used. The PAIGC established its hegemony and was accepted as the interlocutor with the Portuguese, and it won the elections. The other parties left the stage of Cape Verde politics and their leaders went into exile.[4]

What is important for the current study is to see the strategy of mobilisation adopted by the PAIGC with regard to the peasants organisations. Associativism appeared to be an important way of organising people for the party's political aims. Co-operativism appeared, in this context, as "a way of freeing oneself from the domination and exploitation exercised by landowners and traders" (Monteiro 1991). Although a few co-operatives existed before independence, their number greatly increased from 1974-1975 onwards.[5] The co-operatives were seen as a way of overcoming the traders' control and also as a way of beginning the process of collectivising land and the means of production. The slogan 'the land belongs to those who work it' began to become a reality. The basic laws regarding co-operatives approved in 1979 emphasised that co-operativism 'is a way of freeing ourselves from domination and exploitation'. As a result, a whole political philosophy was connected with them, as well as a view of the world and of society.

However, peasants organisations have existed time and again during the history of Cape Verde, as we have already shown. But we must not forget that formalisation/legalisation is seen as a means of social and political control. Since the colonial power wanted to control associative movements, it was normal for their members to resist. In addition, and especially after the establishment of the new government in Portugal in 1928, organisations and associations that were not connected legally and virtually umbilically to the state, in the logic of the corporative state, were considered as potentially subversive.

Other associative movements were very early transformed into strategic points for political mobilisation. Traders and landowners were seen as the great enemies of the peasants and of development itself. Moreover,

> everyone knows the poverty that arose at the time; the problem with the supporting groups, non-payment of salaries, etc., in other words an extreme situation. And so the cooperative emerged as a way for people to organise and free themselves from this state of dependency (Monteiro 1991:24).

The co-operaratives that were activite and technically and commercial supported by the Federation of Co-operatives remained informal until 1979, when the first legislation specifically addressing them was passed (the law laying down the general bases of cooperatives through the decree-law no. 80/79), despite the fact that the Portuguiese commercial code had a chapter on cooperatives.[6]

> They emerged in order to resolve concrete, immediate problems, and ceased to exist when the situation to which they responded disappeared, without having the formal need to provide accounts, and retaining in any case the increased possibility of coming together again (Dasilva 1991:56).

This view that it was not important to organise co-operatives is reflected in the fact that from 1897 to 1975 there were only five co-operatives, the first dating from 1897, the second from 1935, and the rest from the 1960s and 1970s.

One notes from after independence, initially, an interpretation of the co-operatives and of the local structures of the party in power, and later on, of the INC. This is so in Cape Verde and also in other countries on the continent, where

> many African governments have seen the cooperative as an instrument for development, and have devoted enormous financial and human resources to it. But it is a question of creating structures controlled by the state, usually for agricultural exports, with a view to bringing in currency for the state or to political goals (Kamdem 1991: 36).

This way of seeing the associative movement as a political instrument or as a way of executing 'development projects' does not mean, however, that the members' specific interests disappear, or that the institution is reappropriated, even under the apparent domination by supervisory or other kind of institutions.

The autonomy of these organisations was not seen by the political system or by many of the peasants organisations themselves as possible or even conceivable, since they constituted the "natural" continuation of political party or governmental structures".[7] As a result, the very invisibility of these organisations was a fact.

It should be emphasised that this process of enforcing peasants organisations occurred chiefly in those that tended to be formalised in a bureaucratic and administrative manner, with formally defined structures, operating rules, etc. Organisations of an endogenous nature, that is, those that emerged from the internal peasant dynamics, or from the rural population, tended not to be formalised.

According to information provided by several members of these kinds of organisations, "the possibility of formal organisation never came up"; "we don't even know the administrative processes of legalisation and we are not interested in them. We don't see what we would gain". Others say: "Being recognised by the authorities doesn't interest us. The authorities want to control us. The further away we are, the fewer problems we have". Lachenmann, referring to S. Amin, notes that the latter considers.

> African regimes as having been at best populist; they do not really raise the social question but attempt only to attenuate the negative consequences of peripherisation, while the

"masses" about which they speak at such length are kept in a passive, amorphous state, "mobilised" in order to "support" the regimes, but not allowed to organize as an autonomous force vis-à-vis the existing power structures (Lachenmann 1993:71).

Nonetheless, even for those that followed the path of bureaucratic formalisation and those that were activated by the new political actors that came on the scene with independence, we see an authentic strategy of reappropriating their organisations. Moreover, in many cases, supervisory institutions come into play in order to draw profits from them to the advantage of the organisation to which they belong. One of the first co-operators, the leader of an agricultural co-operative, says that:

formalisation has its advantages. The possibility of gaining access to financing for projects is great. Our co-operative obtained a great deal of financing. Members received training in many areas. The problem is that there are only a few people who took advantage of this training because many of them were illiterate. I have travelled a good deal, and I have visited other islands in Cape Verde. Now, things are no longer the same. The INC's technicians came here several times. Sometimes, excessively. They drew up inventories, did the accounting. Now they no longer come and many co-operatives are in decline."

In what this co-operator says, we can clearly see a strategy of taking advantage of the opportunities offered by supervisory institutions, even if he pointed out that the low level of education did not allow all members to take advantage of training programmes. We can also note that the INC's abrupt disengagement, particularly in the area of technical assistance, has led to a decline in the functioning of co-operatives. Moreover, this is what Kamdem says concerning Africa: 'it is feared that this sudden departure [of the state], without having prepared the way, will resemble the "rush" for independence in the 1960s without the trained people to take over from the colonists."[8]

Speaking of the co-operators' reappropriation of the co-operatives, I. Evora, in a study on the co-operative imagination, notes the following situation as regards the members of a dissolved co-operative:

the dissolution of a co-operative was experienced by the co-operators as a loss, and the process of elaborating this loss was painful for everyone, for it mobilised unconscious aspects connected with the question of each person's attachment to the place of their existence; losing the land that they were cultivating together meant not only the loss of the co-operative—and of the goal for which it existed—but also the perception of the weak side of attachment to the earth (they were finally told that it did not belong to them), with the place where one lived and where one sees oneself in the real world (family, partner, children, church, community…). For the co-operators, ceasing to engage in a productive activity that required a good deal of initiative, devotion, and organisation meant, in addition to the current material (concrete) difficulties, a return to small-scale, subsistence activities, and the disintegration of the group that had been a local reference point (Evora 1991: 14).

The individual and collective importance of the organisation, a reference point for the community and the link between members of the collectivity is fairly evident. Discontinuity of this process leads to a rupture within the community, a social disintegration, which indicates the importance of the organisation not only in production but also in social terms.[9]

The data in table 1 (see appendix for all tables) show the significant growth of the co-operative movement in Cape Verde, which began in 1975, at the time of independence, and especially from 1979. In fact, registration of co-operatives[10] did not begin until after the creation of the INC, which obliged co-operatives to carry out this formal process if they wanted to obtain legal recognition and gain certain privileges, such as access to credit. We can conclude, on the basis of the data in the table, that in eighteen years the co-operative movement increased by 34 times compared with the figures for the period between 1897 and 1975, which is the equivalent of a growth of 3000%. It is important to recall that the data concern registered co-operatives. There are in fact many organisations that are co-operative in nature but have not registered because of the bureaucratic and legal formalities involved or because the members did not want to register.

According to the same source, more than two thirds of the co-operatives have peasants as members. However, the activities of co-operatives were multiple: agricultural production, providing factors of agricultural production and/or marketing, and sales of consumer goods. In the same way, associations were not included if they were short-lived or already existed and regularly operated without any legal basis, since the law which regulates the creation, functioning, and legal recognition of associations was not approved until 1987.[11]

Another thing to keep in mind is that the co-operatives tend to be concentrated in the main islands, and in particular in rural areas. This fact is due not only to a stronger concentration of population in these zones or to agricultural and pastoral activities, in which co-operatives are involved, and which are limited to rural areas, but also to the strategy of co-operative promotion and revitalisation. Strengthening and improving supplies of essential goods at moderate prices for the poorest strata of the population and mobilising this same population for the objectives in the programme of the political party in power and of supervisory structures[12] were among the reasons mentioned.

The revitalisation connected with peasants associations appeared as a need at the end of the 1970s and the first half of the 1980s in the perspective of a political party strategy, and in the light of the gradual decline of peasant support for the party in power (PAIGC/CV) compared withthe level of support it enjoyed during the early years of independence. Thus a few projects began to be developed, particularly in the framework of the activities of the Amilcar Cabral Institute, the school that trained party leaders for power and of mass organisations that were close to this party.[13] Research on the agrarian question led to a publication entitled *The Agrarian Question*. At the same time, work was begun on the island of Santiago, and in particular in Santa Catarina, with a view to revitalising the peasants associations.

It should be recalled that access to land in Cape Verde was gained, in most cases, through share-cropping or leasing. The data indicate that 48.9% of the farmers cultivate their own land, 21.5% are share-croppers, and 15.2% lease their land (Teofilo and Machado 1993: 47). However, the landowners' discretionary power, especially in the case of tenant farmers, allows them to take possession of the property at any time, despite the approval of legislation prohibiting this practice (Furtado 1993). E. Teofilo points out that sharecroppers, tenants, and those who hold usufruct see

> peasants organisations as the last line of defence of their rights, and what they fear most is losing access to the land. This is justified, because although they do not really own the property and do not draw from their little piece of land what they need to live, they see themselves as being in a better situation than those who depend solely on salaried work. (Teofilo 1993:40).

The existence of peasants organisations is due to the commitment and the interests of their promoters and organisers, within the framework of the definition of common goals and the interrelation of this commitment with the environment, whether political, economic, or social. This means, in our opinion, that the dynamic of peasants organisations, whatever these organisations may be, includes a dialectical relationship with the community (understood here not only as a place or a space, but also as social actors in process, in action) and the outside world.

While in a first stage, after independence, co-operativism was the primary way in which peasants organised, in a second stage associativism (peasant groups) also underwent an evolution. Moreover, if in the period that immediately preceded independence, and in the first years after it, it was the PAIGC that revitalized the process, and after the government took it over, peasant associativism was also encouraged by political party structures, that counted on the technical assistance of the governmental agency responsible for rural development.

This shift in terms of strategic action regarding peasant associativism took place in 1982. Then, within the framework of activist research, "the peasants organisation (...) was supposed to be at the centre of attention, apart from the party's internal organisation." These other questions involved the search for new support for the PAIGC among the people, since there had been a growing erosion of support for or apathy towards the party, especially with the tensions resulting from the discussion and approval of the basic law regarding agrarian reform and from the tension that still exists between the PAIGC and the government supported by the party, on the one hand, and the Catholic Church on the other.

A revitalisation of peasants organisations had started in the commune of Santa Catarina and more precisely in the Tabugal hydrographic basin, supervised by the political party that was then in power, and relying on the technical and financial assistance of governmental and municipal institutions. For the PAICV, this was a strategy to strengthen the defenders of the agrarian policies that had been established and to attract new adherents to the party. From this point of view, the peasants organisations and associations that were activated and led by this party's "militants" were its chosen instruments for political activity in the countryside. The autonomy of these organisations was not an issue in this context, insofar as the party's political doctrine regarded the interests of these organisations to be the same as those of the PAICV.

In nearly all the hydrographic basins of Santa Catarina, therefore, commissions to activate peasants associations (ACTA) were developed, and alongside them, CARA was set up. There was a very obvious interconnection between support for agrarian reform and for the peasants organisations that were intended to "advance the process of agrarian reform." In the same way, the two forms of associative peasant organisations—the cooperatives and the associations—intersected. Indeed, the associations were regarded as a privileged way of bringing together the forces of struggle that were fighting for their specific interests. The co-operatives would move forward with the support of the associations, in the sense that they would ensure the continuation of the results that had been obtained. In other words, once the associations had succeeded in carrying out the agrarian reform and had assumed complete ownership of the land, the agricultural co-operatives would take over and ensure collective production and marketing. It was no accident that the strategy of mobilising the associations was based on the hydrographic basins and therefore on production. In this way, one can also understand how the National Co-operative Institute was involved in this process. This perspective is clear when it is recalled that :

> This resulted in the fact that Tabugal could be regarded as a potential productive "unit", and that with the carrying out of the agrarian reform, it would gradually be transformed into a sort of co-operative, either as a single whole, or partially or by sub-divisions.[14]

Attention thus turned towards the structures empowered by the PAICV, and not to traditional or endogenous structures. Thus we can readily comprehend that the peasants organisations were not really viable. One tends to forget completely the associative dynamics that occur either in terms of the collective organisation of agricultural and community work, or for the defence of certain interests, even though these may sometimes be temporary. Again, the peasants' non-involvement in some of the associative initiatives, even though expressed in an individual fashion, may be a way of challenging certain practices of institutions outside the community. Moreover, the reaction to external involvement may take two forms: agreeing "to play the game", and to respect its rules, afterward reversing the order of things so as to derive the largest benefits, or else not participating in the game at all.

Institutionalised associative practice after independence experienced both sides of the coin. Many associations functioned to the extent that there was an opportunity to make real gains. That explains why, of the various peasants associations revitalised within the framework of the party's activity from 1982 onward, only one is still operative. This is an association whose members have benefited from access to land, to the *trapiche*, for the industrialisation of sugar cane production, and to irrigation pipes, reservoirs and rainwater-control projects in the hydrographic basin, in the framework of a project financed by international co-operation, without cost to the peasants. In our view, the fact that community or developmental activities were linked to a party organisation was from the outset the factor that made it possible to continue these activities over an extended period of time. This remained clear during the transition in the political system and the regime in 1990.

In the group dynamics achieved with a group of members of the sole peasants association on the island of Santiago that dates from the first phase of independence, we see a strong instrumental reason for the adoption of positions both by individual members and by the association as a whole.[15]

> We have a history of resistance dating from 25 April. We broke all ties to the landowners and we are working the land, the *trapiche*, ourselves. We obtained many things when we were united. We even succeeded in establishing a school. La Ribeira de Tabugal has had many projects completed, many investments. Now we have no problems with reservoirs and rainwater control. The only problem we have to resolve with the farmers who are not members of the Association is the water problem. The MDR's technicians had already talked to us about that, ten years ago.
>
> With the change in the system and the fall of the PAICV, the association was somewhat shaken. Some people were in favour of it and others were against it. In the end, we saw that if we did nothing but argue with each other, the situation would get worse and worse for us. So things have not got worse. But neither have they remained as they were before.
>
> Before, we were treated differently, we had more support form the government and the party. The engineers came here more often. Now, just from time to time. But it's better like that... [Interview].

This clearly indicates that the members of the association ultimately created strategies suited to the group in relation to outside agents and institutions, through an apparently contradictory process of accepting the rules of the game that were imposed on them, while at the same time appropriating these same rules to serve collective interests. These interests involved above all the opportunity to mobilise resources (financial, technical and material) for the activities planned. In this way, the collective succeeded in more easily opposing the agents and social categories. In this specific case, it was a matter of peasants who broke, de facto, their contractual relations with the landowners. Those who leased their land stopped paying rent and the rest no longer shared their production. Legally, the situation of these peasants still remains precarious, so long as they have not bought the land and that the state has not taken land from them in connection with agrarian reform. In this way, government support is fundamental for the peasants in their struggle against the landowners.

The Political Transition and the Place of the Peasants Organisations

The political transition to a multi-party system led, as might have been expected, to powerful social and political changes It also accentuated cleavages and tensions. The new positioning of the political actors made itself felt not only in the political domain but also in other spheres and structures. This was to be expected. All public institutions in the context of a single-party system were, to a greater or lesser extent influenced by the party's standards. It was, after all, what we call a one party-state.

In the domain of the peasants organisations, the co-operatives were the first to be affected by the political divide. The search for supporters for political and social projects led to co-operators being "obliged" to take a clear political position. The political struggle was superimposed on associative, organisational, and economic interests. Division, bankruptcy, and crisis occurred in many peasants organisations. The commissions that had been created to encourage peasants organisations in the hydrographic basins ceased to function. The party's immediate concern was no longer the organisation in the party of an organic search for defenders for certain political measures, but above all to win over voters and their votes.[16]

The peasants organisations (co-operatives or associations) have significantly decreased in number, and in some islands and municipalities they have disappeared altogether. According to the data in table 2, the island of Brava, which formerly had two co-operatives, now has none at all. All the islands have experienced a decline in the way these organisations function. In general, among the registered cooperatives and those that are truly functional, there has been a decline of about 60%.

The emergent political parties found fertile ground in peasants organisations for recruiting supporters for their political and social projects.

For the PAICV, it was a way of consolidating its political base in the context of a pluralistic political system. A whole set of arguments was elaborated to show what the post-independence government had done to improve the people's living conditions, and to seek ways of ending the excessive exploitation to which the peasants had been subjected. On the other hand, the opposition and new parties, particularly the MPD and UCID, sought to show how the peasants and their organisations had been used for the needs of the party in power and for the interests of certain individuals, and at the same time sought to emphasise the authoritarian—not to say totalitarian—nature of the PAICV (the party in power).

The excessive politicisation of the peasants organisations, which was linked to an absence of a *habitus*[17] of pluralist politics in Cape Verde, has led their members to give the party's interests precedence over those of the peasants organisations themselves. A situation–opposition dichotomy had been noticeably absent in almost all the peasants organisations, with the exception of those that were politically homogeneous, in one way or another.

The change in the political system, which was to culminate in the legislative elections held in 1991 and won by the MPD, also led to important transformations in the social fabric. A new economic model was put in place, based on neo-liberal economics. The role of the private sector in the economy was emphasised, and political measures were taken in this area. The state's role was to be limited to promoting and facilitating private intiatives. This involved the progressive withdrawal of the state, in order to ensure economic and financial stability, leading to privatisation implemented with the support of the Bretton Woods institutions.

The existing peasants organisations found it necessary to readapt their organisations and mentalities. In the case of agricultural and agricultural-woodland-herding co-operatives, the impact of the state's disengagement was significant. The supervisory institution (the INC) gradually stopped providing free technical assistance, while fiscal incentives and subsidised credit, sometimes interest-free, and the absence of taxes, all ceased to exist. Similarly, training programmes became less frequently available, and against an international background of declining public development aid, there was even a withdrawal on the part of international non-governmental organisations that had been providing technical and financial assistance to the national peasants organisations. Some co-operators mention that:

> Our co-operative was not prepared to find itself suddenly deprived of the INC's support. Our leadership was not competent to handle the accounting, and contacts with foreign organisations existed only through the co-operatives. We became orphans. It is only now that we are beginning to get back to the situation we knew before.

Some peasants organisations were able to adapt rapidly to the new social, political, and economic reality, however, by taking advantage of the political proximity of the new leaders. But new peasants organisations were also created in a process that extended to the whole of Cape Verdean society, and affected various social levels. From 1991 on, many new peasants associations and agricultural co-operatives were set up.[18]

The data resulting from the census of the co-operatives reflect the tendency seen at the beginning of the political transition and the implementation of new economic policies. In addition, the significant decline in co-operatives had an influence, and the new co-operatives that did emerge were found more often in the productive sector. The INC's records, which show that 'after 1991 the new co-operatives created were oriented primarily toward the production of goods and services, particularly in agriculture, livestock-raising, fishing, and industry' (Recenseamento Cooperativo 1995).

The same document adds that after 1990, with the choice to adopt a market economy, the promoters were interested above all in the production of goods and services, since the cooperatives now belonged to the private sector, particularly because they make it possible to generate profits from small amounts of capital (Ibid).

Peasants associations clearly tended to grow after 1990. On the one hand, this was due to the need, in the context of economic liberalisation, to associate, not merely in order to defend specific interests, but also to gain easier access to certain public goods and services and to improve their effectiveness as a pressure group. On the other hand, the co-operative form of organisation had in the meantime been partly eroded during the transition to a new system, and the need for a different form of organisation was felt.

By analysing the implications of programmes of structural adjustment on the dynamics of peasants organisations, and without forgetting their social, economic, and labour consequences, E. Kamden notes that

> the privatisation of the marketing of export products, to the detriment of the famous "price control boards," which in other countries had become "the marketing boards", thus (provided) an opportunity for cooperatives of a certain size to export their products directly and to eliminate the intermediaries that increased their costs without increasing the products (INC 1991: 34).

The data in table 3 show that at the national level, there is a significant change in the appearance of peasants associations over the period from 1990 to 1997. Before 1997, the existing peasants' associations were limited to the commune of Santa Catarina on the island of Santiago and to be more precise, there is only one that is still operating. In other areas of the commune, there were revitalising commissions; however, these did not lead to the creation of new associations.

These data once again show clearly that peasant associativism is concentrated on the islands of great agricultural importance, such as Santiago, Santo Antäo, and S. Nicolau, which have respectively 67%, 19%, and 6% of identified peasants associations. It is interesting to note that the island of Fogo, although important from an agricultural point of view, seems to have no peasants association. This can be explained by the fact that this island had a high level of cooperativism. It is also interesting to note that the island of Boavista has 6% of the total number of peasants associations, even though it is a sparsely populated island of little agricultural importance in the national context. On this island, stock-raising connected with agriculture must have led to the appearance of these associations.

If we analyse each island internally, we also see that it is the municipalities of great agricultural importance that have the highest numbers of peasants organisations. On the island of Santiago, the municipalities of Santa Cruz and Santa Catarina contain a large number of peasants associations. On the island of Santo Antäo, the municipality of Ribeira Grande has the largest number of peasants associations.[19]

Recent developments in peasant associativism in Cape Verde, especially on the island of Santiago, cannot be analysed and understood without taking into account reasons that are endogenous to the rural communities themselves. With a large part of the rural population entrenched in poverty, associativism appears both as a better way of fighting against difficulties, and as a way of gaining better access to sources of income.[20] We also see a growing willingness on the part of public institutions to entrust farmers' associations with the task of operating, maintaining, and managing infrastructure, in particular those connected with soil conservation and rainwater control, by contracting the infrastructure to farmers' association, financed by public institutions with the help of certain donors.

An important process in the emergence of peasants associations, and in particular those that are legally constituted, can be seen in the 1990s. As we have already pointed out, two interests intersect here. One is the endogenous nature of the associations that have turned toward development either in the interests of their community or in order to respond to the individual needs of their members, and the other is exogenous, driven either by state supervisory structures or by national and/or foreign non-governmental organisations. Even though a significant number of the existing associations in the country are not legalised, both because of ignorance of the procedure and bureaucratic bottlenecks involved, table 3 shows that more associations are becoming legalised. The fundamental reason, as pointed out by the leaders and members of the associations investigated, lies in the fact that the legislation facilitates access to public and private funds, including those from bilateral and multilateral aid agencies. We are witnessing the setting up of a National Platform for peasants organisations, bringing together all the peasant organisations in the country. To join this Platform, however, the organisations have to have a legal form or status.

Peasants Organisations and How They Are Regarded

The description given by their members to their organisation and the importance they attach to it, the internal dynamics of the association, the relations between its members and between them and outside agencies are all important to better understand the changing reality that we are analysing. This kind of methodolgy, taken in conjunction with a knowledge of what the institutions that maintain relations with the associations think and expect of them, and enables us to follow the sense of movement and of the social dynamics, and the possible divisions that exist in social, political and economic relationships that grow up between the various social actors. As far as research techniques are concerned, this methodological approach gives better results when it is applied to relations between the groups—group dynamics or focal groups—given that it enables the rapid drawing up of a sociogramme of the organisation under study.

Ten peasant organisations have been selected for this kind of analysis, five from the island of Santiago, two from Santo Antao, two from S. Nicolau and one from Fogo. Two of the organisations are agro-pastoral co-operatives, one emerged before the transition period and the other during the political transition. Of the eight associations chosen for study, one was already in existence before the change of regime and the political system, the rest appeared after the changes that took place in 1991. We have, however, only been able to consider the organisations on the island of Santiago.[21]

The Agricultural, Woodland, Pastoral Production Cooperative in Tentativa

This co-operative, which was founded in 1991, arose from an attempt to organise an association of alumni of the training school for agricultural-pastoral 'supervisors',[22] who are scattered over several areas and islands of the country. From the idea of an association of supervisors desperate to remain active, it evolved into a co-operative that would make it possible to capitalise on the members' experience for specific economic activities without neglecting the associative aspect that gave birth to the project in the first place. This is an organisation that seems to combine the entrepreneurial aspect with the associative and communitarian aspect. It emerged as 'a way of solving certain employment problems, but also of benefiting students from the former S. Jorge school and taking advantage of their abilities. At the beginning, we wanted to create an association, but it was difficult, and so we decided on a co-operative', explained a member.

Other members observed that

> at the beginning we considered several possibilities: an enterprise, an association, a co-operative. Then we decided that *A Tentativa* was more social in nature and was closer to the group's idea of forming a co-operative. We wanted to show that we could work in arid and semi-arid zones and that it was possible to use the forest areas for certain purposes. In this way, a co-operative would have the most impact, and it could also gain some international aid.

Institutional support seems particularly important given that the co-operative intended not only to work in a forest area, but also to work on the forest perimeter, which was not welcome to the Forest Service. "Despite the high level support we had, it was very difficult because the Forest Service administration opposed everything we proposed. Even though we had bought the land, we found it difficult to enter the area and establish ourselves there".

This co-operative started in mid-February 1990, with six members, and an area of 13 hectares of agricultural land, acquired through contributions made by members of the co-operative, amounting to 80,000 escudos (about $1,000) each. On that basis, they got bank loans for installation and start-up costs. At the beginning, they asked for a loan of 17 million escudos (about $200,000), but obtained only 12 million escudos ($141,000). "The rest of the money was provided by members' contributions while the project was being carried out".[23]

The co-operative is composed solely of men, since it includes only former students of the S. Jorge school, which trained only men. The co-operative functions well in organisational terms. According to one of its members, even though no women were members, members' wives ended up indirectly becoming members. He adds that "We also employ women, especially for work connected with aviculture, thus helping to create jobs for the area's population".

In financial terms, the co-operative is healthy, and is able to meet its financial commitments. Even though it cannot be said to be well off, it is able to make new investments. Within the co-operative, there is a clear division of duties and responsibilities: the members take turns to work during weekends, in order to ensure that clients can be served and to take care of the animals and the whole infrastructure all week.

Relationships with outside institutions are said to be good. "All the institutions we contacted acknowledge the worth of what we are doing, they are always open to supporting the Tentativa co-operative by solving the problems that come up".

For members of this co-operative, the political transition that took place in Cape Verde was positive for the co-operative. "If there had not been a political opening-up, it might not have been possible, in fact, to establish A Tentativa".

The A Tentativa co-operative is concerned to establish healthy and fruitful relationships with other peasants organisations. Thus one of the members of this co-operative is directing the process of setting up the National Platform for peasants organisations, which will be the sole representative of the various peasants organisations in dealing with public and private institutions, both national and international.

During its short existence, this co-operative has performed very well in organisational, economic and financial terms. Similarly, it has managed to integrate itself well into the community, particularly through the jobs created for young people and women in the area, which combine employment for these groups with professional training. The members' stock of knowledge and social relationships has greatly facilitated its development. In fact, this is a group that has specific technical abilities in the area in which the co-operative works, and has solid contacts and useful sources of information both inside the country and outside it. Its members also have "high-level" contacts, which explains how they can develop the forest perimeter and gain access to bank credit, as well as the financing made available (sometimes as grants) by national and foreign institutions.

The co-operative's activities include: aviculture, livestock-raising (beef cattle, pigs), and market gardening, in addition to the conservation and development of the forest perimeter, particularly in order to obtain firewood and charcoal, as well as fodder for animals. This latter activity has not recently been pursued, because the people living near the forests have invaded them to get firewood, which may make the perimeter non-viable.[24]

This co-operative benefits from privileged relationships with national public institutions and even with international institutions. In fact, the social and political capital, as it is defined by Pierre Bourdieu, has made it possible—and still makes it possible—to mobilise financial and other resources rather easily. This is a modern peasants organisation that operates within the neo-liberal economic policy now being pursued. Its members have a good grip on the political, economic and social environment, and take advantage of it. This is the ideal type of organisation for the period following the political opening-up.

The Thirteenth of November Agricultural Co-operative

Another co-operative, known as The 13th of November (*13 de Novembro*), was officially legalised between 1983 and 1984. It is situated in the commune of Santa Cruz. Like most of the co-operatives that arose after independence, the interest in organising it came from a group of militants and people close to the PAIGC, who requested the support of the national co-operatives' headquarters at that time in order to realise their initiative. At the beginning there were around 28 members. However, during a preparatory period that some people considered too long, eight persons had withdrawn before the co-operative began operating in 1978;[25] though at that time, the co-operative already had a warehouse and had prepared the land for agricultural activities.

According to one member of the co-operative, much was achieved with the financial help of international organisations. "In 1982, we had an office and a water reservoir as well as a 700-meter long water pipeline between the well and the reservoir. It was inaugurated in November". At that time, the co-operative had gained legal standing, and it included 20 members, of which eight were women and twelve were men. According to two of the founding members, the co-operative's activities began with the help of the INC and the Oecumenical Church Council (Conseil œcuménique des églises/PSDS).

The co-operative now has sixteen members, including nine women and eight men. It works 4 hectares of land, practicing both agriculture and aviculture. According to its members, they have been supported by the INC, the FAC, the Ministry of Agriculture, and the FAO to carry out various infrastructural projects. "We have a reservoir, irrigation pipes, and recently we have installed a drip micro-irrigation system, in order to make more effective use of the water". Regarding the process of establishing the co-operative, the accountant states that 'the land was completely arid and had not been cultivated. We did all the work of restoring the land so that it could be used for agriculture. We had many losses. It was hard but rewarding work'.

During the first period after independence (from 1975 to 1990), they had no problem relating to the land they were occupying—which did not, however, belong to them. Nonetheless, after the political transition some landowners claimed a right to the property. Negotiations are currently being pursued in order to purchase the land, even though the plaintiffs have not succeeded in proving their ownership of the land. The price asked by the plaintiffs for the land is considered exorbitant, and a technical audit has been made to determine the monetary value of the land. Moreover, the members add that "the question of who the land belongs to might have been damaging for the co-operative. We succeeded in acquiring ADF financing[26] to extend our activities. Fortunately, the financing was granted even without the problem having been finally resolved". They add that the question of the ownership of the land came up when the alleged property owners found out that the co-operative was about to obtain financing. Better still, they stated that the president of the municipality supported them and said that the land belonged to the municipality and that there was thus no problem. "In 1992 he even gave us a document confirming that the land belonged to the municipality".

The change in the political system and the political parties after 1990 affected the co-operative and divided its members. Consequently, there has been a decline in the associative dynamics, mistrust among the associates, and a disaster in the accounting and organisational system for production. The projects, whether at the level of production or upstream or downstream from it, were carried out by three completely autonomous groups, without any production control so far as cost, the necessary investments, the results of production, etc., were concerned. Certain members held back some of the co-operatives' property, as the co-operative disintegrated.

Other problems and conflicts experienced by the co-operative over the past few years are related to a few landowners in the area; "we even lost a whole apple orchard because a landowner denied us access to irrigation water. We lost about 2.5 million escudos ($29,000), and up to this point we have not yet been compensated. The well that feeds our reservoir is on this man's property and he decided to cut off our water supply, but the water belongs to the state". This situation has forced the INGRH, the INC, and other institutions to intervene in order to find a solution. "We spent 90 days without irrigation water and we lost everything. The co-operative almost disappeared because of that. The ADF's financing was of great help to us in restoring production".

In organisational terms, the co-operative is in a phase of restructuring the functioning of its social organs, its production planning, marketing, and its whole accounting system. The main problem, according to the older members and founders of the co-operative, has two aspects. First, most of the members have little or no education. In their opinion, this prevents a proper rotation of the leadership positions which require the ability to read and write short documents. Second, the advanced age of many of the members. At present, only four remain of the twenty founding members. Of the 16 who are no longer active, two have died and the others are ill, old, and cannot take part in the active life of the co-operative. Some have emigrated. They have been replaced by their children. The latter "do not have the co-operative spirit and or even any desire to work in agriculture. They want to emigrate or move to the city". In the elders' opinion, this may put the co-operative's continued existence in jeopardy.

The political and economic stakes involved in a context of economic liberalisation have seriously affected this peasants organisation, not only in the way it is run but also from the economic and financial points of view. Sometimes the survival of the organisation seems to be at risk.

The problems this peasants organisation has experienced demonstrate the political and economic stakes involved in the political transition period.

The eighth of February Peasants Association

The 8th of February Association brings together about 105 farmers from Santa Cruz. It was created in 1993, after the change in the political system. Its leader has been a municipal councillor since 1992. Almost 15% of the members are women. Although it was set up in 1993, its origins date back to 1985-87. According to its president, "at the beginning there was a group that was formed to obtain a motorised pump with the help of the FAC. The group's need to grow, along with the interest shown by new farmers, led to the idea of an association".

It is worth pointing out that at the beginning, the group was moving toward an organisation of a cooperative nature, because the loan obtained from the FAC was based on that alternative. However, according to the president the members had a stormy debate about the most suitable form: "we had to vote to decide whether to form a co-operative or an association." One must keep in mind here that the association was established in the final phase of the political process in which co-operatives, in particular those in Santa Cruz, had been seriously jolted by the invasion of co-operatives by groups with diverse affiliations or political interests. Thus it was natural that co-operatives were seen as potential sources of friction, especially when, as some members of the association said, "we have members who belong to different political parties. We are not here to practise politics, but to work".

This association first sought to defend its members' interests with regard to access to factors of production and credit for agricultural activity. The work of production was carried out individually, because one member's land is sometimes far away from another's. This made difficult any attempt at joint labour and production on the land. However, they practise Djunta Mon in the activities they undertake. Similarly, certain factors of production are purchased jointly, thus diminishing the cost for each individual. The storage and marketing of certain products, such as potatoes, are handled by the leadership of the association itself, which waits for the right moment to put the product onto the market, thus allowing improved results. According to members of the leadership, the association plans to negotiate with the INFA the promotion of seed potatoes for sale to farmers on the island of Santiago. Members of the association work about 60 hectares of irrigated land.

In the area of training, the members have already benefited from projects in plant production and protection, irrigation, and the installation and maintenance of a drip irrigation system, in close collaboration with the

DGASP and the DGAR. Many peasants have not taken much advantage of these projects because of their very low level of education. "Training activities with illiterate peasants have been conducted more in practical terms and in the field, which requires more time and availability".

According to one of their leaders, the association has already succeeded in mobilising a certain amount of support for credit, training, and technical assistance: "members of the association have received credit for the purchase of factors of production in the area of market gardening, thanks to a project financed with Swiss co-operation. They have also received credit for the installation of systems of micro-irrigation, in addition to technical assistance for putting them in place". The new irrigation systems seem to be important for the association, since "the drip irrigation system makes more effective use of the water and allows the surface irrigated to be increased. Before, we used a flood system of irrigation, which used a lot of water. Now that less water is available and its saline content is increasing, new systems of irrigation are important".

The resources in social relationships that have been accumulated by members, or rather by the leaders of the association, seem to have facilitated contacts and access to information and requests. In fact, the leader of the association, even though he is a politician at the municipal level, is an agricultural technician who knows the ins and outs of the Ministry of Agriculture and of the institutions and projects that finance rural development plans.

For two members of the association's leadership, the fact that it appeared after the political change is significant: "Before, we didn't see the importance of an association. Also, we'd had more experience with co-operatives that were very politicised. Many people did not want to join for that reason. Now that is no longer true. The climate has changed and people are beginning to see that they can make more progress if they unite. It is not only the farmers that are forming associations. The landowners, young people, carpenters, et al. have also chosen associativism.[27] The government itself supports the creation of associations. Here, for example, associations are wanted to manage the irrigations systems and wells, drilling and irrigation equipment. The government says it won't need many bureaucrats and the farmers will end up with more responsibility for preserving what they use. In fact, this is true. But it is sometimes difficult to reach an agreement".

The Tabugal Farmer's Association

A peasants association, the only one that still exists since the political change, has 36 members, including one woman, and has 1.4 hectares of irrigated land, 1 *trapiche*, and 1 still. It has constructed a primary school in the locality and another in the residential area in Achada Tossa where most of the members live. The association was created in 1983 in the framework of the PAICV's activity in Santa Catarina and its commitment of support for implementing the basic law on agarian reform (LBRA). It should be noted that this law was passed in March 1982 and went into effect the following year. As we have seen, it was at this time that the PAICV, through the Amilcar Cabral Institute and in the framework of a research-action project, carried out a study on rural life in Cape Verde, in particular in Santa Catarina, and on the promotion of peasants organisations in the commune's various hydrographic basins (Pires and Stockinger 1986; Stockinger 1990). That is how the Tabugal Farmer's Association was established.[28]

The 1.4 hectares of land occupied by share-cropping peasants were finally expropriated by the government in the context of the process of agrarian reform, and they became the property of members of the association.

Although it continued to exist after the political changes, the association experienced problems in managing its property and capital. The decline in the support it was used to receiving from the local offices of the Ministry of Agriculture or from its central offices seems to have created problems for the association's finances. We should add that the relatively advanced age of the members (on average, over fifty years) and their low level of education (between illiterate and elementary school) may explain in part the current situation, without forgetting the impact of political arguments and differences inside the association.

In the framework of the reconversion of the FAIMO, the association has recently been involved in soil and water conservation projects. At the same time, a reorganisation is currently being carried out with the help of the Ministry of Agriculture. According to information provided by members of the association, the process of legalisation has almost been completed. In this way, they added, they would be better able to take advantage of the benefits offered by international organisations.

The advanced age of many of the association's members, as well as the high rate of illiteracy, represent limits to the revitalisation of the association. Thus there are some members who use their leadership positions to take advantage of the resulting lack of supervision. Only since a new member was able to get training in accounting management, has the organisation begun to be better managed. This new situation has led to a reorganisation of the leadership, and the members are more involved in the association's life. Nonetheless, two technicians from the Ministry of Agriculture continue to make weekly visits.

The Santa Cruz Farmers' Association

The Santa Cruz Farmers' Association was established in the context of the privatisation of the Justino Lopes company, a state enterprise concerned with agriculture and livestock-raising. Thus it is an association that emerged in the process of implementing the economic reforms undertaken by the government after the change in the political system, which involved economic liberalisation and state withdrawal from the entrepreneurial sector. According to the association's president, knowing that the government was going to privatise the firm, the workers had wanted to form an association in order to take over at least the agricultural part. "The government showed interest in our idea and we moved forward with it. At the beginning, in 1993, there were 105 of us workers, but many withdrew. In August 1994, when we were working the land, there were only 79 of us left, 18 women and 61 men".

In the context of the privatisation of the firm, the associated workers elaborated a project intended to help the agricultural component recover and become viable. Similarly, in the negotiations with the government and in the contract signed, 59 hectares of land belong to the association, and are to be amortised over 35 years. The price of the land was set at 195 million contos ($2,283,000). In return, the compensation due to the workers on the sale of the enterprise would be converted into supplying electrical energy for irrigation, among other benefits. However, members of the association say:

> The municipality did not accept this principle, and charges nearly 22 escudos per kilowatt hour. In this way, the price of the land was reduced to 164,000 contos ($1,930,000): the 30,000 contos corresponded to the compensation due to the associates as former employees of the Justino Lopez company.

The association is at present experiencing problems over obtaining financial resources for making necessary investments, particularly in the area of irrigation. The association has developed a project that may find funding. However, the financial results, particularly the last year's have not been positive. The association had a deficit of 3 million escudos (around $35,000); it is able to pay its operating expenses, but is not amortising the capital and can free up even fewer reserves for investment. In the banana production sector, the association belongs to the Association of Banana Producers, and in this framework it has received aid from the STABEX programme and the European Union, which have allowed it to make investments in this production sector.

The association's other problems lie in its relationships with the community.

> We have been the victims of several attacks. Agricultural products, tools, and equipment have disappeared. The police have not been able to resolve the situation. This has created an atmosphere of tension, not to mention the loss it has caused the association.

> The former employees of the Justino Lopez firm who preferred to receive compensation are now unemployed and without income. Many of them said we were crazy because we would never get the money they were going to get in compensation. I know only two who have managed to invest the money they got and are comfortable. The others are unemployed and have nothing. We have problems, but we have managed to pay salaries, and we now employ over fifty people, in addition to the members. For that, we need additional aid. We have no institutional aid.

The leaders of the association and some of its members show a certain dissatisfaction with public agencies, and in particular those connected with the lack of aid and technical assistance on the part of the Ministry of Agriculture. They had agricultural equipment that had broken down but was urgently needed.. One of the Ministry's agencies had the kind of equipment that was needed, but did not lend it to the association, which had to buy it in Portugal. Too late, however, because the whole crop was ultimately lost. This case makes it clear that the association does not have the capital necessary to give it access to decision makers or even to the institutions that could support the association. It must be emphasised that of the four peasants organisations analysed, only this one had not received an operating budget from either financial and credit institutions or the international organisations active in this sector.[29]

Its president emphasised the lack of institutional support when he stated that"The association is the one that has the greatest scope at the national level and employs many people. When the President of the Republic visits us, together with the Minister of Agriculture, we will raise all these problems."

We can see, on the basis of the analysis of the careers of these various peasants organisations, the internal dynamic of the rural world in Cape Verde, which is reflected in the peasants' organising movement.

We can also see that these peasants organisations lack visibility to some extent, even though the statistical data and the social-agrarian reality demonstrate their presence in the rural world. The context in which such models (cooperatives and associations) are activated ends up by making these organisations part of a complex organisational blind spot. This results in other political and governmental organisations being more prominent. It seems to us that the paradigm of participation underlying the policies and programmes of promoting peasant associativism resembles that of the single party regime; S. Amin notes that these regimes tend to see the population as "the masses", as something homogeneous and amorphous.

Even taking account of the ambiguity contained in the expression "participation," it appears as an institutionalised form of "popular participation" in the country's development, and therefore as support for the regime in power, by transforming itself into a bureaucratic and bureaucratised mode of participation. Participation as "Empowerment of the people to effectively involve themselves in creating structures and in designing policies and programmes to serve the interests of all as well as to effectively contribute to the development process and share equitably in its benefits" (Economic Commission for Africa, African Charter for Participation, 1990: 19) does not seem to have been a leitmotiv of the whole process, at least at the beginning.

Although there may be a certain adoption of peasants organisations by governmental and non-governmental organisations, it is certain that the peasants organisations have succeeded in taking advantage of the fissures in the relationships that are established, and in putting their own interpretation on these relationships, seeking to maximise the benefits. Thus, we cannot say and the results of research demonstrate so, that a given project is unilaterally imposed by a determined action, or whether the actors in a relationship exclusively determine the meaning of the project. One of the peasants said that "We know how to obtain financing from certain organisations. So when they have nothing to give us, they don't interest us". He meant that communication and social relationships are interactive, even though sometimes one can show that one of the poles of the relationship is dominant. In addition, the interests of the various actors, even though diverse, can sometimes give rise to common denominators. While on one hand "empowering" the population and members of associations in particular tends to be seen as a factor in development, through these organisations the population itself tends to show that what is required is an organisational strategy that involves dialogue, or better still, negotiation with other institutions and partners.

Gender Relationships in the Dynamics of Peasants Organisations in Cape Verde

Legally and formally, men and women had equal right of access to land ownership long before independence. After independence, legislation strengthened the principle of equal rights. In real terms, however, we can see an inequality between men and women, especially with regard to access to land, and a clearly gender-determined division of labour in the countryside. It is also true that the strong migratory movement affects the structure of the social and gender division of labour, which means the structure of power relationships within the family. Emigration, which basically involves men, thus transforms women into heads of families and makes them responsible for the family unit as an element of economic production.

Here we must emphasise one of Cape Verde's specificities regarding the structure of the family and its implications for gender relationships. About

36% of rural families are headed by women. More than two-thirds of these women are unmarried, whether because their marriages have ended or because they are cohabiting with a man. Their children often have several different fathers, something that can be explained by repeated efforts to find a permanent companion or a husband who is not only able to guarantee stability but can also increase the family income.[30]

In demographic terms, women constitute about 52% of the population of Cape Verde, and are thus in the majority. As for the agricultural population, 52.7% are women. Thirty-six percent of the family farms are run by women. The municipalities of Santa Cruz Catarina and Tarrafal, both located on the island of Santiago, have a higher percentage of women running farms, with 47% and 48% of the total respectively.

Although many women are involved in running farms, it should be emphasised that these are generally on not very productive land, which is highly subject to the vicissitudes of drought. This land is used for rain-fed agriculture. For instance, 41%of rainfed agricultural operations belong to women, whereas only 20% of mixed (rainfall-dependent and irrigated) agricultural exploitations and only 23% of irrigated agriculture belong to women. The lands and farms run by women are also smaller than those run by men, and are usually farmed by sharecropping or renting. We should, however, note that women are involved in agricultural production and livestock-raising at every stage, from the preparation of the land to harvesting and marketing, the latter being mainly women's activities (ICF 1993).

Women play an important role in peasant associativism, though it is not so great as that of men, not only because of possible inequalities related to gender, but also because they have less available time, given the multiple involvement of women in the social and biological reproduction of the unit of household production (the family) and in economic production, as well as other activities.

In peasant cooperatives and associations, though we do not have data classified by gender,[31] we can see that there are fewer women than men, though their numbers may still be considered significant. The data in table 8 show that in the sixteen associations, women constitute 47.5% of the membership, and men 52%. These numbers are significant when we recall that each member usually represents a family unit, and that in rural areas 36% of the families are headed by women. This leads us to conclude that in some families headed by men, the women participate in associative movements.

One of the two co-operatives selected for analysis has no active female members, for reasons related to the way in which it was established. However, in the other co-operative, the majority of the members are female, even though no women are currently involved in the the leadership. Nonetheless, according to the members of the co-operative interviewed, one of the women previously served as the president.

The membership of the associations includes far fewer women than men; women constitute 15% and 23% respectively in the two associations. Moreover, in both kinds of organisations, the number of women members corresponds to the number of women involved in agricultural production, especially with regard to running family farms. We mentioned this in connection with our analysis of the ways of gaining access to land. It is usually the heads of families and/or farms who participate in peasants organisations. According to the data provided by the agricultural census, 36% of the total number of farms are now run by women. If we analyse the number of women involved in peasants organisations, it tends to reach this same level. In addition, the data in table 8 show that in the associations listed in the table, women constitute 47% of the membership.

Nonetheless, following a certain logic of the division of labour in this society, we find associations in which all the members are women; for instance, associations that are usually connected with sewing, lace-making, and embroidery, activities. Currently, the sale of hand craft products to tourists and to Cape Verdean émigré communities abroad has become more prominent in the life of the community. Thus the development of women's associativism in this kind of activity reflects, on one hand, the importance of this sector, and on the other, the existence of a desire to seek new sources of income for women.

However, we must draw attention to the gradual growth of associativism among women or among women and men together in the domains of agriculture and especially of family operations devoted to raising livestock. These are domains that Cape Verde women know very well, since they are involved in the agricultural workforce. Generally, women devote themselves to the children and to raising the family's livestock, especially small ruminants and poultry.

In the area of soil, water, and forest conservation, women's role has increased. Activities related to reforestation, including the preparation of the land (low walls and dikes), nurseries for new trees, and planting as such have been dominated by women. For that reason, we should not be surprised by the important part played by women in the agricultural associations working in the various hydrographic basins on the islands of Santiago and Santo Antäo, in the framework of reorganising areas of labour intensive activities and of efforts to change it.

Conclusion

The bulk of studies on the rural Cape Verde tend to emphasise that peasant organisations are low (Stockinger 1990; Pires and Stockinger 1986; Durand 1997). As a rule, these studies cite only the bureaucratic-formal ones as forms of organisations that fit into the perspective of modern impersonal, structured organisations, as characterised by Max Weber (Weber 1964). In this sense, organisations that are traditional, spontaneous, or that have arisen only as an emanation of forces "outside" the rural world, tend to be omitted or just not seen.

It is true that there are cases in which experiments with associations have failed, just as in cases of the individual or family search for solutions to problems encountered in everyday life. Teofilo (1993) suggests that one of the reasons for this situation is the imposition of organisational solutions external to the socio-cultural reality of the rural Cape Verde. Similarly, Thomson *et al.* (1992) mention that the way the government is structured, historically, has not permitted the creation of spaces for participation and decision-making at the local communities, which could produce a participatory dynamic for promoting the emergence of the community's own organisations, and which could be transformed into interlocutors for other agents and institutions. We should also add that the droughts and famines that periodically grip rural Cape Verde lead to an immediate concern with survival and a search for individual solutions, a dimension fostered by the requirement of the immediacy of the issue. At these times, signs of the traditional solidarity of the countryside—manifested in the practice of Djunta Mon, in funeral associations (Boto and Mitin), in the exchange of meals and food among neighbours—tend to disappear, and to re-emerge when normal conditions return. We must also remember that for centuries, the colonising power not only opposed and prohibited peasant associativism, but also derided it as a sign of primitive culture. Even the so-called modern forms of associativism were conceived only within a corporate framework, in the service of the colonial state.

With independence, we see a new attitude with regard to associative movements in general and to the peasants in particular. The traditional organisations were encouraged, especially in their public rituals rather than in their socio-associative aspects. Nonetheless, in accord with the organisation of the rural worldview and the proper economic structure, new organisational forms were conceived and promoted—cooperatives. Even though they were conceived and structured in a wider organisational framework (the whole organisational network of the state was under construction, and in a certain way occupied the social space, superimposing itself on the latter). These new organisational forms ended up creating spaces of autonomy and approval/disapproval with regard to the state (Mafeje 1993).

With independence, despite the emergence of peasants organisations, they were still seen and considered as such, and especially those that fitted into the defined organisational schema. Thus all other forms of organisation remained invisible, and even the stimuli for establishing them were largely unseen.

While in an initial stage the organising form was, par excellence, the cooperative, in a second stage the association became the preferred form, and underwent a major revitalisation after the political transition. Currently, new spaces of intervention in the rural world are being created with the appearance of new peasants organisations and with significant promotional activity and visibility in the public sphere.

New ways of organising the economy and production, together with new ways of organising social relationships, are appearing in a new political context in which spaces for dialogue and the defence and imposition of certain interests have been established. The network of interests at stake allows more negotiation and flexibility in determining the directions of social action. When the latter is combined with greater empowerment of the organisations at the base, the social dynamics is greater. In this way new spaces are emerging that allow greater visibility for these organisations. In reality, peasants organisations have always existed, despite various restrictions. Their invisibility was connected with the absence of expression in a public space and not with their own absence.

Notes

1 By analysing peasant organisations in Cape Verde, we can differentiate between co-operatives and associations, as stipulated by Cape Verde's legal framework.. The census of co-operatives did not take into account the variable of gender, which makes it difficult to analyse it in terms of gender. This gap will be filled in by interviews centered on the key informants and in-group dynamics involving members of peasants' organisations included in the sample.

2 These rural organisers are officials of the Ministry of Agriculture. They work at the community level, and are especially active in assisting farmers and livestock raisers. They work in all the country's islands, and particularly in the ones that are most important from the point of view of agriculture.

3 In connection with the definition of Djunta Mon and its difference from Djuda, see Silva, Tomé Varela. " Djuda " et " Djunta Mon " : formas informam (marjinal ?) e di Koperativismo in: INC – Encontro de refelexäo sobre o Movimento cooperativo caboverdiano. Praia : INC, 1991.

4 For a more detailed analysis of this period, cf. Furtado, Claudio, 1997, *Genèse et (re)production de la classe dirigeante au Cap-Vert*, Praia : ICLD.

5 According to the available information, before independence there were only two agricultural co-operatives in the Cape Verde Islands. However, we must remember that many peasants' organisations existed in an informal way.

6 Eurico Monteiro points out that at independence, an attempt was made to break with the earlier system. He says: "Around 1975-1976, the mere idea that cooperatives would be regulated by the commercial code was already sufficient reason to be able to discipline in a different way, because it was as if people wanted to conflate the cooperatives with traders merchants, and with the associations... The idea of traders tended to convey a certain pejorative feeling in relation to the nature of the cooperative itself, and from which it was felt necessary to escape, whatever the cost. Curiously, the idea of being a trader merchant was something almost negative; one could notice the desire to be free from the conditions of the commercial law at the base of the cooperatives, but more clearly, I think, in chapter 3, article 14, which stipulates that a trader cannot be a member of a cooperative, and that one of the conditions for being a member of a cooperative is that one must not be a trader". (Monteiro, Eurico, Intervention..., INC, op. cit., p. 23).

7 Concerning cooperatives, Nindi points out that " As a rule, two chief tasks were assigned to them, both of which are revealing to the instrumental role reserved for the peasantry in most African national development plan. First on the political plan, rural cooperatives were supposed to organise the villagers with a vierd to facilitating the transmission, towards rural masses, of political orientations and instructions decided at the top. Second, on the economic level, their planned function was to serve as relay between state comries, parastatals, and government departments on the one hand, and the mass or pretty producers on the other" (Nindi op. cit, 49).

8 Cf. Kamdem, Emmanuel "L'OIT et les coopératives en Afrique," in INC, op.cit, p. 3. Later on, we will analyse the consequences of the state's disengagement and co-optation for peasants' organisations in the post-independence period.

9 With independence, and in the framework of political mobilisation, the slogan "the land belongs to those who work it" became widespread. Thus many sharecroppers and farmers broke their connection with the landowners. From that point on, a few agricultural co-operatives or associations of agricultural producers appeared and took de facto possession

of lands without any legal basis. Many landowners resorted to the courts in order to take possession of their lands. And that is what happened to this co-operative. We should emphasise that in Cape Verde the land is privately owned, and thus access is gained through purchase or share-cropping (cf. Furtado, Claudio, *La transformation des structures agraires dans une société en mutation, Santiago du Cap-Vert*. Praia: ICLD, 1993).

10 The National Institute of Co-operatives is a government agency entrusted with promoting and supporting the co-operative movement, and also carrying out inspections of the co-operatives in order to verify that the legal requirements of the law on the general bases of co-operatives is being respected. At the INC, a fund has been created, whose goal is to give financial support to the co-operative movement by granting credits.

11 We should have to mention the fact that the goals of the associations did not include political activity. The political regime in power, which was based on a single-party system, did not allow political action outside the framework of the single party and the mass organisations connected with it. The law on political associations was approved in 1990, in the context of the political transition to a multi-party system.

12 We note that at the beginning, the initiators of co-operativism preferred to conduct their activities in consumers' co-operatives, because the suburban and rural markets were completely without basic products. It was only in a second stage that there was a preoccupation with promoting co-operatives for agricultural production and livestock-raising, especially in areas where fields were irrigated. The opportunities for promoting co-operatives for agricultural production have always had physical limits. Arable land is in short supply, and there are few irrigated lands. According to the data provided by the Ministry of Agriculture, Cape Verde has 42,500 hectares of land used for agriculture, and of these only 2500 hectares are irrigated, with the possibility of extending these up to a limit of 5,000 hectares.

13.There were basically three mass organisations: The Women's Organisation of Cape Verde (OMCV), the Amilcar Cabral-Cape Verde Youth Organisation (JAAC-CV), and the Abel Djassi Pioneers Organisation (OPAD-CV). Some analysts see the existing Labour Union—The National Worker's Union of Cape Verde (*Union nationale des travailleurs du Cap-Vert-Centrale*, UNTC-CS) as the PAIGC/CV's fourth mass organisation. With the transition to a multi-party system, the OMCV and the OPAD-CV were transformed into non-governmental organisations. The JAAC-CV disappeared. The UNTC-CS had to be reorganised in the framework of the new trades labour union legislation.

14 Pires O. et Stockinger, op. cit., p. 59. In connection with CARA, Stockinger said that it "had no bureaucratic foundation. It depended on the practices that it carried out. It consisted, at that time, in the "proliferation" of the Tabugal experience in the other "ribeiras" of the commune, that is to say in setting up peasant committees. Peasant leaders actually joined in this work. Up to the beginning of 1984, the frameworks of associations (the mobilising committees) were set up in several of the more important "ribeiras" of Santa Catarina." Cf. Stockinger G., Cronicas de Campo, Praia, ICLD, 1997, p. 46.

15 In the framework of the research project, we planned, "in addition to the inquiry, open, unstructured interviews that will be carried out with members of the organizsations, for which focus groups of less than ten participants will be organised" (Furtado 1995).

16 The political transition to a multi-party system in Cape Verde was very rapid. The public announcement of the political opening-up was made in February 1990 and legislative elections were held in January of the following year. In eleven months, the constitution had to be revised, the laws regarding political parties and elections had to be approved, etc.

17 The concept of *habitus* is understood here in the sense defined by Pierre Bourdieu, according to which it is " system of enduring and transportable arrangements, structured structures that are predisposed to function as structuring, that is, as principles generating and organising practices and representations that can be objectively adapted to their goals without assuming a conscious direction toward goals and express mastery of the operations necessary for achieving it, objectively "regulated" and "regular" without being the result of obedience to rules, and, being all that, collectively orchestrated without being the result of the organising action of an orchestra leader." Pierre Bourdieu, *Esquisse d'une théorie de la pratique,* Geneva: Groz, 1972.

18 It should be emphasizsed that the emergence of peasants' associations was accompanied by a movement on the part of landowners, who created an association for the defense of their own specific interests.

19 From an administrative point of view, Cape Verde currently has 17 municipalities. The fact that it is an archipelago does not itself determine the administrative division. In a few cases, islands have only one municipality. That is, the island and the municipality coincide. In contrast, the larger islands are subdivided into several municipalities. Santiago, the largest island, has six municipalities, the island of Santo Antäo three, and Fogo, two.

20 Even though it may seem radical, we must remember that the sometimes extreme poverty in which a significant part of Cape Verde's population lives can lead to an individual and familial quest for survival that is superimposed on associative and communitarian initiatives. A study made by the World Bank in 1993 shows that 30% of the population of Cape Verde is poor, including 14% who are very poor. H. Proença and A. Rocha, *Cabo Verde, 1993, A Poverty Study*, World Bank.

21 The island of Fogo, which was initially chosen as part of the sample in fact, according to our list, has no association at all. We were unable to visit the islands of S. Nicolau and Santo Antao for practical reasons, despite making several attempts. We believe, however, that the results of work based on Santiago apply, *mutatis mutandis*, to the country as a whole. We have to remember that Santiago contains the bulk of the agricultural and rural population of the country, as well as the majority of peasants organisations. According to the figures in the co-operative census, 12 out of the 20 agro-pastoral co-operatives existing in 1993 were on Santiago. And according to the list we drew up, 55 out of the 82 peasant associations that exist are also on Santiago.

23 The co-operative also benefited from financial support provided by the ADF, amounting to about 8,000 contos (around $94,000) for extending its activities and acquiring equipment.

24 Despite the existence of forest rangers, it is difficult to control these incursions, especially in a drought year when families are having problems surviving. In this case, firewood represents an important source of income. In general, firewood obtained in the forests is used only for cooking and in the bakeries in Praia. The incursions usually take place at night, when the forest rangers' activity is limited.

25 Its activities aiming at cooperative promotion are supposed to have begun in 1976, according to a technician who was working at the cooperatives' national headquarters, and who was assigned to supporting the group. We should also emphasise that this technician is still helping the cooperative with its organisational restructuring, acting as a contract consultant in the framework of ADF financing for the cooperative.

26 The ADF (Fondation africaine de développement) is a North American organisation that finances the activities of associations and non-governmental organisations in developing countries.

27 According to data provided by the Directorate for Youth Affairs (*Direction générale de la jeunesse*), there are currently almost 400 youth associations in Cape Verde, and these have very diverse aims. This fact reflects the dynamics observed in the associative movement in general.

28 As we have already said, in 1974, after the April Revolution in Portugal, the workers in Tabugal, like many others who work the land in a system of share-cropping and tenant farming, stopped maintaining any sort of tie with the owners of the lands they farmed, even though these ties remained *de jure*. Since that time, these peasants have set up a system of common use of certain equipment, such as the *trapiche* and the still used in the production of grogue and honey. The approach on the part of the PAICV that took place in 1983 in the sense of creating an association did not represent anything new. It was primarily a continuation of the process of mobilising the peasants in an area that was easy to penetrate politically, in the context of agrarian reform, making it possible to obtain rapid results that served as an instrument of persuasion and marketing in dealing with the peasants of other communities and hydrographic basins in the commune of Santa Catarina and perhaps in the country at large.

29 With the exception, in this case, of the aid received from the European Union. This aid was granted to the Association of Banana Producers and Exporters, and the Santa Cruz Farmers' Association, as a member, received part of the financing.

30 Cf. *Analyse de la structure des familles à partir de l'analyse du Recensement de 1990*, ed. Claudio Furtado. Praia: DGE, 1995; Ministère de la Santé/UNICEF, *Cap-Vert, Estudo sobre a saude das crianças menores de cinco anos, Estudo Etnografico*, Praia: Minstério da Saude, 1996.

31 As we have already said, the cooperative census carried out by the INC does not break down data by gender, and making any analysis by gender. As for the census of associations we carried out in the course of the present research project, we were not able to obtain complete information regarding either the number of the members of the associations or their gender distribution. Nonetheless, as we have shown in the present study, this information is available for the associations and cooperatives selected for deeper analysis through focused interviews and for the associations that work in the area of soil and water conservation.

Abbreviations

ACTA: Commission dynamisatrice des associations des agriculteurs Commission for Farmers Associations)
ADF: African Development Foundation
CARA: Agrarian Reform Support Commission
FAC: Cooperatives Support Fund
FAIMO: Labour Intensive Front
ICF: Women's Issues Institute
INC: National Institute of Cooperatives
INERF: National Insitutue of Rural Works and Forests
INFA: Agriculture and Stock-Raising Development Institute
INGRH: National Water Resources Management Institute
JAAC-CV: Amilcar Cabral African Youth - Cape Verde
MCE: Ministry of Economic Coordination
MDR: Ministry of Rural Development
MPD: Movement for Democracy
OIT: International Labour Organisation
OMCV: Organisation des femmes du Cap-Vert (Cape Verde Women's Organisation)
NOG: Non-governmental organisation
PAICV: Parti africain pour l'indépendance du Cap-Vert
PAIGC: Parti africain pour l'indépendance de la Guinée et du Cap-Vert
GNP: Gross national product
UNDP : United Nations Development Programme
UCID: Union capverdienne indépendante et démocratique (Cape Verdean Independent and Democratic Union)
UDC: Union démocratique capverdienne (Cape Verdean Democratic Union)
UNTC-CS: Union nationale des travailleurs du Cap-Vert-Centrale syndicale (National Workers' Union of Cape Verde - Central Branch)
UPICV: People's Union of the Cape Verde Islands

References

Albuquerque, Luis and M. Emília Madeira Santos (coordinators), 1992, *Historia Geral de Cabo Verde*, vol. 1. Lisboa/Praia: IICT/DGPC.

Almeida, Germano, 1991. *O dia das Calças Roladas*. Mindelo: Ilheu Editora.

Biaye, Mady, 1996. *Perspectives démographiques du Cap Vert à l'horizon de l'an 2000*. Praia,: MCE.

Bourdieu, P. 1972. *Esquisse d'une théorie de la pratique*. Geneva: Groz.

Cabral, A., 1978. *A Arma da Teoria*. Lisbon: Seara Nova.

DGE, 1992. *Recenseamento Geral de População e Habitação*. Praia: DGE .

DGE , 1995. *Análise da Estrutura das Famílias à partir do Censo de 1990*. Praia: DGE.

Duarte, Dulce Almada, 1994. "Identidade Cultural Caboverdiana," in *Pré Textos*, special issue, *Encontro de Escritores Caboverdianos*, June, 62-67.

Durand, P., 1997. *Développement social urbain : identification de projets de quartier*. Dakar: MFCA/ AFVP.

Furtado, Claudio Alves, 1993. *A Transformação das Estruturas Agrárias numa sociedade em Mudança*. Santiago de Cabo Verde: ICLD.

Furtado, Claudio Alves, 1997. *Génese e (Re)Produção da Classe Dirigente em Cabo Verde*. Praia: ICLD.

Hopffer Almada, D., 1991. *Caboverdianidade e Luso Tropicalismo*. Recif: FJN.

ICF, 1993. *A Mulher Caboverdiana e o Desenvolvimento*. Praia: ICF.

INC, 1991. *Encontro de Reflexão sobre o Movimento Cooperativo Cabo Verdiano*. Praia: INC.

INC, 1995. *Recenseamento Cooperativo*. Praia: INC.

INC, 1992. *Plano de Desenvolvimento Cooperativo 1992-1996*. Praia: INC.

Lachenmann, G., 1993. "Civil Society and Social Movements in Africa : The Case of the Peasant Movement in Senegal." *The European Journal of Development Research*, vol. 5, no. 2, December, 68-100.

Machado, P.A.L. and P.E. Teofilo,1993, *Lei Rural e Estruturas Agrárias*. Rome: FAO.

Mafeje, A., 1993. "Organisations paysannes en Afrique : possibilités d'échange entre économiste et sociologues, quelques remarques théoriques/méthodologiques," *Bulletin du CODESRIA*, Dakar, 1, 14-18.

Mariano, Gabriel, 1991. *Cultura Cabo Verdiana (Ensaios)*, Lisbon.

DRP/GEP, 1988. *Recenseamento Agrícola*. Praia: MDRP, vol. 1.

Ministério da Saude/UNICEF, 1996. *Cabo Verde-Estudo sobre a Saùde das crianças menores de cinco anos, Estudo Etnográfico*. Praia: Minisério da Saùde/UNICEF, vol III.

Ndiobo, M.N.D. *Réalités rurales et animation pour le développement : les lieux/moments dans la communauté rurale de Coumbacara au Sénégal*. FOCSIV, Centro Studi e Ricerce Africa, s. d.

Nindi, B.C., 1993 "Dilemmas of Development in Rural Institution in Sub-Saharan Africa." *Journal Eastern African Research & Development*, 49.

Pires, Olivio & G.O., Stockinger, 1986. *Partido e a Questão Agrária*. Praia: IAC.

PNUD, 1993. *Relatório sobre o Desenvolvimento*. New York.

Rahmato, D., 1991. *Organisations paysannes en Afrique : problèmes et perspectives*. Dakar: CODESRIA, Document de travail, 1/91.

Roch, Adão, and H. Proenca, 1993. *Cabo Verde : A Poverty Study, Praia*. World Bank.

Semedo, J.M., M. Trujano, 1997. *Cabo Verde- O Ciclo Ritual das Festividades da Tabanca*, Praia: Spleen Edições.

Silveira, Onésimo, 1990. *A Tortura em nome do Partido Unico*. Mindelo: Terra Nova/Ponto & Vírgula.

Stockinger, G., 1990. *Crónicas de Campo*, vols I and II. Praia: ICLD.

Thomson, et al., 1992. *Descentralização, Democratização, Provisão e Administração de Servicos Pùblicos e Gestão de Recursos Naturais-Iniciativas e Dilemas em cabo Verde*. USA, ARD.

Appendix

Table 1: Co-operatives registered with the INC

Island/Commune	Number of co-operatives
Santiago	118
Praia	80
Santa Catarina	10
Santa Cruz	22
Tarrafal	6
Santo Antão	27
Ribeira Grande	12
Porto Novo	9
Paul	6
S. Vicente	11
Fogo	22
Brava	2
S. Nicolau	3
Sal	3
Boavista	5
Maio	13
Total	204

Source : INC 1994.

Table 2: Co-operative organisations existing and functioning in 1993

Island/Commune	Number of co-operatives
Santiago	40
Praia	24
Santa Catarina	2
Santa Cruz	13
Tarrafal	1
Santo Antäo	9
Ribeira Grande	1
Porto Novo	6
Paul	2
S. Vicente	10
Fogo	15
Brava	-
S. Nicolau	6
Sal	1
Boavista	2
Maio	10
Total	93

Source : INC 1994.

Table 3: Existing peasants associations in Cape Verde

Islands	Number of peasants organisations (associations)
Santiago	55
Santo Antäo	16
S. Vicente	1
Fogo	-
S. Nicolau	5
Brava	-
Maio	-
Boavista	5
Sal	-
Total	82

Source: 1997 census (adaptation).

9

Women Peasants Organisations in Cameroon: The Case of the Department of Lékié

Jean Paul Komon

Introduction

In Cameroon, the 1980s and 1990s witnessed a deepening crisis and the implementation of programmes of structural adjustment. These two phenomena would be the source of the decline in income and the deterioration of the purchasing power of most families, to the point that most families had enormous difficulties in meeting their basic needs. Being the most deprived, women and their children were the most affected and consequently were active in seeking new strategies for life and survival. Under a government whose authority and legitimacy have been strongly challenged after being forced to embark on strict austerity measures (Bratton and Walle 1997; Mamdani and Wamba Dia Wamba 1995), women in both towns and villages were quick to devise various solutions to attenuate the severity of their living conditions. Some of these solutions were in the form of coming together as members of associations and organisations with the common aim of taking care of their interests. These peasants organisations represent new hopes. They are social innovations, through which women, at their own risk, collectively organise their modest means and abilities, in order to increase their productivity and income, achieve a degree of autonomy and social equality, and contribute to transforming the productive system and the way in which the rural world is structured.

In addition to their diversity and number, however, a crucial question remains: to what extent is the level of women's power and their participation in decision-making centres, or—their role in the restructuring that is currently going on in the rural world?

This question is all the more important because, despite the significant contribution women have made to socio-economic life, phallocratic domination usually leaves them unequal partners in family structures, in the political hierarchy, and in development. They are thus excluded from the planning and implementation of policies for managing rural areas, where they are the centre of all social life. Therefore, whether peasants organisations will proffer real opportunities for women to overcome some of these challenges will depend both on the way in which women peasants organisations are created, organised, and are run, and on the kind of relationships women's groups have with other social actors (men, aid donors and especially the state).

Studying the operation of women peasants organisations in the Department of Lékié, Cameroon, the most important agricultural area in the country, will enable us to judge the true ability of women in confronting the current challenges. These involve the re-ordering and control of the rural world.

In this study, we will first offer a few general observations on the concept of peasants organisations in Cameroon and on the dynamics of organisations connected with colonial history and changes in the institutional, political, and economic environment. The second part of this study will analyse the dynamics of women peasants organizations; their operation and their relationships with other social actors in order to see whether they are capable of breaking with the logic of the existing socioeconomic structures and redirecting the politics of the rural world toward the worthwhile objective of satisfying the needs of a social stratum that has always been exploited, excluded, and discriminated against.

1 - Peasants Movements and the Development of the Institutional, Political, and Economic Situation in Cameroon

According to Grabowski (1994), a proper distribution of power among social actors in a social space will produce better cooperation and good economic results.

This has been confirmed elsewhere (Taiwan, Korea, etc.), and may have encouraged the authorities to set up an institutional framework favourable to the growth of peasants organisations.. But despite the advantages that this movement should have produced, the results of the multiplication and diversification of peasants organisations still remain limited by certain factors. Unless care is taken, these limiting factors might lead to social confusion and uncertainty in rural areas.

1.1 - The concept of a peasant organisation in Cameroon

Generally, a peasant organisation federates associations of peasants. This association is almost an enterprise, whose members, all have the same status and seek to take advantage of the activities they undertake together, chiefly for the benefit of the members.

This definition has three essential elements:

(i) An endogenous social consensus: this is what holds together peasants who are free to make their own choices and whose basic goal is to live together in a continuing spirit of cooperation and social justice.

(ii) An aspiration on the part of all the peasants who are members, to meet the same needs and produce joint results by taking shared risks together.

(iii) A collective of decisions or initiatives whose objectives and programmes are compatible with the most urgent interests of the peasants.

The constitution and promotion of peasants organisations cannot be decreed. They stem from the manifest, endogenous will of the peasants, instrumentalised by charismatic leaders, to construct a community. This means that each member, constrained as he or she is by the normal factors of everyday life, must necessarily take on both rights and obligations. The creation of a peasants organisation also depends on whether triggering factors exist.

In Cameroon, many of these triggering factors are at the origin of the emergence and especially the increase in the number of peasants organisations since 1920. While in the West, endogenous and essentially economic or technical factors underlie the process of creating peasant groups, exogenous factors have made a significant contribution to the development of the peasants movement in Cameroon. Before these factors came into play, there were indeed already regroupings of peasants in the rural world. Many associations, established on a clan basis and with a complex system of reciprocities and obligations, had only a local and limited socio-economic ambition: to preserve the communal bond or the group (Badouin 1982). They were conceived by and for the members, and perceived as a vehicle whose aim was to provide various economic and social services to the rural population. The lack of national and international ambitions, however, restricted these associations to activities that were often not very productive, and required rudimentary means of production and little utilitarian calculation.

Nonetheless, the participation of Cameroon in a monetarised economy that was increasingly globalised, together with the failure of the policy of import substitution pursued in the 1960s and 1970s, led the all the different authorities and regimes to take a greater interest in agriculture. This was reflected in an attempt to restructure the rural world to adapt to current challenges. The peasants organisations, which provided its foundation, reflect the will and social aims of those who set them up.

1.2 - The Structure of the Rural World or the Reflection of Colonial History

The history of organisations in Cameroon resembles in many ways that of other African countries south of the Sahara (Thelen and Sven Steinmo 1992; Braverman, Guadch, Huppi and Pohlmeir 1991). It is, however, specific in certain ways that are connected with the human and cultural geography of Cameroon, and especially with the country's colonial history. The current structure of the rural Cameroon dates back from the country's political history, particularly the socioeconomic changes resulting from its insertion into the global economy in the colonial and postcolonial periods. This is reflected in a juxtaposition of a dual logic of cooperation, imposed by the twofold colonisation of Cameroon by Great Britain and France.

Thus the forms of peasants organisations dating from 1920 and corresponding to the two logical systems just mentioned, survived even after Cameroon gained independence.

Whether it is the British or the French model of co-operation, the role of the state as protector and unifier is primordial, at least in the genesis of most peasants organisations. In the anglophone area, the state based its co-operation policy on pre-existing local community initiatives and used them as the basis for structuring the rural world, and later on, as an instrument for controlling rural areas (Khaled 1997). In the parts of Cameroon under French occupation, the state played a more active role.

In fact, the state created peasants organisations and intervened directly in their internal workings. Since it appointed their leaders it also influenced the socio-economic orientations of the peasant groups. In reality, the state created co-operatives without any co-operators. At first glance, the two concepts seem different. The British concept seeks to preserve and encourage the traditional community spirit, by deeply involving the pre-existing associations that held and managed land in its process of rural restructuring. The French approach, on the other hand, distrusted autonomous initiatives, and from the outset excluded the peasants from participating in the definition, execution and management of the programmes of the associations. Moreover, the structure of these associations, their objectives, and their leaders were defined or appointed by the state. It was in keeping with this logic that zones of priority projects and credit unions, mutual associations and co-operatives were set up, with the help of aid donors who were convinced of the role that such groupings had to play in rural areas. They were all put in charge of people controlled by the state.

The two concepts did converge, however, in the objectives that the peasants organisations were assigned. These were: to modernise agriculture, to dominate and control export products, with a view to guaranteeing public revenues and supplies of raw materials for local and foreign processing industries.

This orientation imposed by the public authorities corresponded to the model of development adopted by the colonial metropole and later by the national elite that came to power after independence. It assumed that the traditional associations were not interested in economics. It led peasants organisations to give priority to economic considerations over social ones, just like their European counterparts. They were concerned with dealing with the external economy rather than the endogenous and durable development of the rural world.

In order to achieve this exogenous objective, the colonial state, and later on the post-colonial nation-state, encouraged the promotion of peasants organisations that were under the control of recognised traditional managers of land, family workforces, and the social product which resulted from the sale of cash crops.

While the men, because of their status, were socially involved in the centres and formal associations for production and sale in the cash crop economy, the women, who had no power over the factors of production or the social product, were concentrated in auxiliary activities and in informal associations that had no national or international ambitions, and they were thrown back on themselves.

The harmonisation of legislation on peasant co-operation in Eastern and Western Cameroon carried out in 1972, after the unification of the country, did not change the logic on which the outward-looking and monopoly structure of the rural world had been based since 1920.

Although the idea of community development contained in the British model was explicitly recognised and implemented through specific actions begun under a specialised office of the Ministry of Agriculture, it was the

French model of co-operation that would be imposed during the development of the peasant movement in the 1970s and 1980s.

The law passed in December 1973 on the status of co-operative societies and the decree of October 1974, which put it into effect, confirmed the control of the state over the structuring of the rural world. It was the Ministry in charge of agriculture and parapublic development enterprises, which gave impetus to the movement. It was they who nominated the co-operatives' officials, removed members of their supervisory councils, and inspected and regulated their management. Through the presence of their representatives in general assemblies and supervisory councils, they controlled how power was shared out.

This privileged position of the state gave it the opportunity to direct rural development policy in accordance with its wishes and contrary to the interests and expectations of the rural population, and also to dilute or divert the power of those who were socially excluded because of their sex.

All the same, an examination of peasant movement development strategies in the 1960s, 1970s, and 1980s, which were driven and controlled more by the state than by the peasants themselves, does certainly reveal the presence of women in these associations (55% of the members). Unfortunately, however, this massive presence was not reflected in any significant representation of women in the decision-making bodies of the associations.

The movement was built on men, on local ideology, and on state power that was inspired from outside and was patriarchal in nature. It postulated masculine authority from the outset.. Even when women were included in the decision-making bodies of the associations, their participation was imposed from the top of the political hierarchy or of traditional institutions, which maintained the inequality of the sexes. This was only to be expected, however, since in contrast to countries like Tanzania, the place and the role of women are not governed explicitly by any basic law (the law of the family).

As a result, women continue to make a small contribution, because when they belong to associations or to their decision-making bodies, they do not play a proper role as partners, with obligations and rights equal to those of men.

As Courade et al. (1992) have emphasised, the system which increased the complexity of society, and contributed to the structuring of the rural world in the 1960s and 1970s, helped to set up a general right to protection, security and assistance and thus to a right to interfere in structures that were dominated more by the managers than by the true producers of the resources.

Instead of being endogenous and self-developing structures, they became semi-public institutions that were used by the state more for redistribution and thus for domination than for the actual creation of wealth and the promotion of a democratic society.

Unfortunately, at the end of the 1980s, the fine structure of the co-operative movement was severely shaken and almost destroyed by several factors, which led to a process of social disintegration. This was reflected in the multiplication and diversification of peasants organisations. Alongside the peasants organisations under state control and development enterprises and village associations and groupings, we find more peasants organisations that are private and based on religious distinctions. These, moreover, seem to be diversifying their activities.

The structures that have arisen from this change shows a twofold ambition: to help the small peasants, who are for the most part poor, without fundamentally changing the distribution of economic power, and to strengthen the traditional associations or groups that play a significant socio-economic role in the rural world (Attwood 1988).

2 - Institutional Economic and Political Development and the Differentiation Between Peasants Organisations

For a long time, the lack of any political culture among peasants, their ignorance of what was at stake in the rural world, and their downtrodden social situation allowed actors external to the rural world—the state in particular—to control the restructuring of this world. These actors organised the peasants with the sole aim of controlling production. But economic and political institutional developments resulting from the vast movement of globalisation that began in the early 1990s have led to profound changes in the peasants' world. Thus we find both a multiplication of and a differentiation between peasants organisations, which produces a worrying outlook, all the more so because the process, which has a political dimension, is occurring at the same time as the state is restructuring itself and the demands of women and donors for freedom are being sharpened.

2.1 - The growth in peasants organisations and the competitiveness of the rural areas

The changes in the rural world that took place during the 1980s and 1990s were due to two basic factors: the economic crisis and the great wave of socio-political liberalisation that washed over central Africa in general and Cameroon in particular.

The decline in government revenues that resulted from the crisis led to a reappraisal of the relationships between the state and the peasant associations that were kept afloat in return for services rendered.

Various budget cuts were imposed that deepened further the poverty of the people, and made them invent, strengthen, or broaden local forms of solidarity (Rahmato 1991). This was possible only because the politico-institutional environment had changed. The various donors, who were aware of the stakes involved in the restructuring of the rural world that was currently going on, sought at first to bring the peasants and their organisations back into the economic and socio-administrative decision-making centres, and into the centres concerned with the definition of development policies. They would then be given an opportunity to take up a more advantageous position in a social space that was becoming increasingly liberalised, homogeneous, and dominated by the new culture of negotiation and compromise. Finally they would be able to influence the politics of rural development, in order to increase the democratisation of the rural world by developing independent associations (Hanak and Loft 1988).

The many peasants organisations created in a decentralised way during this period function with a double objective: to bring together the maximum number of peasants, in order to fight together against poverty, backwardness, and dependency with regard to the state, and to provide responsible and powerful interlocutors for the government.

The need to resist and survive, and the need to help families live in difficult conditions "have taught women to emerge from their traditionally submissive role and to confront family and community pressures." Later on, they began to work together to improve the conditions under which they and their communities lived. We can thus see a strong participation by women in the peasant movement. They represent 51% of the members of peasants organisations and 39% of the membership of management committees. This participation in mixed organizations does not, however, exclude the establishment of specifically female organisations to deal with problems specific to women, such as their economic and social marginalisation.

But will the conditions under which they are created, the low level of peasants' education and political culture, the polyvalence of peasants organisations and the relations they maintain with other social actors really allow the establishment of a different social climate?

2.2 - The diversity of the peasant movement, amid increased social competition and inequality

The objectives of peasant associations, following the laws of 1990 and 1992, have been determined largely by the conditions under which the associations were created. The kind of activity to be promoted is connected with agro-ecological conditions, with the ability of members to manage and regulate the associations and to finance them to a reasonable degree (Bottomly 1989). The achievement of these objectives also depends on the level of the surplus derived from selling their products and from financial aid and training received from other actors.

With regard to the factors mentioned above, we can distinguish two broad categories of peasants associations active in Cameroon: those devoted to agriculture and essentially economic activities, and polyvalent or multifunctional associations, which pursue social and economic activities simultaneously.

The second category constitutes more than 80% of the peasant groups and associations in Cameroon. This pre-eminence is connected with cultural factors (tradition) and also with the diversity of social problems that emerged with the collapse of the system of social participation implemented in 1960 (Berry 1992; Mahieu 1990; Glazier 1985; Guyer 1981) and shown up by the crisis of the 1990s.

Established with a small basic income and with little chance of becoming viable, peasant associations in Cameroon generally suffer from four problems: segmentation, fragmentation, tribalism, and sexism, which prevent them from meeting their objective of bringing together "the exploited people of the rural world".

• *Segmentation*

The various donors have found an opportunity in political-economic liberalisation to assist peasant associations that are deprived of training and resources, to make them more effective and boost their expansion. Their promises of aid have led to a massive creation of rural peasants organisations. The actions of non-governmental organisations do not stop at this, since they also actively participate in drafting and submitting proposals for the associations they sponsor. Through this activism, non-governmental organisations would like to attract the peasants' goodwill, and in the long run, their allegiance.

Unfortunately, the various kinds of aid, more competitive than complementary, have turned out in some cases not to serve the material interests of the peasant leaders and their associations.

In general, this aid has two types of negative effects. On one hand, it exacerbates frustrations when it is insufficient for the associations' needs. This is evident during the early years of their existence. On the other hand, it further impoverishes the associations, because it stifles all endogenous initiative, particularly when peasant associations seek only "to capture outside aid".

Because this aid, especially financial aid, is insufficient, as the associations multiply and diversify, competition for it increases. The leads first to a destruction of the internal coherence of the associations, rivaly then to an increasing gap between the associations and their original objectives, and finally to a hardening of relations among members of the associations or between different associations that are active in the same sector and activity.

The dismantling and dismemberment of the UGAPAB and ATNA associations, together with the tribulations of the leaders of PCIDRK and BOSAPPAL in recent years, are examples which indicate the risks run by an impoverished peasant movement that is dependent on external aid. Dissension among members regarding the management of gifts and aid provided by non-governmental organisations contribute to the breaking up of associations which were originally small in size.

• ***Fragmentation***

At first glance, one might think that a fragmentation of existing peasant organisations since 1991 indicate their desire to reshape themselves, and to seek a new social foundation and an optimal size for effective operation. Unfortunately, this is not the case. Among other things, it results from the lack of compatibility between the liberal institutional framework and statutes, on the one hand, and the patriarchal, autocratic culture of the peasants leaders on the other, besides external interference in the internal functioning of the peasants organisations, and the refusal of the leaders/creators of peasants organisations to co-operate with donors (Njongang and Dikoume Matam 1996).

All these elements are obstacles to redefining the roles and functions of all the associations' partners. This lack of consensus lead to a weakness of peasant organisations, and to their being exposed to external influences. Indeed, the fragmentation of peasants organisations increases their social instability, and at the same time deprives women of structures through which they could express their sufferings and fully assume their social responsibilities for feeding and caring for their families.

The excessive fragmentation of peasants organisations leads women to engage more in individual than in community activities. In the Department of Lékié, we find that nearly 62% of the income of women belonging to an organisation come from individual activities, and only 38% from community activities.

• ***Tribalism and sexism***

Given Cameroon's history, cultures, and inadequate means of communication and information, the dynamics of creating peasant organisations have remained concentrated in the central, southern, western, and north-eastern regions, where there is a long tradition of working together and pooling resources. Nonetheless, with the help of partners outside the rural world (the state, non-governmental organisations, donors and churches), this dynamic has become stronger in the rest of the country, thanks to study visits, education and training projects. The effect of training and imitation has now made possible a relatively balanced spatial distribution of peasants organisations in Cameroon and in the Department of Lékié.

Because they are part of a precise cultural environment and of a social space that for the most part remains little imbued with the political dimension of restructuring the peasant and rural world, peasants organisations are still closely attached to those who have set them up and to the place in which they were established (the clan, tribe, ethnic group, etc.). This is, moreover, instrumentalised by the fact that peasants organisations are given a name in a national language, which indicates the place where they were created, their space, and their sphere of influence.

In addition, peasants organisations are marked by phallocratic predominance in the social sphere, a result of the influence of the traditional social institutions which consecrate male primacy (Meillassoux 1982). As a result, women, who constitute the largest cohort of members of these groupings and the locomotive for their socio-economic activity, are not represented in the decision-making centres of the peasants' movement. This social inferiority does not always allow them to take advantage of all the outside aid and of agricultural reform policies and socio-political liberalisation. Hence the vital necessity for women to organise themselves.

Segmentation, fragmentation, tribalism, and sexism limit the horizons of the peasants organisations' ambitions. They do not prepare them to deal with the challenges of a globalised economy, whose effects in rural areas can already be seen and are revealed by remarkable social differentiations.

The multiplication and diversification of peasants organisations at a time when the state is reorganising itself can lead to two phenomena whose future consequences cannot yet be discerned. First, there is a contradiction between the recognition of peasants' political rights and the state's inability to guarantee their economic and social rights. Secondly, there is a social differentiation within the peasants organisations and among them, that is, between those that would profit from socioeconomic liberalisation and others that would suffer from it, because of their inadequate or poorly adapted initial endowments.

Overall, the restructuring currently being carried out in Cameroon is going to create dramatic social situations, reflected in the persistence of poverty, the aggravation of inequalities among peasant organisations and within them and the skimming-off of the social fabric. The latter phenomenon is all the more sensitive, because the regional or national regrouping of associations and groups remains weak: less than 5% of the organisations of women peasants are connected with a peasant federation, and less than 12% have regular contacts with other structures in the rural world.

The relative social isolation of organisations of women peasants raises the problem of their ambitions and their contribution to the reorganisation of social relations and the rural space. This is marked by an increasingly extensive involvement of international and national institutions financing development and of the participation of peasant leaders in the decision-making structures of the political parties.

3 - Women and the Rural World of Lékié

In the past, one rarely heard the voices of women. Their contribution to economic life was less seriously regarded, because they were active in less monetarised sectors. In more recent years, however, their role has been more recognised. This has taken place through a silent process of women's emergence, either alone or more often in associations, in all sectors of the agricultural and non-agricultural economy. Thanks to the role they play in the material and moral maintenance of families, they are negotiating their participation in power and are strengthening the spheres of power, which the culture and the state's policy have marked out for them.

But what explains their current dynamism? What forms does their coming together in associations take, and why? What are their speci[illegible] objectives and their contribution to the perpetual negotiation that they conduct between their status on the one hand as heads of families and their very heavy responsibilities on that account, and on the other hand, the prerogatives conferred on them by statute, which used not to be accorded to them? What are their relationships with other social actors, notably with men, with the traditional authorities, and with the state?

We shall try to answer these questions by focusing on the basic factors in the emergence of organisations of women peasants in the essentially agricultural Department of Lékié, on the forms and problems that this emergence raises for the peasant movement's cohesion, and on the relationships of power in families and in society.

3.1 - Women in Lékié in a context of economic and political crises

The structural adjustment and budget tightening measures, the democratic uprisings, the conflicts between men and women in families and in society over dwindling resources, and the influence of a Cameroonian system, in which the rich become richer and the poor poorer, have all led to enormous imbalances that affect women in specific and significant ways.

The difficulties being faced by women to access to material and financial resources, and to take advantage of the economic and technological inputs that confer power, the gaps in training and education, the obstacles to women's participation in economic decision-making centres, and the silence or impotence of laws in face of injustice have imposed important constraints on women's development. Their marginalisation has prevented a sustainable and balanced development of the rural space, and triggered a revolt, which took shape in the creation of peasant associations specifically for women.

The spread of these associations was directly linked to attempts to reduce women's poverty and thus the poverty of families, and to the recognition of women's social roles in production. Thanks to the peasants organisations, women are hoping no longer to be seen solely as family labour involved in agricultural production, but as social, moral, and economic organisers of the family, which is still largely based on subsistence crops mostly raised by women. Their peasants organisations appear as a reaction to adapt to socioeconomic changes in an increasingly risky, uncertain,[1] and restrictive environment.

The many women's associations reveal their dynamism and imagination, and reflect the desire of women to intervene directly in areas affecting their lives and the long-term future of society. This concerns equity, efficiency, and autonomy.

Because they represent nearly 90% of the agricultural labour force, producing more than 90% of subsistence crops (World Bank 1990), and represent 51% of the members of peasant associations, and have formed more than 40% of the associations active in Lékié, women want to be treated as well as men are.

The material and moral sustenance of families, which they provide, calls for their productive capacity to be strengthened, and subsequently for them to acquire financial autonomy and greater control over their own destiny.

Whereas during the 1960s and 1970s, most peasant associations were led by men, and women were only concerned with production, in the 1990s we see the rise of associations led by women. More than 40% of the peasant associations active in the region were established by women and 59% of the leaders of these associations are women.

The major role of women in the peasant movement is contributing to a gradual transformation of the rural world. Previously involved only in subsistence and traditional cash-crop agriculture (cocoa beans), thanks to the imaginative activity of women's groups, Lékié has witnessed the appearance of new structures since the end of the 1980s and a new agrarian capitalism. On the marginal and swampy lands left to them by men, women are developing a dynamic and diversified market gardening operation. Its products today bring in more than 30% of the agricultural income of peasant women and their associations. The extent to which women's associations have spread depends on cultural traditions, the economic activities of migrant workers, and the structure of markets (Bratton 1986), and also on how close Lékié is to the political capital of Cameroon.

Greater integration of community action is a way of fighting against difficult economic conditions. It still seems, however, that the groups derive most of their income less from community activities than from individual ones. Only 14% of the groups derive their income from community fields or activities.[2] This observation also shows the way agents behave with regard to associations and the way they are run.

3.2 - Forms and functioning of women's associations connected with systems of farming and development

Peasant associations and the way they are run vary, according to whether the farming system used is simple or mixed, and whether the system of development is egalitarian or not.

The differentiation is all the clearer because in farming systems dominated by cash-crops and in systems of development in which male pre-eminence is accepted from the outset, exploitation, exclusion, and discrimination against women are a permanent and unacceptable reality. In the culture and society of Lékié, women do not have access to land or control over its products. Scarcely 5% of the women have a right to own land, and no woman has the full status of a head of a family (Sow 1995). This situation of being a social inferior is less marked in a system of mixed agriculture with a preponderance of food crops, where women enjoy a right to the temporary use of land.

In view of the everyday restrictions imposed by various systems, and by the burdens of domestic and productive activities, in terms of space, assets, and time, women are diversifying the form of groupings and the way they work, in order to escape from the vicious circle of poverty. Their groupings in Lékié can be classified into two categories: formal structures and informal structure

• ***Formal structure***

We call them "formal structures", because to gain recognition by the authorities, they have to fulfil the conditions required by the 1991/1992 law relating to the number of members (at least 25), the type of activity, and the sources of financing for their various activities.

In Lékié, these structures are more numerous and diversified because of the natural and structural constraints of the traditional economy. Their development since 1991 is exactly the same as that in the whole peasant movement in Cameroon. From 1990 to 1993, their numbers rose form 0 to 35, with a total of almost 1,000 members, but in 1994 their numbers began steadily to decline, until by 1997, only 15 such associations were still actually functioning.

This development can first of all be explained by the arguments that underlay the setting up of these associations. When the 1991/1992 law was promulgated, non-governmental institutions and national and foreign donors came into the rural world and promised enormous amounts of financial and technical support for the associations, with a view to modernising women's production. These promises thus led to an initial spread of organisations; but when most of the commitments made by donors were not fulfilled, many associations folded up. This development is also explained by the limited viability of the associations. This was due to the mismatch between the associations' democratic statutes and the(autocratic) way they were run in practice, and also to the specific problems that these associations had to face because they are run by women. The life of women peasants' associations and the way they were run in Lékié are closely connected with their promoters. Most of these —90% at least—had completed primary and even secondary school education, and they had acquired experience with associations elsewhere in politics (the one party) or in traditional and religious association movement. Many women leaders were, moreover, newcomers, who had recently moved to rural areas after their male partners had lost their jobs in the city (55%).

An investigation involving 20 leaders of these associations shows that 25% of them had been grassroots level officials of the single party, 65% came from mixed-sex and religious associations, chiefly Catholic, and were migrants, and 10% had been trained by non-governmental organisations. Of the 20 leaders interviewed, 20% were widows or women abandoned by husbands who had become migrant, urban, or lower-level workers, after being sacked because of the steep decline in the price of cocoa beans. No unmarried woman ran an association, but 10% of the women leaders had husbands who were either rural notables or directors of mixed-sex peasants organisations, and thus acted directly or indirectly in the associations.

The past and the social or matrimonial status of the leaders of peasant associations, together with the system of farming, influence to varying degrees the personality and the character of the organisations these leaders had helped to create and the means at their disposal. Although more than half had not completed primary school education, these leaders were able to inculcate a breath of democratic culture into the running of peasants organisations and to raise women's consciousness of their socio-economic situation. This was done by drawing up a definition, a census of those responsible for this situation, and re-establishing a consistent correlation between the ongoing changes in the rural world in general and in Lékié Department in particular, and the policy of an effective distribution and social allocation of resources and power. The latter is defined, moreover, in terms of the dynamics of social relationships between men and women and between women and the rest of society. It is clearly implicit in the ability of women and their associations to participate in authority, to achieve their objectives, and to regulate their various products.

The women's co-operatives and associations that one finds in Lékié were established at the level of production, processing, marketing and support activities for food crops, which are necessary for families' food security.

The dynamism and the diversification of their agricultural activities, which ought to be facilitated by the ecology and the environment, is however hobbled today by the difficulties women have in gaining access to the means of production. It is hard for them to gain access to land, which is owned and managed exclusively by men, and even when they do gain access to it, their use of it is limited in time (by the year), its area is small (4 to 28 ares per person), its quality is poor (swampy or exhausted land), and it is allocated by village associations (for 600 women, an area varying between 2 and 5 hectares is allocated).

The difficulties involved over permanent grants of land to women, which derive from cultural, demographic, and economic factors, explains why the activities promoted by their associations are devoted to annual food crops and to market gardening, which do not tie up capital for an extended period of time.

Though these activities certainly do not cover cash crops in terms of areas compared to food crops, they add to the burdens of work for women. As a result of their social status, women are supposed to be involved both in the production of perennial and food crops and in domestic activities. This double role for women requires on average 14 to 16 hours of work per day, and decreases their productivity, defined in terms of production and area, unless they can delegate some of their tasks.

Because of this, formal women peasants associations cannot count on the work factor to improve their viability. The free labour given by their members does not completely belong to them, since it is divided among the individual "priority tasks" depending on male authority and on other supporting tasks, which are done exclusively by women or by their associations.

In theory, the relative absence of labour under the control of women's associations might be offset by making more use of children's labour[3] and that of paid employees, or by mutual aid among women, or by technology and mechanisation.

Unfortunately, the legal restrictions on child labour, the poverty of the associations, and the low level of education and training of their members (80% are illiterate), makes access to these factors of production and to technical services difficult and unequal (Ventura-Dias 1973).

This difficulty weakens the formal associations.To carry out their diverse activities, they are forced to resort to a reduced number of donors and not to classical financial institutions, which have little confidence in them (for lack of credible guarantees). We must add that the help provided by these donors is made conditional on the democratic functioning and good governance of these associations.

An analysis of the statutes and formal operation of these associations shows that democratic practices have been established. The election of members of boards of directors, the management of funds, and decision-making regarding expansion and diversification are carried out in accord with the norms of negotiation and compromise. The statutes also provide for possible changes in the leadership of these associations.

Democratic practices in the choice of projects or personnel are respected by more than 70% of the members of women peasants organisations. It is exactly this which attracts an increasing number of members. Democratic practices impose on leaders and members a new culture of authority and new power relationships, without however changing the most pernicious and ancient practices connected with unequal relationships between men and women.

In spite of the aid provided by donors, which remains inadequate to meet real needs (hardly 10% of the investment or operating costs) the situation of associations and their members has not greatly changed and has not fundamentally altered the traditional social hierarchy.

At the cultural level, women's associations have not changed the concept, held by men and by traditional institutions, that women are inferior beings who cannot manage and use resources, despite the indispensable socio-economic roles they play.

Women can gain access to resources only through men (through marrying and producing male children) or, to a lesser extent, through the market, especially the real estate market. This latter alternative is increasingly used in Lékié, where land is relatively scarce, because of the large population and competition between food and cash crops. As regards their activity in the real estate market, a few women's associations now rent lands to use them, as the rental contract stipulates, solely for food production and market gardening activities. This remains an isolated and novel case in Cameroon, since women have access to resources only by following rules defined by traditional institutions, which are largely dominated by men and the traditional authorities.

The inequality over initial endowments and access to resources increases women's inferiority. This inferiority is also due to the limited use by women of technology and economic inputs, which greatly reduces the level of both production and productivity, and affects the choice of activities to undertake. Women take up activities (food crops, craft work, etc.) where income remains largely dependent on a market, which is small, little-diversified, distant, and demanding, and which does not provide them with means to gain access to technology and to modernise women's production in rural areas.

In addition to this economic and cultural inequality, there is also another connected with the dependency created by donors for training and educating women, for popularising agriculture and for financing projects.

Non-governmental organisations and donors exercise a monopsony power on the market for support or assistance to groupings. This monopsony dominance, which restricts social welfare (Blair and Harrison 1994), depends on the elasticity of the supply of the product and on the control of the groupings and of the rural area, the donors' market share (the number of associations they advise or support); and the elasticity of demand of their competitors or rivals—that is, the non-governmental organisations, and the classical governmental financial institutions.

The risk run in restructuring the rural world of Lékié is that it might fall under the exclusive control of the donors and others who provide finance for it. With strategic withdrawal from the peasant world of the state, which might have been able to provide some resistance to the donors and other providers of finance, these latter are adopting "predatory" practices, which consist in proposing high levels of diversified aid to the groupings, in order to bring the majority of the groupings under their control and eliminate their competitors, and ultimately to obtain a monopolitic position in the rural world.

This strategy is closer to reality, because the donors and finance providers are perfectly aware of the stakes involved and the political dimension of the social restructuring currently in progress in the rural world. By controlling the rural world, they want a monopoly over the levers of power, which would allow them to have a say on the allocation of resources and the manipulation of structures in a project that is not always related to or compatible with women's interests.

But this attempt at monopoly, which is becoming more obvious and clearly defined in the region is too expensive and too risky to be applied widely and be generally accepted. It is perhaps to escape abuses of this monopsony power that women have created other forms of groupings and organisations.

• *Informal Structures*

Informal structures do not result from socio-economic and political liberalisation, but from the community constraints characteristic of traditional society.

These structures have always existed, but nevertheless their number and visibility have grown steadily since the beginning of the 1990s, with the tightening of constraints and increased rationing of assets and time (Azam 1987; Berthelemy and Gagey 1987, Bevan, Rigsten, Collier and Gunning 1987). Between 1990 and 1996, the number of informal structures rose from 60 to more than 100. As each structure includes on average 60 to 80 members, that means that more than 8,000 women are involved in them.

These structures represent more than 25% of the groups active in Lékié. There is a double challenge involved in setting them up: how to pursue personal gain while at the same time maintaining the balance between each member's rights and obligations in an uncertain and repressive world. As a result, every individual choice to join or to promote an association is determined by the evolution of the productive system and by community pressure. This varies depending on the degree of openness to modernity or to the market and the development of the state's role.

Community pressure, diffused through a dense network of very personalised vertical relationships (family lineages) and horizontal relationships (mutual aid), calls for associations to be created less on ethnic than on multi-ethnic bases. Their members could thus belong simultaneously to several groups. There is no exclusive attachment to a single given structure, but rather to several such structures, in which repeated transactions among the same members produces loyalty and decency, which are guarantees of equity.

This multiplie sense of belonging to several associations goes with a diversity and range of cost in time, assets, and services, which are extremely high. The costs, which amount to a multiform endless tax or social debt, which individual members owe totheir various communities, lead women and their groups to engage in multiple activities: producing food crops, education, caring for the health of children, etc. The association appears to be an insurance service that can provide its members with benefits in cash and in kind.

Women's multiple membership activities, considered necessary to increase the general well-being and to meet the pressing requirements of the community economy (Mahieu and Odunfa 1988), raise problems of co-ordination and risk-management both within families and in associations. Because the production of these associations has long-term benefits and depends on a unified view including the life cycles of the members, the incentive is strong not to sacrifice investment in the associations for short-term gains.

The emergence of informal associations in an uncertain environment is dependent on the existence in the rural world of ambitious, generous, available, and altruistic women leaders. In Lékié, most of the leaders of informal associations have had little education or training, and are more than 30 years old. More than 80% of them are between 30 and 40 years old. Nearly 90% are married and have children whose average age is from 14 to 15 and who can, if need be, replace them during repeated or prolonged absences. The domestic activities that occupy a large part of these women's work-time can, in fact, be more effectively handled by their daughters. For this reason, the latter are likely to leave school early or to begin their schooling late, thus creating the conditions for their future socio-economic inferiority.

The freeing-up of this amount of work-time—on average 6-8 hours per day per woman—ought to further productive community activities. While the older women (at least 45 years of age) devote these extra hours to the moral and material maintenance of the community and groupings, the younger women, who are still seeking to consolidate their union and their participation in their community, devote these hours to domestic or individual activities.

The differentiation in the use of the freed work-time shows that peasant associations interest older women (30-45 years of age) more than younger women. This explains why the leadership of these groups is composed of more experienced women who are well integrated into their various communities and motivated by an ambition for autonomy and responsibility.

This observation does not alter the fact that the members of these groups are increasingly younger women. More than 60% of the members are between 25 and 29 years of age. This is due to the increase in women's poverty resulting from population growth and the deterioration of the environment (Cecelski 1987; Jacobson 1993).

Their youth, their commitment, their energy, and their attachment to the communities from which they come or with which they have become connected lead toward an inevitable social restructuring and a readjustment of the system of reciprocities and obligations peculiar to the associations based on a community. This restructuring is all the more necessary because the socio-economic crisis has put social ties under increasing pressure.

Informal groups are finding it increasingly difficult to meet the condition *sine qua non* for their existence: in order to be a force for power or conquest, they have to achieve a completely unprecedented alliance among three socio-cultural groups, which have different and even opposing interests.

The first of these groups consists of women who are wives or sexual partners and whom power of social regulation have been denied them by traditional social institutions. These women are responsible for 50 to 55% of Lékié population, but they lack resources and power to control it (Mafeje 1991). They dream of having land, of cultivating it and of using what they produce.

For a woman to belong to an association of women peasants would disturb the peace of family relations in this area, where conflicts have always been managed by force and violence. This can be more readily envisaged, because the young women who belong to associations are less and less imbued with the patriarchal culture. Their education, no matter how brief, and their openness to the market and to modernity makes the current social regulation regarding them increasingly irrelevant. These women aspire to own private property, to be free and to change society. But because of their multiple activities and their belonging to many organisations, they do not have the capital to undertake any long-term activities, nor the capacity to undertake them. The mass of disinherited women who live in Lékié cannot by themselves transform their revolt into a political movement.

The second social group needed for informal women peasant groupings to exist is composed of lower middle-class women in the countryside (older wives of notables or local political leaders, and widows who are *de facto* managers of lands and family plantations). They suffer from not having sufficient access to the local socio-political system, because of the fact that they are women. As the *de facto* leaders of production units that provide most of the domestic goods and income, without their being able to exercise the role of head of the family, these women are not revolutionaries who want to overthrow the traditional social hierarchies. On the basis of their status, age, and their ties with the traditional authorities, they simply want to share power peacefully with those who have always held it.

They are not inclined to demand any abrupt break with traditional institutions, but rather want to attract a wave of sympathy from the mass of disinherited women, whose interests are, in the long run, in conflict with their own. These leaders use all kinds of seduction (altruism, accessibility) to gather around them as many women as they can.

The third group is the one that produces the ideology of the peasant women's movement. These are women who for the most part have been to school (primary or even secondary school), where they have learned both the rudiments of modern techniques and about certain notions of freedom, democracy, socialism, etc., values subverted by the men in power, even when they proclaim them. They are women who are involved in religious or political associations and are open to the influence of urban culture, either because they have lived in the city for a while or because they maintain regular relationships with agents or with their "sexual partners" who are still working there (formally or informally).

Aided by non-governmental organisations, donors, churches, and associations of urban women, these women leading peasant women's movements are in no way the heirs of a long associative tradition. In order to create an ideology that serves their ends, they draw on the immense democratic corpus. They reduce it, instrumentalise it, in order to find something in it to justify a challenge to male power and the construction of a new society in which the equality of the sexes is acknowledged.

One might at first think that the emergence of these informal structures, instilled by the leaders of the third group, might disturb relations within families and society. This view is supported by the fact that the management of conflicts, particularly between men and women, in an area where phallocratic predominance is a reality, is carried out through force and violence.

The work of family economists (Becker 1974-1981, Bourguignon *et al.* 1994, Lundberg and Pollack 1996), however, describes models of negotiation and collective models of co-operation or of non-co-operation, which remove violence from social regulation. For these authors, violent conflicts can only arise among men and women, especially over the use of common resources, if using them collectively is less productive than what they could hope to achieve by using them separately. A unilateral social separation, imposed by one party alone, would lead only to small gains, especially since property rights in traditional Cameroonian society, those concerning land, for instance, are still governed by inequitable unwritten laws.

While aware of this reality and despite their original social inferiority, women are forced to associate with men, at the same time that they belong to specific groups that would increase the diversification of risks and benefits. Women's multiple membership of associations would allow associations to avoid or at least to attenuate the effects of property rights involving relationships between men regarding material and immaterial things. If there is opposition, it is more to the traditional authorities than to men or to the state. The confrontation becomes polarised over land.

But how do these associations work? What are the power relationships inside them? In these structures, who holds the power to allocate resources to members or to activities? What are their relationships with men? With the emergence of these structures and their underlying responsibilities and obligations, do women enjoy a new status and new prerogatives, and do they contribute to changing the distribution of power in society?

3.3 - The Emergence of Women Peasants Organisations and Power Relationships in Lékié

In addition to the diversity of their names, organisations of women peasants are set up and run on a communal and even altruistic basis. The associations are composed of interrelated people. These relationships are lasting ones, because they provide a common body of rules that are more moral than economic, and are based on an altruism that functions in accord with a system that integrates individual decisions into social decisions. It is, moreover, altruism that makes it possible to build a bridge between the family bond and the social bond. In an unbalanced environment that is one of generalised inequalities, the negative effects of altruism become even more serious.

The concern of women's groups about supporting families and those of their members who are in difficult socio-economic situations inevitably attracts a number of members and persons who might be altruistic as well as selfish. In Lékié, more than 6 women out of 10 are members of an association. This involvement results less from a long associative tradition than from the activity of the many non-governmental organisations established in Lékié. Associations of women peasants represent for them an effective lever or support for policies of rural development.

In peasant associations, which include both altruistic and selfish individuals, the behaviour of each of these individuals will influence the association's final results, that is, it will contribute to either improving or decreasing the overall income. In fact, if all or most of the members are altruistic, they will not undertake any action that might harm the others. This will produce an increase in the overall income and serene social relationships within the associations.

It is not impossible that selfish people in this movement might disguise themselves as altruists, in order to take advantage of the contributions made by the altruists. Their expectations of altruistic reactions by other members oblige them to pretend to be altruistic and to allow the maintenance of the whole community's standard of living.

Nonetheless, they may behave in an overtly selfish way that could lead to a decline in the general standard of living, of transfers and gifts, and to their social exclusion. Such an attitude will harden social relationships and weaken the association, which would find itself forced to appeal to outside aid in order to survive and carry out its multiple socioeconomic activities. This choice seems all the more likely, because women peasant associations start off with limited resources. In Lékié, fewer than 5% of women are landowners, and more than 70% have no training or education in the area of agricultural development.

A general survey of women peasant associations in Lékié shows that they contain both altruists and egoists. This coexistence, which is often unbalanced because there are more egoists than altruists, and which seriously affects the associations' overall productivity, is facilitated by the fact that women belong to several different associations.

If we take into account the fact that very large numbers of women belong to associations, each of them acting altruistically in order not to be excluded from the system of mutual assistance, we see that in reality, the altruist who founded the association can, without any violence or coercion, lead several egoistic members to pretend to be altruistic in the association and to show affection for each other in order to please the altruist. The latter could eventually become the leader of the association.

This leads in this case to a woman becoming leader of an association because she has resources and is seeking to satisfy her own needs, while taking into account those of the other people in whom she is interested.

This altruistic behaviour, which is explained by the motives of seeking a reputation, of asserting status, of a desire for reciprocity, and concern about participating in a collective action and in life in society, gathers around the altruist a large number of members who benefit from the association.

In order to take advantage of the material gifts (land) and non-material gifts (advice) of the leaders and their associations, most of the members are "spoiled children" (Becker 1971) who will make use in any way they can of friendship, brotherhood, unity, discipline, commitment, and submission. The result is the leader's absolute power over the members who adopt economic behaviour that is rational in the sense of methodological individualism. Moreover, this power is strengthened by the leader's status and especially her age: in Lékié more than 34 years of age on average.

The love offered by these leaders is therefore not disinterested. It is balanced by a very egoistic satisfaction: absolute control over the association. It is reflected in anti-democratic practices, notably in the choice of the officials entrusted with the everyday management of the group, the allocation of resources, and the orientation of the group's programmes. Most of the officers are people closely connected with the leader. They cannot challenge the leader's strategic choices, even if they are contrary to their own interests. This phenomenon can be seen in all the associations, and is due to the fact that these structures are constituted on a clan or family basis, and increasingly led, because of family circumstances (widowhood, divorce, celibacy, male emigration, etc.), by women heads of family.

The grouping of women around altruistic leaders who have gradually acquired the status of heads of families might at first lead one to think of reassessing how generalised altruism is, and considering more the establishment of personal property rights that are exclusive and freely transferable. In other words, one might imagine that with the emergence of these associations, property rights might be made clearer and the retrocession of resources to their main users clearly envisaged.

The multiple activities women in Lékié are engaged in, particularly in developing market gardening, have certainly allowed them to increase and diversify their income: from 1988 to 1996, we find an increase of 3% in the average income of women producing food crops. This increase in their income and therefore of their economic power has allowed them to begin negotiations with traditional land-holders in order to obtain land with a view to engaging in more diversified activities. In fact, the impoverishment of men that followed the decline in the price of cocoa beans, and coincided with a slight improvement in women's economic situation, might have accelerated the process of negotiation. This would have been assisted or encouraged by providing for private property and freedom in the laws of December 1990, guaranteeing socio-economic liberalisation. Unfortunately, the advent of liberal democracy or the rule of law, heralded by these laws, has not yet taken concrete form. Consequently, the unequal relationships that underlie the permanence of traditional society, particularly between women and men and between women and the state, have not fundamentally changed, even if some evolution is perceptible.

Despite the fact that the market and the state have gained a foothold in Lékié, established social and cultural male predominance still constitutes the foundation of social regulation in the region. It has a significant influence on the allocation of resources and power among social agents. It still guides the community's ambitions.

Nevertheless, the blossoming of an independent and enterprising spirit among women, which has strengthened their contribution to socio-economic life in recent years, has forced society to make a relative adjustment in social relationships. The confirmed role of women as purveyors, producers, distributors, and managers of wealth in particular circumstances (widowhood, celibacy, etc.) has coincided with a role as head of a production unit or as an almost exclusive provider for the family.

Even though men have retained the symbolic title of head of the family, part of their power has been devolved to women, who control a large part of wealth, which they use both to meet the need for food and to keep the community going.

Confronted by the need to survive and carry on, society as a whole and families in particular have to make choices or rather negotiate new contracts over the allocation of resources and spheres of power, which are compatible with the status and the heavy burdens that every person carries, and with the prerogatives conferred by status. This continuing negotiation must take account of population growth, the development of the price differential between cash and food crops, and the dynamism of the women peasant movement. In addition, it will try to avoid social alienation and rupture between the sexes.

In this model, an alternative to common preference models, it is acknowledged that women and men, have distinct preferences, and participate in working out families' consumption patterns and thus in the general welfare. These consumption patterns, whether in a situation of overt conflict or even rupture, or in an easier relationship (marriage) will depend not only on prices and overall income, but also on the "danger point" determined by the preferences of each spouse and especially on the income separately controlled by each spouse.

The negotiation in co-operation, observed in families in Lékié, no longer necessarily implies that household or family income is held in common. Socio-economic evolution has resulted in partners (men and women) behaving and acting in spheres that are relatively distinct, even if one partner must include within his or her scheme of preferences the resources of the other spouse, represented by the consumption of collective goods which they pay for.
In this case the co-operative gains are greater than those in the common preference model, and depend on the danger point. The incomes respectively controlled by men and women affect the behaviour of households and the relative well-being of each of their constituents, if this control bears on the danger point.

This result implies that if one changes the rules of access to resources for all and if this change affects the income controlled by each spouse, there will be effects on allocation within families and therefore on individual and general well-being.

If the allocation benefits women, the result will be, as we see in the case of Lékié, an increase in medical care, and a sharp reduction in expenses connected with the consumption of alcohol and cigarettes.

We should, even so, add that these effects will be a function of agreements made between men and women within the framework of traditional institutions. Unfortunately, since the latter have always given priority to restrictive agreements, which completely depend on allocation and are agreed before the marriage, spouses, and in particular women, retain little room for manoeuvre in any future negotiations over resources and spheres of power.

In fact, despite the clear contribution made by women to economic activity in Lékié, their access to resources and technology is limited. This limited access for women, who are caught between the subsistence economy which they sustain and the market economy whose grip on domestic production is increasing, results not only from socio-cultural factors but also from economic and political ones (the monetarisation of the economy, the structures of markets, low rates of growth, and unequal distribution of wealth).

The traditional place of women, which has resulted in their having difficulty in gaining access to resources, education, professional training, and to development in general, confirms how marginalised they are. This does not seem to bother the traditional authorities, who regard women's groups as rivals for the management of land.

The marginalisation of women is further accentuated by the fact that the women peasant movement is very segmented and fragmented. More than 62% of the members of organisations of women peasants are unaware of the existence of federations of peasants organisations, and fewer than 42% of their leaders have contacts with other organisations, and even fewer with federations (Barbedette 1993).

The isolation of the women peasant movement and its autarkic functioning do not increase the clout of women peasants organisations or their negotiating power with other actors, who are more aware of the stakes involved in the current restructuring of the rural world.

The marginalisation and even the exclusion of women from technological policies, especially in agricultural development, combined with the segmentation and fragmentation of women's associations, reinforces the effects of a sexist policy of development, which has been implemented since independence. It also raises the crucial question of the power of women or their associations and their participation in decision-making centres.

For the time being, the spread of organisations of women peasants has not fundamentally changed the social hierarchy. At the local level, their leaders have certainly gained a certain notoriety, but this has not resulted in a reassessment of the distribution of power. How could one achieve that with associations that are relatively poor, whose survival essentially depends on the support and gifts of non-governmental institutions, donors and the state?

In addition, the state has recently acknowledged the contribution made by women to socio-economic activity, and requires the basic political-administrative structures (district, community, village) to include women or their associations in the milieus in which policies of rural development are conceived and decided upon. This participation certainly enables women's groups to put forward their demands over the control and management of resources. But since their economic and political power is small (because of their segmentation and fragmentation), they have real difficulty in challenging the local political structure and challenging erroneous concepts of rural development, even though decisions and orientations arising here nonetheless bear heavily on their destiny. On the whole, women's organisations do contribute to the gradual transformation of the classical rural milieu, by developing a new culture, but they remain unequal partners in the social structures and in development.

Conclusion

The state's withdrawal from productive activities in rural areas has contributed, along with other economic, historical, cultural, and political factors, to the emergence of peasant associations in general and women's associations in particular.

The latter's objectives were diverse: to replace the state in its role of struggling against poverty, which so severely affects women, to release women's energies, and to lead women to decide their own destiny by themselves and together in groups.

One of the direct consequences of the emergence of these associations is the reassessment of the traditional and the modern systems of social regulation dominated by men and by the state.

One can see that in Lékié, despite the energy of the women peasant movement, which is characterised by a multiplication of their associations and a diversification of their activities, the relationships of social forces have not fundamentally changed. Women are still confined to the traditional roles of maintaining and supporting the family.

Through their inventiveness, shown here in the diversification of their productive activities, they have certainly augmented their contribution to the local economy. Unfortunately, they have not yet been able to obtain freedom of access to resources and freedom to manage these ressources for their specific interests. If they could now gain access to them, according to current social rules, they still could not make use of them in accordance with their own desires and aims. Even the social and economic liberalisation, undertaken by the public authorities since 1990, has not yet allowed this evolution to take place. As a result, women remain in large measure dominated and excluded from centres of decision-making. Even their organisations, which are sadly impoverished and financially and technologically dependent on outside sources, have not allowed them to challenge an anachronistic social hierarchy, the logic of land, the constraints relating to labour or technology, which have kept them in a marginalised and poverty-stricken condition that is increasingly unacceptable.

The emergence of women peasant associations at the same time as the state was reorganising itself raises the problem of the social ideal in Cameroon in general and in the rural world in particular. Does the laissez-faire philosophy that underlies this process make it possible to reconcile equity and efficiency? To put the question more clearly, can the market economy and the system of freedom that are at the base of the restructuring of the rural world today help improve not only the conditions under which women live but also their status in the family and in society?

While the multiplication of organisations of women peasants is a positive step forward, it must be emphasised that if their effectiveness, especially in the equitable distribution of power, the strengthening of autonomy, and the promotion of women's freedom, is to be strengthened, it must be supported by access for women to civil rights, education, health, training, and information, and to financing targeted to their individual and collective activities.

All these activities are part of a coherent political system, which is democracy.

Notes

1 The uncertainty results from the fact that the economy is dependent on a few crops for export whose prices fluctuate and whose markets are shrinking in the present conjunctural situation.

2 This phenomenon can also be observed in the southern part of the country. See the study of the rural organisations in southern Cameroon, "Projet Appui aux stratégies paysannes et à la professionalisation de l'agriculture", (ASPA), January 1998.

3 It is important to recognize the contribution of children's labour in family agricultural work, especially in women's fields. It is almost the sole labour women have at their disposal. However, this contribution raises questions relative to its importance, the quality of its contribution, its evaluation in economic terms, and the constraints it puts on education.

References

Attwood, D.W. and Baviskar, eds. *Who Shares? Cooperatives and Rural Development*. Delhi: Oxford University Press.

Azam, J.P., 1987. *Repressed Inflation in a Semi-monetary Economy, The Collapse of Cash crops in Mozambique 1981-1985*. Mimeographed. Clermont-Ferrand: University of Clermont-Ferrand/PERDI.

Badouin, R., 1985. *Le développement agricole en Afrique tropicale*. Paris: Cujas.

Bakker, I., 1994. "Introduction. Engendering Macroeconomic Policy Reform in the Era of Global Restructuring and Adjustment." In I. Bakker, ed., 1994, *The Strategic Silence—Gender and Economic Policy*. London: Zed Books and the North — South Institute.

Banque mondiale, 1995. *L'état de pauvreté au Cameroun*. Washington DC: World Bank.

Becker, G.S., 1981. *A Treatise of Family*. Cambridge, Mass: Harvard University Press.

Berry, S., 1992. "Social Institution and Access to Resources in African Agriculture." *Africa* 59: 45-53.

Berthelemy, J.C. and F. Cagey, 1987. "The Agricultural Supply Price in Africa." *European Economic Review* 31: 1495-1507.

Bevan D.L., A. Bigsten, P. Collier, and J. W. Gunning, 1987. "Peasant Supply in Rational Economy." *World Development* 15: 431/449.

Bottomley, T., 1989. *Centred Entreprise for Agricultural Development*. Oxford: Punkett Foundation for Cooperative Studies.

Bratton M. and N. Van De Walle, 1997. *Democratic Experiments in Africa— Defining Transition in Comparative Perspective*. Cambridge: Cambridge University Press.

Braverman A., J. L. Guasch, M. Huppi, and L. Polhmer, 1991. *Promoting Cooperatives in Developing Countries: The Case of Subsaharan Africa*. Washington D.C.: World Bank Working Papers.

Cecelski, E., 1987. "L'émergence et le travail des femmes en milieu rural: la crise, les réactions et les solutions." *Revue internationale du travail* 28, no. 1 O. I.T Geneva.

Courade, G., 1988. *Organisations paysannes, sociétés rurales, État et développement au Cameroun (1960-1980)*. Paris: ORSTOM.

Glazier, J., 1985. *Land and Users of Tradition among the Mbere of Kenya*. New York: New York University Press.

Guyer, J., 1991. *Representation Without Taxation: The Essay on Democracy in Rural Nigeria*. Boston: African Studies Center Working Papers, 152.

Khalid, El Amin, 1997. *'Peasant Organizations, Participation and the State:* Two Case Studies from Rural Sudan'. Dakar: CODESRIA, Working Papers.

Jacobson, J., 1993. "Women's Health: The Price of Poverty." In Koblinsky, M., T. Timyan, J. Gay, *The Health of Women, a Global Perspective*. Boulder, San Francisco, and Oxford: Westview Press.

Lundberg S., and R. A. Pollack, 1996. "Bargaining and Distribution in Marriage." *Journal of Economic Perspectives*. Vol 10, pp. 139-158.

Lynch, S.G., 1990. *Income Distribution Poverty and Consumer Preferences in Cameroon*. Ithaca NY: Cornell Food and Nutrition Program, Working Paper no. 16.

Mafeje, A., 1991. *Ménages et perspectives de relance de la production agricole en Afrique au Sud du Sahara*. Dakar: CODESRIA.

Mahieu, F.G., 1990. *Les fondements de la crise économique en Afrique*. Paris: L'Harmattan.

Mamdani, M. and E. Wamba Dia Wamba, 1995. *African Studies in Social Movements and Democracy*. Dakar: CODESRIA.

March K., and Taqqu, R.L., 1986. *Women in Formal Associations in Developing Countries*, Boulder CO: Westview.

Ntonang and Dikoume Matam, 1996. *Cheminement démocratique des Organisations paysannes, jeux de pouvoir, démagogies et influences extérieures*. coll exp. Des fédérations/Saild.

Sow, F., 1995. "Femmes rurales chefs de famille en Afrique subsaharienne." Rome: FAO.

Thelen, K. and Sven Steimo, eds., 1992. *Historical Institution in Cooperative Politics in Structuring Politics, Historical Institutionalism in Cooperative Analysis*. Cambridge: Cambridge University Press.

Ventura Dias, V., 1985. "Modernization, Production of Rural Women in Kenya." In I. Ahmed, ed., *Technology and Rural Women: Conception and Empirical Issues*. IL: George Allen and University of Wisconsin, World Employment Program.

Part Three

Peasant Association in the Development and Economic Reform Process

10

Peasant Organisations in the Development Process in Kenya: Opportunities and Constraints

Mary Njeri Kinyanjui

Introduction: Rural Development and the Peasantry

Rural development has been largely informed by non-leftist paradigms of which modernisation induced by the state and donor agencies has been the most dominant. This paradigm views agrarian reform as evolutionary rather than revolutionary in which agrarian transformation of peasant producers was envisioned in terms of shifting from subsistence production to market-based production. In the process, governments and donor agencies are regarded as the main actors in the process of change. Under this arrangement, the people involved were silent recipients of the benevolent change. The people commonly referred to as *wananchi watukufu* belonged to the Kenya nation regardless of class, race or religion therefore, the nation has a responsibility to take care of them. It is perhaps because of these ideological stances that social class differentiations such as peasants, middle class and upper class became uncommon in the development planning literature. Class analysis of the Kenyan social fabric[1] seems to have had no impact on main stream development policy.

In the 1980s, the development euphoria and optimism waned, as most of the development projects initiated were a fiasco when the pendulum of socio-economic indicators—literacy, health, food security, income and employment swung towards grey areas. The reality dawned that African countries were experiencing increased poverty and unemployment while famine and food shortages were becoming more intense. In addition, gender inequalities were also increasingly becoming conspicuous. Not to mention that the debt burden became unbearable for majority of African countries. The effect of these deteriorating economic conditions on the peasantry was devastating.

To address this situation, the government wrote the 1986 Sessional Paper No. 1 of 1986, which adopted a liberal approach to development.[2] This approach de-emphasized the role of the state in development but proposed the use of non-governmental organisations and the private sector in initiating change. The paper also endorsed the adoption of market-centred structural adjustment reforms proposed by leading lending donor agencies—The World Bank and the International Monetary Fund (IMF), which, considered the approach as efficient in resource allocation (World Bank 1989). The paper also made attempts to incorporate people-centred approaches proposed by philanthropic donor agencies with particular emphasis on community mobilisation and human resources development. Attempts were also made to incorporate gender concerns in development policies. People centred development is premised on the centrality of human beings in development. People are both producers and consumers in the development process.

The way forward for bringing both women and men into the development agenda is still marred with uncertainty. Market-centred approaches target individuals or specific institution but under the circumstances, the number of poor people has been increasing. Moreover, as markets are selective in resources and inputs allocation, the benefits have accrued to the more entrepreneurial, competent and well-endowed with technology and resources; consequently, disfavouring the impoverished majority.[3] The critical issue is the markets' ability to operate efficiently and be open to all people including the rural poor and the peoples' ability to exploit the markets efficiently. Non state actors have been pivotal in filling the gap created by the negative impacts of implementation of economic reform.

Therefore, regarding peasant organisations, there is need to understand the nature and extent of people's level of participation; the formation, composition, financing and fundraising procedures; and their organisation dynamics and governance structures, and linkages with agents of change—the state and non governmental organisations—as well as with other organisations.

The State and Peasant Organisations

The "state" in Africa is currently seen as detached from the people, with its hegemony built upon the dismantling of indigenous formations and institutions. In early post-colonial Kenya, the first regime was seemingly tolerant to peasant associations as long as they engaged in social reproduction, and to some extent, property buying. No effort was made to incorporate these institutions into mainstream state institutions or incorporate them into "politics" and decision making.

The state used the "provincial administration" as the avenue for determining political action and furthering its development agenda for the people. The provincial administration was based on a hierarchical classification of the territory into administrative provinces, districts, divisions, location, sub-locations and villages. District boundaries have recently been changed, with new districts being created by subdividing the older ones. Ostensibly, these new spatial forms are being created to boost the efficiency of local administration, but have been criticised with increased protests over the location of headquarters.[4] These protests are symptomatic of the state's failure to consider the grassroots people's view, and the state's indifference to indigenous institutions and the impotency of these institutions when juxtapositioned against the state. It is on the basis of this that one of the research questions here is with regard to the relationship between the state and the peasant institutions and how these institutions have responded to the changing political space.

Various researchers have carried out studies on peasants and peasantization in Kenya; among them Kaplinsky (1978), Kitching (1980), Anyang Nyong'o (1981), Cowen and Kinyanjui (1977) Njonjo (1981), and Leys (1975). Other works like that of Rahmato (1991) define peasants as the poor, underprivileged and excluded individuals, the majority of whom are found in the rural areas. These people are: cultivators, pastoralist, farm and rural craftsmen. There is no general consensus in these works on the definition of the Kenyan peasants. Peasants are defined on the basis of their mode of production, which involves reliance on agriculture. Peasants also rely on family labour, family farm for wage rather than paid employment and use simple technology.

Mafeje (1995) negates the characterisation of African peasants by classical peasant analysts. Mafeje points out that in his earlier works on social organisations and economic production of African agricultural producers he "discovered that whoever the African producers were, they engaged in more than one economic activity and entered into a variety of production and exchange relations" (1995:4).

On the basis of evidence provided in these works, African peasants are supposedly characterised by poverty and live in rural areas were they are engaged in agriculture. Some of them are farm labourers producing for own consumption and dispose surplus in order to be given passage by the state. Most of the their actions are motivated by survival instinct and they are engaged in more than one activity and a variety of production relations. Peasantry in the African setting has evolved within the context of inequality and exploitation. Peasants are rural cultivators whose surplus is transferred to a dominant group of rulers.

Rahmato (1991:1) defines peasant organisations as forms of associative activities through which cultivators, pastoralists, agricultural labourers and rural craftsmen recreate conditions of their existence. Such organisations are either formal or informal bodies involving peasants both men and women, with the prime objective of pursuing mutual benefits in return for mutual obligations. Mafeje (1993) disagrees with this view of peasant organisations. He argues that peasant organisations must be understood in the context of peasant societies and the peasantization process. He defines peasant organisations as a generic mode of social organisation associated with peasantry of classical antiquity (Mafeje 1993:17). The *raison d'être* goes beyond mutual benefits and obligations for the member. Peasant organisations tend to be resistant movements or organisations against domination by the state, landlords, merchant capital, or men in case of women's organisations (Mafeje 1993:17). They perform a political function involving mitigating the effects of certain self-imposing structures in their society (Mafeje 1993:17). Despite the conflicts between the works of Rahmato (1991) and Mafeje (1993) on peasant organisations in the African context, there is convergence on the phenomenon of peasant associative behaviour or the tendency of peasants to form alliances. The major source of controversy is with regard to the ideological and political function of these associations. In Kenya, peasant associative behaviour has been documented in the works of Maluki (1971) and Parker and Torres (1994). Cooperatives, traditional self-help groups, business and entrepreneurial associations being the main forms of organisations identified.

Most of the peasant studies are by and large gender neutral. The use of masculine phraseology for example in the work of Cowen and Kinyanjui (1977) could be viewed as a matter of style. Nevertheless, it appears that the process of peasantization within the context of transition or in the form of relations of local and international capital exhibited a strong masculine tenor

Maluki (1971) states that self-help organisations are a political process involving people in organising and planning for actions to meet community-defined needs and problems. Funds are derived from government and non-governmental agencies. Although initially based on the clan principle of the Akamba people, women's clan organisations are a new form of social development prompted by the exigencies for better living conditions. They are a social innovation necessitated by collective demand for modernisation. Muluki views the performance of women's clan organisation as being threatened by administrative politics, the patriarchal structures of the Akamba society and lack of policy support. Faaland and Jansen (1977) view co-operatives as vehicles of rural development, that could improve people's quality of life by increasing productivity and incomes through improved farming methods and by introducing essential social amenities, recreational and educational facilities.

Owen (1995) documents success stories of women who have come together to work for their own welfare and change. Using examples from India, Uganda, Ghana and Kenya, she demonstrates the need for women to unite. In India, the Shramjivi Samaj succeeded in developing a collective awareness of corruption, injustices and the paralysis of the *panchayats* (village councils) which were male-dominated and rarely responded to their complaints. She demonstrates the success to TASO (a Ugandan widows' association) in helping members cope with death as well as prepare themselves as female heads of households. She also reports a success story of Ministry of Widows in Northern Ghana that rescues women banished from their homesteads and assists them initiate income-generating activities

A comparison of these associations shows that organisations in the three African countries are welfare and relief-oriented. In contrast, the widow associations in India have become consciously political and give real meaning to the overused but little understood word 'empowerment'. Westergaard (1982) argues that women co-operatives could play an important role in poverty elimination. Through co-operatives women can pursue economic activities and can be trained in productive skills, and could be involved in disseminating knowledge on health, nutrition, family planning and literacy. Parker and Torres (1994) found that a fifth of small-scale entrepreneurs belong to a business support group, providing financial services to members including men.

Maluki (1971), Owen (1995), Faaland and Jansen (1995), Westergaard (1982) and Parker and Torres (1994) seem to be in agreement that associations are important forms of social organisations among different categories of people. Associations have potential that could be tapped to initiate change through mutual support, collective action, knowledge and skill transfer, leadership training and savings and credit. Saving and credit associations usually referred to as solidarity groups have recently attracted attention as a means of mobilising resources for enterprise development (Berenbach and Guzman 1994; Ryhne and Otero 1994).

Relations between the entrepreneur and his clients are a major asset for business development, as such relations are built on informal control, professional collaboration, interdependency and trust. Despite this recognition, very little efforts have been made to investigate the applicability of these principles of social, political and market relations in the development agenda. It would be therefore interesting to find out how associations and self-help groups fit into network models in the development process, particularly in mainstreaming people in the development process. However, what is lacking in these studies is the theoretical basis for social, economic and political change generated by such institutions. The studies adopt approaches which atomise in a frequently changing pattern the development requirements of peasant agriculture into a series of discrete technical fixes such as extension, credit, technology and markets (Moyo 1995).

Methodological and analytical framework

Linking mainstream economic models that emphasise the role of markets as the key forces of change and political economy change within the framework of democratisation and civil institutions underlie this research. Associations involving women are important in the development process, as they deal with the social, economic and political spheres. In this study, we operationalise the role and impact of peasant associations on the basis of change, structural impediments, domination and exploitation, trust and ideological relations. The first issue involves an examination of the origins, participation levels, and organisation dynamics of peasant organisations. The second involves an investigation of forces underlying women participation in peasant organisations and finally the third investigates the participation of peasant organisations in the rural development process. The underlying factors that inhibit women's participation in peasant organisations is also discussed and the general role of peasant organisation in the development process.

The study was carried out in Gatundu North division of Thika District, in five locations namely: Ndarugu, Chania, Mang'u, Kiganjo and Karatu. The population of the area is 239,000 people with about 120,000 men and 119,000 women. The majority of the people are small-scale farmers engaged in the production of foods, fruits, horticultural products, animal husbandry and poultry. The small-scale farmers form a variety of organisations which many be grouped as follows (Table 1).

Table 1: Grouping of Small Farmers

Organisation	Basis of Relationship	Space
Family clan	blood or marriage	Village
Religious		
Small Christian Communities		
Prayer groups		
Fellowships		
Women and Men Organisation	Spiritual	Village
Religion		
Parish		
Economic		
Marketing Co-operatives		
Farmers' societies		
Land buying companies	Economic	Village Location
Division		
Age groups	Coming of age	Village Location

Gatundu division, Thika District, has varied socio-economic and ecological conditions. The individuals in associations were investigated by use of a household survey. A hundred households were randomly selected for the survey from the five locations, and six organisations were selected as well. In each of these associations, group leaders were interviewed as a team using a questionnaire comprising structured and open-ended questions. Focus groups, each consisting of ten members, were used to discuss key aspects of the study especially regarding the choice of group's economic actions, leaders, sustainability and future programmes, state association relationship and the constraints facing their association. The associations that were studied include the following: Gatukuyu Coffee Farmers Cooperative Society, Age Group, Mbari ya Ikang'u, Mbari ya Ikang'u Githaku, Group ya Kurimia Chania and Igamba Itega Group.

Characteristics of Peasants and Peasant Organisations in Gatundu North

In Gatundu North, 61.5% of the married men respondents went to school up to primary level, 23.1% secondary level, and 3.8% A-Level. Uneducated married men were 11.6%. Most of married women 44.8% attained primary education while 31% of them had no education at all. Only 24% of the married women had attained secondary education. Slightly less than a quarter (22%) of the single women had no education at all ; 66.7% reported having primary education while only 11.1% reported having secondary education. Majority (50%) attributed their low levels of education to lack of school fees, 33.3% failed examinations while 16.7% could not continue with their education because of pregnancy.

The married men advanced several reasons for attaining their respective levels of education. The respondents without any education at all indicated that their parents refused to take them to school or were given household responsibilities to take care of, such as farming or herding animals. Lack of school fees was another reason which contributed to the low level of education attained. Failure to pass examinations was also key in determining the level of education an individual attained. Some respondents stated that they decided not to continue with school after finding school uninteresting while a few got employed before they could pursue further education. The number of children for married women ranged from 1 to 9 with a mean of 3.8 children.

Land ownership is based in the Swynnerton Plan of 1954 which led to the individualisation of land on the basis of free title lease. Land became a commodity for sale, implying that one could not only acquire land for sale but could also buy it from a willing seller. In addition land could be mortgaged in order to obtain bank loans. At least 96.2% of the married men respondents owned land and the size of land varied between 1 and 12 hectares. In most instances the land owned, inherited or purchased and those respondents who purchased land did so because the land they had inherited was small or it was bought for their sons. Although most respondents stated that they owned land, majority did not have a title deed to the land as only 30.8% of the respondents reported having a title deed.

Reasons given for not having a title to land were that land had not been subdivided by the office of the registrar of lands, and family disagreement among brothers or one of the two parents was still surviving meaning that she or he is still the trustee of the land. According to custom, land is given to male children but the decision of when and where to give children land rests on the parent. Respondents indicated that the aspects considered when a parent transfers land to his children are age and marital status. In order for one to be given land by a parent, the land recipient is supposed to give the father a goat, beer and honey. Majority of the respondents were third generation owners of the land as land was first owned by grandparents. Others were second generation owners since their parents were the first to own land.

Majority of the married women respondents did not own land (75.8%) and thus only one quarter of the group owned land. None of the women had purchased land but half of them have access to land for cultivation. The women who owned land inherited from parents, none of the women had a title deed to the land they claimed to own. The reason for not having land title deeds was because land belonged to the husband, had not yet been subdivided or the title deed had not been obtained from the District Land Board. Majority of the women were ignorant of the procedures of land acquisition. A few stated that land is only given to sons. The few women who owned land were third generation owners since land was first owned by grandparents. While majority of them were second generation owners, having acquired land from their own parents. Only 38.9% of the single women owned land and out of that two had land title deeds and the size varied between one hectare and six hectares.

Many respondents had no money for land transfer or did not know the procedure of acquiring land, particularly single women. The duration they had stayed on land was between 18 and 45 years and most of the single women were engaged in full-time farming for subsistence with a few engaged in farming on a commercial basis. Crops grown in Gatundu North include maize, beans, arrowroots, potatoes, bananas, vegetables, yams, coffee, tea, french beans, avocadoes, macadamia nuts and passion fruits. In addition to farming the peasants also keep livestock and poultry.

The peasant associations identified in the study area were mainly cooperatives, family, clan, age group and women organisations and land buying companies. Membership in land buying company, however, was low: only 3.8% of the married men and 7.4% of married women were members of land buying companies, namely, Mangu Investments and Gwa Gakai Land Buying Company. Most of the married men were members of the Gatukuyu Cooperative Society which is involved in coffee management and marketing. Membership in family and clan organisations was also common in the area. Few of the respondents were members of more than one association. Married women registered a fairly large proportion of women (85.2%) who were members of peasant organisations, married men followed with 76.9%, single men 42.2% and single women 41.2%. Peasant organisations in Gatundu North are institutions geared towards change and overall improvement of members' social and economic life. The organisations consist of both married men and women as well single men and women. The household survey revealed that the longest duration a married man has been in a peasant organisation was 51 years. For married women, the longest duration was 22 years while for single women it was 7 years. The longest duration for single men was 8 years.

Most peasants covered in the survey joined their respective associations between 1990 and 1997. Social and economic shocks played a key role in motivating association formation and in reactivating old ones. The association start-ups were initiated by the members themselves. Indeed, 20% of the married men and 11.8% of the single men were involved in the founding of their respective associations. The others joined associations which had been started by their parents, age mates, elders, clansmen, teachers, and the church or other women in the area. Except for the respondents of the associations who were involved in the formation of their organisation, several were introduced into the association by friends, parents, clan, spouse or church member. The need for unity and solidarity also motivated single men, married women and single women to join associations.

Politics was insignificant in the start-up of the associations. Only 13.6% of the married women respondents associations were started through political influence. In the case of married men, the proportion was five percent while that of single men was 5.9%. With regard to whether membership was closed or open, majority of the respondents in all the categories indicated that membership in their association is open. The proportions of married men who are members of association with open membership were 73.7%, married women 65%, single men 94.7% and single women 66.7%. Three factors were considered in order for the respondents to become members of the associations. These are age, sex and geographical location. Geographical location was however considered for single men, married women and single women. Factors such as religion, political affiliations, profession or economic status were not considered. The associations are also open to both urban and rural people. However, there were associations where membership was only for rural people and all the single women in the survey were members of rural-based associations. The respondents indicated that the associations required membership fee of between Ksh 10 and Ksh 50.

In the preceding sections, we saw the various reasons that motivate individuals in Gatundu North to form alliances with others. Needless to say these alliances are not promoted by affection but are based on a consciousness and solidarity spirit. It may not also be plausible to argue that the communities are developed if they are middle class or striving to become middle and upper classes. Or that they are not part of civil society because they do not have money and power. The relevance of this can be expressed in the popular culture expression that "life is not a rehearsal". Peasant organisations therefore actualise life in rural communities, resist control and domination, accumulate capital, provide insurance policy to members and play the role of civil society in the rural economy. They are, indeed, social movements whose consciousness is based on common experiences generated internally or externally. These common experiences are landlessness, lack of employment, state exploitation and domination, low returns from investments in primary agriculture, limited money supply, lack of social infrastructure including health and education. They also serve as structures of identity and self-respect.

Undoubtedly, these experiences generate helplessness and hopelessness which students of survival strategies over-emphasize. Students of survival strategies narrow themselves to the helplessness and hopelessness of peasant communities and fail to see the insidious consciousness upon which joint actions are undertaken to bring change or address structural impediments within relations of domination and exploitation and those of trust and ideology, Beyond this state of peasant communities' hopelessness and helplessness, a consciousness had evolved which makes people take responsibility for their own community's change process and structural impediments. Since the scope for acting individually is limited, they work together in solidarity.

The collective actions taken bring change for the current generations. For many of their members, peasant organisations are social movements through which they are able to bring about major changes in their society at different levels. These levels are at the individual, household and community levels.

Peasant Organisations and Agrarian Transformation

The role of peasant organisations in agrarian transformation has largely gone unnoticed. Peasant organisations are supposedly passive recipients of change initiated by government, non-governmental organisations and donor agencies. The reality is however different, peasant organisations have been proactive in mobilising members, resources and laying down infrastructure for further agrarian transformation. Since the Gatukuyu Coffee Farmers Society split from the Giant Kiambu Farmers Union, it has been actively involved in promoting coffee growing in the area. One notable achievement has been in locating factories closer to the people. At that time of the split, farmers transported coffee for distances between 5 and 10 km. By locating its factory nearer, the distances have been reduced to 3-5 km. The location of these factories close to the farmers ensures that farmers can spend more time harvesting coffee in a day and spend less time on travel. The new factories have also contributed to the reduction of wastage and improved efficiency in coffee.

The Gatukuyu Coffee Farmers' society has also continued to purchase farm inputs such as fertiliser and pesticides in bulk for sale on credit to farmers, subsequently improving coffee yields for individual farm members. The farmers society also sets standard for quality coffee and has encouraged the formation of committees in each factory to ensure that good practices in coffee husbandry are maintained. The society has also installed electricity, and water in its factories to improve processing and enhance quality of the processed coffee beans. Finally, the society organises training programmes for farmers and participates in agricultural shows. This evidence shows that Gatukuyu Coffee Farmers Society has initiated, designed and implemented projects for its members, the development philosophy being "doing the best we can for ourselves". The association is making efforts to reach and meet the needs of all its members in spite of all odds as we shall see later.

On a lesser scale, we also observed the efforts made to enhance agrarian transformation by the Mbari ya Ikang'u Githaku Peasant Organisation. In this organisation, the members have started a project of buying manure collectively in bulk and then distributing it to the 20 members for use on farms. This phenomenon has been prompted by the fact that the cost of composite manure and fertilisers rose with liberalisation. The farmers have found it uneconomical to buy small quantities of these products for use on farms.

Joint effort in supplying farm inputs is also evident in the Chania Kurimia Women Group. Besides, selling labour, to large - scale farms, these women also help each other to farm on their own farms. This ensures that digging and weeding is carried out on time for early planting, as well as ensuring that the members have spare time to work as farm labourers during the peak periods. These last two organisations are again making efforts to benefit themselves by improving farm productivity in their own small way by purchasing inputs, and rotating labour assistance.

The transaction regimes, money generation and supply are the main structural impediments to agrarian transformation in Gatundu North. The embeddedness of these transaction regimes constitutes a major market failure involving unequitable profit-sharing between the various participants. As early as 1976, farmers in Kiambu[1] registered complaints of exploitation in the coffee industry within the transaction regimes. The farmers called "for the immediate elimination of some unnecessary middlemen who were making huge profits at the expense of the farmers" (*Daily Nation* 23-7-1976).

In response, the then chairman of Coffee Board of Kenya, Mr. Kuria exonerated the board by stating that "Although the CBK was authorised by law to deduct 3% of total seasonal sales, the board only deducted 1.5% of the deductions. After these deductions, the remainder of the money is passed over to five authorised agents down the ladder to the local cooperative society and then to the farmer. He also pointed to the effect that farmers' complaints may not have been genuine because this season the farmers will receive the best returns since 1956. The expected sales were: "Ksh. 15,000 per tonne of clean coffee. The second interim payment to be made in April will be on the basis of Ksh. 13,000 per tonne" (*Daily Nation*, 23-7-1976. He indicated that in London, Kenya was accused of paying the highest returns to farmers, subsequently encouraging expansion of coffee growing. He also stated that he had encouraged the Kenya Planters' Co-operative Union (KPCU) to establish its own mills for threshing coffee in order to avoid policy measures.

The farmers' awareness of exploitation in the coffee industry was heightened by a visit to the London Auction by the chairmen of the farmers unions that constituted the Kiambu Farmers Cooperative Union. According to a respondent, some of the chairmen upon return informed the farmers about the high prices of coffee in the London Auction. However, unfortunately, the earnings of coffee per tonne were not transferred to the farmers by Kiambu Farmers Cooperative Union. In response, the farmers accused the Union of representing their interests inadequately in the coffee industry. They, therefore, demanded to form autonomous associations from the Kiambu Farmers Union.

The call for the formation of autonomous associations spurred tensions and hostility between the farmers who were pro and those who were against the Union. The farmers adopted several measures to deal with the impasse that followed. One of them included a visit to the then head of state who was also the parliamentary representative of Gatundu. The visit comprised representatives from Gatundu, Gatitu, Gatukuyu and Kiamwangi societies. The delegation was headed by Mr. Kahugi Gituro who presented a memorandum on behalf of the delegates. One of the key issues in the memorandum was the societies' demand to split from the Kiambu Farmers' cooperative Union. The reasons advanced for the split were, poor prices paid to the farmers by the Union, expensive farm inputs and delayed payment. The President responded that "there was no law which prevented the societies from dissociating from the Parent Union. So long as you can look after your own affairs, you have the right to choose to belong to a union or not." (*Daily Nation,* 10-1-1976).

Unfortunately, the visit seems not to have had an impact on the volatile situation in the division. The situation was aggravated by the coffee smuggling and thefts that were taking place in the country. The smuggling and thefts intensified the farmers' fears. They did not trust that the union could take care of their coffee already in the factories.[2] They feared that the union management could steal the coffee and sell it to the smugglers. Subsequently, the farmers set up vigilante groups to keep watch of the coffee in the factories and the protests intensified. The *Daily Nation* of 28-1-1976, for example, reported that a farmer Mr. J.K. Mbugua threatened to uproot his coffee and burn about 250 bags of coffee if what he called negative attitude of the Kenya Planters Cooperative Union was not reversed. To placate the top echelons of small-scale producers, the Kenya Planters Cooperative allowed farmers to set up their own factories to process their coffee and sell it directly to KPCU rather than the union. Farmers in Gatundu North such as Gacheha, Gitau and Ndekei were allowed to cease being members of the cooperative union and to start their own factories. Essentially, by removing these farmers, KPCU hoped to diffuse the farmers' riots. Unfortunately, this move did not have the desired effect as the farmers continued to resist. They continued to demand to know the truth about producer prices of coffee, they also refuted annual statements of accounts from the cooperative union management.

Needless to say lack of capital to finance investments is viewed as a major structural impediment in economies of the poor. Modernists and propagators of market economies recommend provision of loans and credit facilities as the panacea for poverty alleviation in communities of the poor and peasants. This is unfortunately done without a full understanding of money generation and supplies within these peasant economies.

Peasants in Gatundu North generate money mainly from the sale of crops, animals and sale of labour to other farmers. Crops grown are perishable and are not good store of value and peasants are therefore forced to dispose their products at the going market prices. Second, the flow of the money is contingent upon the seasonality of harvests, which fluctuate frequently. Money generated is limited in amount, which in turn, limits the size of individual transactions within the economy. The banks that are the official financial intermediaries are located in major urban centres such as Thika, Gatundu, Nairobi and Kiambu. Besides, the banks have set standards which most often than not do not correspond to the generation and flow of money in the peasant economy of Gatundu North. It is argued here that these conditions of money generation and supply in peasant economies and the failure of banks to adapt to these conditions have created grounds for peasant organisations to evolve as financial intermediaries.

The peasant organisations act as financial intermediaries in the peasant economy through the collection and pooling together of money generated in the economy. It also assists in enhancing circulation of money by distributing it to individual members in a rotational manner and promotes exchange transaction within the economy. The transactions made are related to investment and include investment in children education, purchase of household goods, farm animals, seedlings health products, and provision of shelter.

Mbari ya Ikang'u - "Githaku" Peasant Organisation

The third case study is a splinter group of Mbari ya Ikang'u formed by daughters and daughters-in-law of Ikang'u second generation children. It is predominantly a female member's association, although a few males are members. The organisation was founded by a woman called Gatherero and two other women. They started the organisation because their interests were not being represented in the Mbari ya Ikang'u associations. Their other complaints were misappropriation of funds and lack of power sharing between men and women. They felt that women members were not considered in decision making, while it was furthering male members' needs as opposed to those of women. Gatherero in particular, felt marginalised because though a daughter of Ikang'u's daughter, she was separated from her husband. Her membership was held with suspicion. The other two women are widows, and not blood relatives of the Ikang'u. In many ways they felt alienated in the associations' function.

The formation of this splinter group indicates the members' need for autonomy and a joint action of dealing with domination and exploitation from the male gender. The inclusion of daughters in-law in the association points to a local evolution of sisterhood. The sisterhood is built around the need to address household needs and address the problems of domination and exploitation. The organisation was founded in 1993 and the membership is Ksh. 10. At present it has 20 members. The organisation's objectives are to help one another in family matters; to pool savings; to purchase household items; to finance children's education; to improve shelter; and to buy farm inputs.

Gatukuyu Farmers Cooperative Society is a member of the Cooperative Union Bank located in Gatundu urban centre. On another scale, the Mbari ya Ikang'u Githaku organisation collects and pools money for members generated within the peasant economy. The members make a monthly contribution of Ksh. 50. Thus, the organisation pools a total amount of

Ksh. 1,000 every month. Similarly, Igamba Women Group collects and pools money for its members. The monthly contributions were Ksh. 50 per month at the start-up time in 1993. The figure has now been raised to Ksh. 100 meaning that the association pools together Ksh. 6,000 every month.

The agents and export companies are the key intermediaries between the farmers and the international market. They set the rules of the game and reap most of the benefits in the retail trade chain because the commodity chain is structured in a way that a given product value increases as the product moves across one transaction regime to the next. The farmers also lose ownership and control of the commodity once they dispose it to the agent who will earmark the commodity for the international market. The farmers lacked information about the market and the transaction regimes involved. The closest Gatundu North Farmers came to learn about the market for their products was during a visit to coffee auction in London in 1977/1976. Women farmers tend to lose out in most of the transactions as they are marginalised by the male dominated marketing system.

In gender analysis the exclusion of women in development is due to partriachy that lead to their lack of authority and power. Elements such as trust, intimacy, commitment, communication and sexual intercourse, procreation and nurturing impart on relations between men and women, authority and power. The literature has also presented partriachy as static institutions unresponsive to change. Beside, it has failed to recognise that patriarchy as an institution has building blocks, rituals and symbols that make its presence known. By delineating male and female spaces as mutually exclusive domestic and public spaces, we are assuming that men have no role of responsibility in the domestic sphere and that women have no business being in the public sphere. We then move on to characterise the male and female characteristics that facilitate men to excel in the public sphere and which women lack.

From the Mbari ya Ikang'u case study we can draw that objectives of the organisation traverse both the domestic and public sphere. Objectives one and two: assisting each other in times of problems and providing support during family functions are clearly issues relating to the domestic spheres. Conversely, objectives 3 and 4: setting a framework for pooled savings and initiating development projects for example buying land and residential plots in urban areas are related to public domain. It appears that Mbari ya Ikang'u planned to undertake collective action in both the domestic and public sphere.

The splinter group Mbari ya Ikang'u Githaku is indicative of internal conflicts within the PO. The internal conflicts rose from abuse of trust, levels of commitment, perception of autonomy, desire for autonomy and levels of involvement. Their organisational objectives were related to improved human relationship between members while objectives two and three—provision of household items and the need to improve shelter—are indicative of improved livelihood for the families concerned. Their objectives are leaning towards the domestic sphere. Pooled saving objective could lead to primordial accumulation. It may also serve as insurance policy when need arises and improve transaction cost. Over all pooled savings ensures a wider circulation of money within the peasant economy.

Mbari ya Ikanga Peasant Organisation

Mbari ya Ikang'u is a clan-based peasant organisation. It is named after Ikang'u and is based in Ngethu village. It was founded in 1993 by one of his sons. Ikang'u had two wives and always called his family members to deal with shocks and crisis. However, since his death no meeting has ever been called to bring the family members together. On 20-8-1993, Gitau, his only surviving son, called a meeting of Ikang'u's great granddaughters to discuss the issue of Gathoni, who was unmarried and had two children. She was working as a secondhand clothes trader in Nairobi. Her children were being taken care of by her mother. The phenomenon of having unmarried women with children is relatively new in the family. None of Ikang'u's daughters had had children in his household, all his daughters were married and living in their husband's home. Gathoni's unmarried status was therefore unique and a challenge to her nuclear and extended family. The case became even more complex after her death, especially because of the children. The older generation does not seem to understand or appreciate this phenomenon, and it is perhaps for this reason, that Gitau called upon his sisters and step sisters to look into the issue and set an agenda and action plan for dealing with the phenomenon.

The first meeting was held in 1993 and was attended by Gitau, his sisters and step-sisters. However, during the meeting the sisters and step-sisters felt that they were old and could not take on the responsibilities of their children and great grandchildren. They proposed that another meeting for their children, grandchildren and great grandchildren be called to deal with the issues of the family and the changing society. A follow up meeting was held in 1994 and attended by Ikang'u's second generation children and their children. During the meeting Mbari ya Ikang'u's family organisation was launched. Its objectives were to assist each other in times of problems; provide support during family functions; set a framework for pooled savings; and, initiate development projects, for example buying land and residential plots in urban centres.

The Sywnnerton plan which institutionalised peasantry gave men the mandate to own and control land resource. In our household survey we observe that 75.8% of the married women stated they did not own land. One respondent indicated her husband is the owner of the land and can only claim ownership and control to land after her husband dies. The situation of land ownership and control gives one spouse leverage over the other since the title of the land bears the male spouse name only. This implies that in the absence of dialogue and trust, society's checks and balances to ensure fairness between the spouses, the spouse with the title may take advantage of the other. For example in one case a male spouse used the land title deed to guarantee someone else a bank loan without the knowledge of the wife. The loanee failed to repay the loan and therefore the guarantor had to be charged for the loan repayment. According to the female spouse the farm was auctioned without her knowledge. One day while doing her routine farm work a court order was brought informing her that the land has already being sold to someone else and she should vacate. The house was demolished and they were evicted from the land which had been their marital home and means of production.

Exclusion and bias of women, basic tenets of patriarchy need to be treated within the analysis of customary law and African tradition in gender and peasant organisations. In our household survey we observed that (85.0%) of the married women were members of peasant organisation and only a few claimed being denied by their male spouse from attending POs. According to one respondent whose organisation could be said to be one of those enhancing gender roles because it concentrates on the domestic sphere said that initially male spouses may have been opposed to some women from leaving the homestead to attend our meetings but they now appreciate our coming together. A male respondent was also positive about the activity of the women participation in the PO. He stated that "at present the wife does not make as many demands as she used to. He was happy that the household has enough plates, and utensils for the family and when they have guests his wife does not have to borrow plates from neighbours.

The Chania Kurimia group is also indicative that the public sphere is not exclusively male, the women are not tied to production activity in the households only. To show that they are not victims of patriarchy, they gang together to bargain for higher wages, to enhance individual accumulation by pooling resources together. In addition, they also save time. The domestic and public spheres of production in the peasant economic, landscape are not mutually exclusive. For example if a male spouse is not able to appear during a *riika* (age group) meeting, the female sits in the meeting and casts a lot for the daily collection. This further reinforces the role of trust and intimacy in the domestic sphere.

The POs have been playing two major roles in reinforcing or rearticulating partriachy among the peasantry to accommodate the crisis and changing demands of the changing peasantry economic landscape. They reinforce female role of nurturing by providing them with basic needs. They also provide women with a chance of gaining some independence, changing their position from being recipients but also a contributor to the household kitty. The POs also provide women solidarity for bargaining for better wages. They also serve as a platform for exchange of information.

Peasant Organisation and Multiparty Politics—"Democratisation"

The repeal of section 2A of the Kenyan constitution that barred the formation of many parties paved the way for democratisation in Kenya. The introduction of other parties was welcomed by most people covered in the survey. Several parties were formed Democratic Party (DP), Forum for Restoration of Democracy (FORD) which later split into two factions; FORD Kenya and FORD ASILI, Kenya National Democratic Alliance KENDA etc. However, the majority of the respondents did not register officially with any of the parties. Most identified themselves with FORD Asili. A good number identified themselves with the Democratic party, a few associated themselves with Kenya African National Union (KANU), KENDA and FORD Kenya.

Two key respondents reported that as far as they are aware none of their PO members have a political party membership card or have ever seen or read the party's manifestos. A respondent in one of the POs indicated that he does not identify himself with any party. He votes on the basis of the personality standing on a particular post. He indicated that he voted for a DP presidential candidate, a KANU member of parliament, and a KENDA councillor. On further questioning of his voting pattern, he argued that DP candidate is rational and has a vision. His choice for the KANU parliamentary candidate was based on the fact that, his home is within his vicinity and he has played a key role in steering development projects in the area. Example of projects given were contributing money for construction of classes, churches, he also helps the needy students in high school and clearing hospital bills. His argument for choosing a KENDA councillor was because he was mature and had leadership qualities. On the relationship with the PO and what seems to be infatuation with parties and choice of candidates he said that the parties have no solid policies and state what they will do or they have done in the past, but on the ground there is no evidence of activity. Moreover, presence was felt during periods of electioneering after which their representatives disappeared. Another respondent commented that the distance between POs with both the central and local authority is large. He said that the offices are not located close to the people in the area, if you need to see policy makers one had to travel to Nairobi, or see his agents who will convey the information. He however observed that parliamentary representative do make appearances in funerals, or visit specific bars (beer halls) but it is still difficult to access a local representative because they will be ringed by a clout of aides or associates, thus remain distant even if they are in the proximity.

In one PO, I enquired what was the distribution of party membership like and how they worked to minimise inter-party politics. The response was clear that political parties did not bring material benefits to the people, and in policy matters they have not influenced government policy in farm input prices, or taxation. Most of the potential development projects are largely ignored and households had to contribute in the provision of labour in the rural water programmes through digging the trenches as their contribution to the project. This case to a large extent represents the view of both bottom and top down approaches to development that the peasantry is an army of idle manual labourers.

The political distance between the PO and the representatives is intensified by the flow of information and use of opinion sharpers. These opinion sharpers are neither free from subjectivity nor are always right. Information is through world of mouth, mass rallied, songs mobile promoters and inadvently the use of emotive songs to captivate attention. Party rallies are more less like a carnival. After one such carnival leaves, the peasant attend the next one. Political party affiliation in all the POs was mixed. At least in all the POs members covered in the study do not belong to one party. They have mixed party affiliations; asked whether they perceive this as a weakness because they cannot lobby a party to incorporate their interests and needs, the response was that the PO's mandate allows freedom of association, expression and choice. The POs interests come first. They said that they keep off party politics in their POs' agenda. In my view it appeared that POs perceived political parties as distant institution which claim to be with the peasantry only when they need support or votes. Further, it is difficult to extend trust which is the linchpin of the POs survival to institutions that are not only abstract but externally formed.

A fundamental constraint to democratisation is reducing the distance between the PO and the democratising agents. The democratising agents' failure to reach the POs lies in their perception of the social political organisation of POs. The POs have their own internal organisation based on trust, kinship, blood relation, initiation bonds, needs; not to mention that the POs have their interpretation of fairness, responsibility, good governance, risk taking and perception of time. The other constraints lies with the POs themselves because they allow the individual members to participate in the political life rather than in the organisation. However, the POs have learnt from past experiences of resisting public policy from the case of Gatukuyu farmers cooperative society which has been met with a lot of resistance. The POs seem to be recapturing the reality of the present characterised by economic crisis, social and political strife. They are unwilling to let go their solidarity under the guise of being staunch supporters of parties as a sign of being democratised. To them the democratising agent is just another passing cloud.

Conclusion and Recommendation

Peasant organisations are important tools for reaching out to the peasantry in the development process for achieving gender justice and balance, and for democratisation. However, in order to do this the social science has to re-orient itself and realise that human beings economic and ecological landscape are both our specimens and laboratory. Mistakes of the past top down approaches, rather than bottom up, dual economies, surplus labour, and reliance on external borrowing models of development have contributed to a lot of suffering to generations not involved in decision making. Our approach to PO should be open minded appreciating and recognising POs have their own worldview passed over from one generation to another.

Thus, investment in people's health, education, nutrition improves their ability to participate in the development process. Perhaps, the most noticeable limitation in the people-centred approaches is the efficient allocation of scarce resources in the face of growing population. Undoubtedly, these two approaches provide a basis for development planning in the 1990s. But unless they are skilfully directed within a framework that takes into consideration people's initiatives, they are equally bound to fail. Thus, incorporating people in the development process by making them actors in the change process should be the initiative of the people, by the people and for the people's associations and/or initiatives such as peasant organisations, might be the most logical beginning point. These associations include a multitude of groups and self-help associations which rural people tap informally to undertake collective action.

Notes

1 Gatundu North was part of Kiambu District then.

2 The smuggling involved senior politicians and members of KPCU. The *Daily Nation,* 7-10-1977 reports that MP for Embakasi and Makuyu were in court because of stealing coffee worth Ksh. 1.8m.

3 Among them Kaplinsky (1978), Kitching (1980), Anyang Nyong'o (1981), Cowen and Kinyanjui (1977), Njonjo (1981), and Leys (1975).

See Kenya (1986) Economic Management for Renewed Growth Kenya Sessional Paper no 1. 1986. Social Dimensions of Development (1994) Ikiara (1999) Kenya (1997) Development Plan 1997-2001

4 The works of Gibbon (1993) document the advance effects that structural adjustments and economic reforms have on vulnerable groups.

5 See Njehu Gatabaki's view on the subdivision of Kiambu into two. Kiambu and Thika districts on the *Daily Nation*. One of his fears was that Kiambu District was to lose its history as the birth place of the first President of the country. Also see the controversy over the sub-division of Kisii district.

References

Anyang Nyong'o, P. (1981) "What `The Friends of the Peasants' are and How they Pose the Question of the Peasantry." *Review of African Political Economy*, No. 20, January - April.

Berenbach, S. and Guzman, D. (1994) "The Solidarity Group Experience World Wide." in Otero Maria and Rhyne Elisabeth (eds) *The New World of Microenterprise Finance: Building Healthy Financial Institutions for the Poor.* West Hartford: Kumarian Press.

Cowen, M. and Kinyanjui, K. (1977) "Some Problems of Capital and Class in Kenya." IDS Occasional Paper No. 26. IDS, Nairobi: University of Nairobi, Institute for Development Studies.

Daily Nation, Issues 23-7-1976; 10-1-1976; 28-1-1976; 07-10-1977

Faaland, J. and Jansen E.G. (1977) *The Role of Co-operatives in Rural Development in Kenya.* Bergen: The Chr. Michelson Institute.

Gibbon, Peter (ed) (1993) *Social Change and Economic Reform in Africa,* Uppsala: The Scandanavian Institute of African Studies.

Ikiara, G.K. (1999) "Economic Restructuring and Poverty in Kenya." in Njuguna, Ng'ethe and Wasunna Owino (eds) *From Sessional Paper No. 10 to Structural Adjustment : Towards Indigenizing the Policy Debate.* Nairobi : Institute of Policy Analysis and Research.

Kaplinsky, R. (1978) *Readings on the Multinational Corporations in Kenya.* Nairobi: Oxford University Press.

Kenya (1994) *Social Dimensions of Development in Kenya: An Approach to Human Centred Development Alleviation of Poverty.* Vol. 1: Conceptual Framework, Nairobi: Government Printer.

Kenya (1997) *Development Plan 1997-2000.* Nairobi: Government Printer.

Kenya (1986) *Sessional Paper No. 10 of 1986 on Economic Management for Renewed Growth.* Nairobi: Government Printer.

Kitching, G.N. (1980) *Class and Economic Change in Kenya: The Making of an African Petite Bourgeoisie 1905-1970*. New Haven: Yale University Press

Leys, C. (1975) *Underdevelopment in Kenya: The Political Economy of Neo-Colonialism 1964-1971*. London: Heinemann.

Mafeje, A. (1993) "Peasant Organisations in Africa: A potential Dialogue Between Economists and Sociologists." in *CODESRIA Bulletin* No. 1, 1993.

Mafeje, A. (1995) "African Peasants: An Elusive Category." Paper presented at a methodological seminar on multinational working group on Peasant Organisations in Africa: Problems and Perspectives, Dakar, Senegal November 27-30, 1995.

Maluki, E.I. (1971) "The Role of Traditional Structures in Self Help Schemes in Machakos." Draft Paper.

Moyo, S. (1995) "Peasant Organisations and Research Issues on Democratisation in Africa". Paper presented to the Codesria MWG Methodology Workshop on Peasant Organisation in Africa CODESRIA, 21-30 November 1995.

Njonjo, A.L., (1981) "The Kenya Peasantry: A Re-Assessment." *Review of African Political Economy*, No. 20, January-April, 1981.

Owen, M. (1995) "Widows Banding Together : Women a Power for Change." in *People and the Planet* Vol. 4 No. 3. Pg 20-22.

Parker, J.C. and Torres T.R. (1994) "Micro and Small Enterprises in Kenya : Results of the 1993 National Baseline Survey." Gemini Technical Report No. 75. Bethesda: Development Alternatives.

Rahmato, D. (1991) *Peasant Organisations in Africa: Constraints and Potentials*. CODESRIA Working Paper 1/91 Dakar, 19

Rhyne, E. and Otero M. (1994) "Financial Services and Microenterprises : Principles and Institutions" in Otero, Maria and Rhyne Elisabeth (eds) *The New World of Microenterprise Finance: Building Healthy Financial Institutions for the Poor*. West Hartford: Kumarian Press.

Westergaard K. (1982) "Pauperization and Rural Women in Bangladesh : A Case Study." CDR Project Paper A.82.5.

World Bank (1989) *Sub Saharan Africa: From Crisis to Sustainable Development*. Washington: The World Bank.

11

Voicing a Peasant Alternative: The Organisation of Rural Associations for Progress (ORAP) in Zimbabwe

Arnold E. Sibanda

Introduction

In the context of political authoritarianism and repression, economic deprivation and poverty, the erosion of basic services and injection of moral decay, peasant organisations have increased their influence in the public sphere. This chapter examines the evolving relationship between democratisation and development through an attempt to answer the question of whether there is a peasant alternative to contemporary development strategies. The research is based upon case studies of the peasant associations under the Organisation of Rural Associations for Progress (ORAP) in Zimbabwe. The main issue explored is whether the emerging African development paradigm, which expects civil society empowerment to lead to social and economic development, is based on social forces that have capacities for assuming such a leadership role.

The state in Africa was seen in the past as the appropriate institution to lead development work, although its role has become pervasive in the lives of citizens and has not socially uplifted the majority. Because of the failure of the state to achieve national development the focus has shifted to other forces with calls for the state to 'roll back'. The African middle class has failed to develop into a national bourgeoisie capable of revolutionising the productive forces. The framework in which it operates only facilitated their incapacitation. Lack of capital, credit and markets, the imposition of debilitating taxes and corruption among state functionaries all rendered the African 'bourgeoisie' 'a pale shadow of its nineteenth century Victorian counterpart' (Rahmato 1991: 8). The African working class or proletariat continues to be a weak social force with a numerical inferiority and deficient organic structure. The uneven and incomplete development of capitalism has retarded the development of industry, thus leaving the 'working class' as a semi-proletariat with one foot in the urban and another in the rural sectors. As a result, the only hope for working class intervention is in an alliance with poorer sections of the peasantry

As far as the peasantry in Africa is concerned, one recurring problem is the treatment of its differentiation in a cursory and non-strategic manner. Peasants are seen as 'the poor, the underprivileged and the excluded' (Ibid.). The fact is that not all peasants are as such, and it is therefore strategically imperative to demarcate the social composition of peasant organisations in order to understand their cleavage and dynamism, their potentials and constraints. Bratton (1990) in a study of farmers' organisations in Zimbabwe concluded that the social composition of an organisation's clientele affects the nature of its advocacy. In most circumstances it is questionable whether the organisation serves 'the poorest of the poor' or some other better-off group(s). In Africa organisations representing the poorest populations tend to waver over whether to concentrate on policy advocacy or on the implementation of technical projects.

Peasant Organisations and Participation

There are numerous unanswered questions with regard to Bratton's conceptualisation of peasant organisations and what is required of them if they are to meet the challenges of the new millennium. The merits of Bratton's contribution lie in improving our understanding of how the social differentiation among the peasants presents problems for peasant organisations, and how this affects their advocacy for appropriate development policy. The demerit of his analytical framework is the fatalistic assumption that unless movements are 'clean' class movements and unless they represent the poorest sections, there is no way forward for peasant social movements. However, this alleged 'messianic' approach to the peasant organisations is not empirically rooted because in reality most peasant families are resource-poor and weak in the face of the state and other social forces operating in the rural sector. This weakness of peasant households can render it impractical to decentralise power to one pure class of the rural household.

The study of peasant organisations tends to be fuelled in the dominant development literature by the question of promoting popular participation. Popular participation was seen as an essential element in sustainable development, and organisations preaching participation were seen as relevant agencies for social reform and instruments of peasant empowerment. Therefore, the involvement of rural or community-based organisations in the identification, planning and implementation of projects would enhance chances of their success (Katerere et al., 1993). The expected success of popular organisations is based on the fact that when members form and manage them, there is greater commitment to them. The major question for this study therefore was whether the experiences of peasant organisations under ORAP could be said to represent a peasant alternative to hitherto existing forms of top-down strategies of development. Does the ORAP experience represent a pointer to possible viable popular initiatives and participation in development?

Dietz & Moyo (1991) studied the structure of ORAP and found that it was a top-heavy structure because of its preoccupation with technical and financial management efficiency. At the same time the ORAP structures retained a strong gender dominance and representation of the poor. In examining the role of ORAP as a non-government organisation (NGO) in shaping popular intervention in Zimbabwe's rural development, various studies found ORAP to be a pioneer in developmental work. NGOs in Zimbabwe pre-date independence and were founded more on the basis of welfarism rather than development—as they provided palliatives to the dispossessed and oppressed (Moyo 1990). At independence in 1980 and beyond they were initially concerned with drought relief programmes but changed later to emphasise development and self-reliance. They were expected by the new government to complement state programmes, and the state was very open towards relief programmes but uneasy about independent rural development initiatives. Therefore NGOs were to perform a 'gap-filling task' (see De Graaf, Moyo & Dietz 1992; Katerere et al. 1993; Riddell 1990; Muir 1992; Dekker 1994). ORAP was an exception to this model.

The major conclusions from the studies above is that there is a large variety of NGOs in Zimbabwe, many of them aiming their programmes at the rural poor. But most of the time NGOs were unable to reach the poorest of the rural poor and generally had scant knowledge of the diversity of their target group. NGOs emphasised social and non-economic activities and practised few macro and business economic analyses. They tended to concentrate their activities around major cities, particularly the capital city, Harare. Vahlhans (1994) claimed there was no evidence that NGOs reached the poorest of the poor, nor that they were more flexible and experimental in their approaches than government agencies, nor that the poor participated more. It was only in a minority of cases where NGO project benefits outweighed costs of intervention (Ibid.).

NGOs in Zimbabwe have been isolated from each other and from centres of power and authority. 'NGOs in Zimbabwe seem to have evolved as a rather special sector, nurtured with large amounts of donor funds, more or less accepted by the governement and at a friendly distance of politics' (de Graaf et al. 1991: 55). In most instances, the response of developing countries to the growth of NGO depends on the political orientation of the regime and the degree of existence of decentralisation and opposition within the state (Vahlhans 1994). In liberal democracies, such as in India, where corporate relations between the state and NGOs exist, the state reacts with demands for more co-operation. Yet in autocracies such as in Kenya, more control is the response. Ethiopia, Tanzania, Uganda and Zimbabwe are some of the regimes that are noted to have responded to NGO growth with more regulation, co-ordination, co-option and even dissolution.

In spite of intervention of NGOs, the effective participation of beneficiaries in development is crucial in the success of development work. To participate, people need to raise their level of consciousness and to form their own organisations. The rural poor need to become increasingly aware of the socio-economic reality around them, of the forces that keep them in poverty, and of the possibilities for bringing about change in their conditions through their own reflections and collective actions. This constitutes a process of self-transformation through which they grow and mature as human beings.

Mulwa (1988) identified effective leadership that shows high respect for people's ideas and experiences and the value of the human person as the most important attribute in the development of NGOs. The 'grassroots' need to be effectively motivated through in-depth awareness of the social forces militating against their socio-economic welfare and to be able to discover their worth in society and their potential to change undesirable living conditions affecting them. The grassroots have to organise themselves and cultivate their own leadership for effective participation (Ibid.). Grassroots organisations should ensure social homogeneity in membership mobilisation to avoid the possibility of domination by more powerful groups.

There is need for support to balance the interests of the rich and poor peasants in rural organisations. A legally constituted non-government agency could fulfil such a balancing role. The grassroots need to learn to move from the known to the unknown, from small projects manageable within their own limited knowledge and resources to bigger and more ambitious projects (Ibid.). There is a tendency for NGOs to create dependence on their bureaucratised expertise and professionalism among peasant organisations, and for this to tend to frustrate participatory development initiatives (Moyo 1991). Although the expertise of NGOs is capital for peasant organisations, there is a real and tangible need to transmit basic skills to the local personnel of peasant organisations to sustain the initiative (Ibid.). Training in human relations and awareness programmes are an integral part of the participatory process to build the necessary group cohesion and co-operation, thus overcoming underlying social differences (Mulwa 1988).

In this context there is an unresolved theoretical problem revolving around the conceptualisation of the peasantry in Africa. Mafeje (1996) suggested there is need to terminate the use of the peasant concept in sub-Saharan Africa precisely because he considered that no such category exists empirically in Africa, especially in its pure form as it historically occurred in the West or in the East. He also suggested it would be theoretically more sound to focus on the study of popular rural organisations rather than the conceptually stretched notion of the peasant organisation. It must be reported therefore that nowhere does a social analytic category exist in its purest form. This is precisely because of variations in historical experiences and specificities. Social concepts should have a heuristic value if they are to help to denote the major basis or expected basis of existence and social reproduction of a social category. In that context, not in its purest form, but at least as having a fundamental common denominator, the peasant concept must denote a social category that owes its primary existence and reproduction (even if deficient or insufficient a basis, but still its core basis) to working on the land in either cropping or pastoral work, or both. There may very well be a fusion of multiple survival strategies which define peasantries, but the denominator which binds them is their dependence on land.

The conceptual framework adopted here in understanding peasant organisations holds that: organisations are considered 'peasant' or peasant-oriented if they (a) seek to advance the interests of those whose common denominator for survival is attachment to the land; (b) are actually composed in their membership population of significant proportions of the social agents as defined in (a); or (c) are a combination of both (a) and (b). Thus they need not be organisations *of* peasants but they definitely must be organisations *for* peasants. We will take the term peasant organisations to be a useful conceptual shorthand.

The study sought to find how organisations such as ORAP are capable of permitting and stimulating the articulation of a peasant alternative to the capitalist mode of development. This was achieved through examining how they organised the rural people and how they were unified by providing the support which allows rural people to organise work, take back control over their own lives, find their own voice and thus provide an alternative voice.

Study Approach

The research sought to identify such a peasant alternative voice, how it is articulated and what role ORAP members play in its articulation. The study explored whether the members of the peasant organisations understand this articulation. Specific questions raised included: What are their expectations and aspirations? Are these expectations fulfilled or not, and if not why not? What are the development constraints and prospects? How is the peasant alternative being articulated at the political, economic and cultural levels? The crux of the matter was to discover whether the ORAP experience does pose a possibility for a popular-participatory and democratic initiative for development.

The first approach used was a review of documentary research concentrating on written material on ORAP, its origins, objectives, structure and activities through a study of annual and evaluation reports at ORAP head office in Bulawayo. The second major methodological element was field interviews and observation covering agents and processes at the head office of ORAP, field workers and selected members from the organisation's various groups reflecting ORAP structures.

The individual membership of ORAP was found to be around 300,000 in 1991, with some 4,000 family units in 1990 in 1,000 local peasant groups (Dietz & Moyo 1991; Halimana 1992) (see Table 1 in the appendix). ORAP is organised as a multi-tier administrative structure based on a decision-making process starting at the bottom with *family units*. The family unit consists of at least five families, while between 10 and 25 families form a *group*. Five or more groups elect an *umbrella*, and up to five or more *umbrellas* elect an *association*. Each association elects four members (two men and two women) to the ORAP Advisory Board. Each association tends spatially to cover a single administrative district, while in many cases umbrellas tend to cover a single ward. Family units tend to coincide with single villages (Dietz & Moyo).

ORAP currently operates in three provinces, Matebeleland South and North and Midlands. The economic terrain of ORAP is located in Zimbabwe's communal areas. This research covered two provinces, whose ecological precariousness influenced the structure of ORAP's development intervention: Matebeleland South and Midlands. However, a major determinant in the selection of the Midlands province was that during the political crisis in the 1980s, where the army confronted 'dissidents', ORAP activities were banned by government officials in this province. It was therefore crucial to include this province in the sample in order to assess the process there in a comparative manner. Two districts were selected, one in each province. The selection of the districts was done after a pilot study in the field covering the two provinces.

The Filabusi Association was chosen in the Inzisa district of Matabeleland South and the Vusisizwe Association in the Kwekwe district of the Midlands province.

The rationale for selecting the two districts was based on the fact that there were two ORAP associations in Matabeleland South in which the movement was operating strongly. The first, Umzingwane, was relatively nearer the urban centre of Bulawayo and with relatively better infrastructure than the second, Filabusi. The movement started in Filabusi in the mid-1980s but had a longer history in the Midlands where political authorities later harnessed its progress. Besides, in Filabusi, the movement had progressed to include the resettled population on former white commercial farmland. Vusisizwe in the Kwekwe district was sufficiently distanced from town but not isolated. It had reasonably good infrastructure with Kwekwe City in proximity and two growth points, Zhombe and Silobela, in its midst. A comparison between the two associations would therefore be much more useful than if the situations were too similar.

A group-based sampling framework was used because of the large numerical strength of the organisation's membership at family unit level. Sampling above the group tended to neutralise the grassroots approach of the project. It was originally intended that 5 per cent of the total number of groups in each district would be covered. However, problems arose during the pilot survey as we discovered that no data existed to provide a sampling frame. Interviews with the agents at the organisations' offices established that no-one really knew how many groups existed in each district as no records existed. In fact, no up-to-date data on the membership was gathered and stored. The data on record, established through the estimates of researchers, reflected the status only up to 1990.

We opted to use the 1990 database for a sampling frame for the groups. Since it was estimated that there were 1,000 groups and 16 associations in 1990, that would mean on average about 60-62 groups in each association. Granted that there may have been phenomenal growth and perhaps some decline in places, a much higher percentage sample than 5 per cent would be safer. We therefore elected to cover 50 groups in each district (that is, about 83 per cent of the estimated 1990 average number of groups in each association). A third methodological procedure, the participatory rural appraisal (PRA) strategy was also used. This involved having the group members appraise the organisation and their own situation themselves. One hundred questionnaires were distributed, 50 in each district or association. However, this approach was adopted after interviews with head-office staff, field workers, village mobilisers and some group members to ascertain capacity and ability to perform the PRA.

One important limitation of this study is that only a few groups were actually visited and observed at their activities, be they income-generating projects or meetings. This was owing to constraints of time and resources. The road infrastructure, especially in Filabusi, is extremely bad and several appointments with groups could not be met following vehicle breakdowns. Spot visits were very important to assess the life situation of members. Research workshops with some groups or the leaders were also not possible owing to costs and time constraints. Some NGOs did not seem to be willing to talk about the work of others. Although there was a denial of competition, there did seem to exist an overlap which could be avoided by more NGO co-ordination. Little information was gained on how other NGOs in the operational areas co-operated with ORAP except from two who are also donors.

The ORAP Alternative Development Philosophy

In the Filabusi Association, the bulk of the groups were formed in the 1991-95 period. This, incidentally, was the period of the implementation of the first phase of the Structural Adjustment Programme (SAP) in Zimbabwe. At the same time, it was in the 1991-92 period that Zimbabwe suffered the most devastating drought in years, followed by yet another drought in 1994. It could be interpreted that the drought issue and the negative effects of the SAP programme accelerated the pace of formation of peasant groups during this period. One group suggested that 'many people had heard of ORAP before but were lazy to join. But when the drought came and ORAP offered drought relief food, many wanted to join'. The above observation may be true also for Vusisizwe in the Midlands, as Table 2 shows that the situation there also peaked in the same period.

Through the implementation of SAP the state started 'rolling back' some of its activities. This mainly affected the rural sector. ORAP intervened not only with drought relief but in some places by promoting the construction of clinics, schools and skills training centres to cushion the impacts of SAP. Economic reforms were apparently incompatible with an unliberalised political scene. All the groups interviewed indicated they had started their groups to share ideas on development and to start development projects to improve their lives and the future.

ORAP strongly believes that if meaningful development is to take place, rural people have to be masters of their own destiny and, therefore, they need to reverse the 'top-down' approach to development through their own 'bottom-up' strategies. Equally important is the need to search for a development paradigm which is rooted in the people's culture as they live it.

ORAP went through an extensive cultural analysis and came out with some shared values within the communities. These values form the basis of ORAP's philosophy which revolves around the 4*Zs* and a *Q* (in the Ndebele language): *ziqoqe*—self mobilisation; *zenzele*—do it yourself/ fend for yourself; *ziqhatshe*—create self employment; zimele—self-reliance (based on interdependence)/ be independent; *qogelela*—create savings (NKomo 1997: 79). Most of the groups were named after these four Zs.

The above was also emphasised by the president and founder member of ORAP, Mrs Sithembiso Nyoni, who said her vision in starting the movement was to alleviate poverty in Zimbabwe. She saw poverty integrated in four dimensions: cultural, psychological, social and economic. ORAP attacked all these four dimensions in its integrated approach to development. People had to rediscover themselves, assert themselves as liberated individuals and regain their human qualities, which they had lost under the impact of colonialism. The philosophy had been built on the African language and culture, said the ORAP president. The development philosophy, dubbed the ORAP Family Approach, was also said to be rooted in the African culture.

There is a very strong internalisation of the philosophy by members. First, this was gauged from the groups' discourses on development and their hopes. This was said to be found in the names given to the groups by the members themselves. The Chavunduka study pointed out this fact and cited a number of cases of groups' names which indicated elements of the ORAP development approach. These were verified by this study. All but two of the groups responding had vernacular names which had a meaning having to do with the philosophy of the four Zs and a Q. The two odd ones had English names such as 'Precious Melodies' (a singing group of formerly unemployed rural youths who also now engaged in several handicraft production groups apart from agriculture), and 'Dambridge' (a women's sewing group).

Secondly, understanding the ORAP philosophy was gauged from the groups' understanding or definition of the four Zs and the Q concepts (see Table 3). The interview question assessed whether the group members' definitions and understandings of the concepts coincided or were at variance with the official ones. Clearly, a majority of the groups seemed to understand the ORAP philosophy as represented in these concepts. However, fewer groups seemed to understand what *ziqhatshe* or self-employment meant, or the definition seemed to be at variance with the official one. This was also true of *zimele* or self-reliance.

A significant number of the membership also had problems with what *qogelela* or 'create savings' meant. Less understanding of 'self-employment' and 'self-reliance' was found in the Filabusi Association, while less understanding of 'create savings' was found in the Vusisizwe Association.

The groups were also asked to indicate what important things they gained from ORAP, and the results are summarised in Table 4.

Obviously, the philosophy of development was recognised as the overall leading gain. However, it was the most significant in Filabusi (Matabeleland South) and less significant in Vusisizwe (Midlands). Material support was recognised as the leading gain in Vusisizwe. All the dissatisfied groups that claimed to have gained nothing from ORAP meant this in relation to their not receiving material support. They too acknowledged that good ideas or philosophy were imparted to them, but said it was 'useless when words are not translated into deeds'.

This criticism was acknowledged by ORAP official agents. Although most of the groups indicated they had realised positive benefits from ORAP, its president was of the opinion that those who felt dissatisfied should not be ignored. Some of the members, however, made unfair comparisons between ORAP and other NGOs they were involved with, because these gave material handouts. ORAP leaders were expected to investigate complaints genuinely but not to be forced into competition with other NGOs.

The ORAP leadership seemed to be caught up in this crisis of expectations, and sought to make a distinction between ORAP as membership and movement, which mobilised for development initiatives, and ORAP as implementer, which had to be practical and faced resource constraints in the same way as other NGOs did. Whether ORAP was split according to this distinction reflecting idealism and reality, most groups felt this perception represented academic escapism. In fact, one group declared: 'ORAP is led by crooks who go around asking donors for money in our names. They use the money to enrich their businesses. If only they practised what they preached.' There was strong support for the ORAP philosophy. However, this is threatened if it is not substantively translated into action.

Perspective on ORAP by Group Beneficiaries

'ORAP is different from other groups: it teaches respect for every moving being. ORAP doesn't believe in one person, it believes in persons.'
'Before coming to ORAP, I didn't know how development started. Now I know that before development, there must be thoughts in mind.'
'Money is available, the important thing is to develop ideas.'
'With ORAP we have learned to work together. Things can move very quickly.'
'ORAP has given us a lot of knowledge and experience, and a full understanding of how rural people can uplift themselves through unity and working together.'

Source: Chavunduka et al. (1985: 4-5)

Yet ORAP was perceived rather positively by most of the peasant groups. One of the reasons given for joining ORAP by the groups was the help needed to start development projects. The study investigated what projects had been established, how they were performing, and what problems were faced by the projects. The groups were involved in a wide variety of projects.

ORAP's main activity areas were service projects (water and sanitation facilities, food storage, grinding mills and cattle raising); training activities (courses and on-the-spot assistance); drought relief programmes (transport of food from surplus to deficit areas); development education activities (seminars, discussions, courses and exchanges); and income-generating projects or micro-projects such as sewing, poultry-keeping, gardening, co-operative endeavours and carpentry/woodwork All these activities were geared to reducing poverty and to develop a long-term strategy to minimise dependence.

Service projects were seen as important to build the requisite infrastructure, while income-generating projects or micro-projects were regarded as transitory or vehicles for real development. ORAP's view was that micro-projects alone were inadequate as development tools. That is why income-generating projects were tied to the overall approach, which included relief programmes and skills development through training courses and development education. ORAP also sought to revive and develop traditional skills, knowledge and practices in areas such as agronomy, co-operative effort and food storage.

These micro-projects by the groups were distinct from ORAP's commercial undertakings. These included hardware (wholesale and retail) outlets, a fuel filling station and a large-scale commercial farm in its areas of operation, and other strong commercial undertakings beyond the three provinces where ORAP was currently mobilising. These undertakings by ORAP as an institution were intended to generate alternate funding for the association, so as to minimise and eventually end its dependence on donor-funding. Some of the earlier investments by ORAP in its operational areas were criticised for having had a 'welfarist' rather than a 'commercial' approach. The idea was now to turn these into real commercial ventures. The ORAP group members tended to consider the progress of their micro-projects as successful (Table 5).

About 71 per cent of the groups said they were making fair progress with their micro-projects, while 24 per cent said progress was good and only 3 per cent thought it was bad. Fewer of the groups seemed to be making good progress in Vusisizwe (Midlands) than in Filabusi (Matabeleland South), although the general trend in assessment was generally the same in both areas. The micro-projects were regarded as important sources of supplementary income by the peasant families interviewed.

Gender Relations in ORAP

ORAP believes that its family approach recognises that women are major contributors to the productive life of the community as well as being the heart of social development and change. According to Nkomo (1997), the approach does not single out a woman, but recognises her within her family as it faces social, economic and political challenges collectively. Mrs /Nkomo, the executive co-ordinator, felt that ORAP was sometimes criticised by feminist groups and some NGOs for not addressing the women's issue separately. She suggested that ORAP was not feminist as such but that the organisation had a gender-sensitive and holistic approach. Dialogue was considered an important part of creating family-oriented gender relationships. ORAP thus believed that collective family work makes it easier to implement development activities. At meeting places of the ORAP groups, both men and women carry out development dialogue. In fact, of all the groups who responded to our survey, only 6.7 per cent were not of mixed sex. The groups met together, decided on projects to undertake and assigned roles to the sexes. 'This methodology', suggested the executive co-ordinator, 'empowers women by encouraging them to participate in the decision-making process, and this gives women confidence to be in control of their situations' (Ibid.).

Indeed, many women were found to have taken up leadership positions in most of the groups. ORAP structures had given women avenues of social participation, economic empowerment and decision-making on ORAP's development agendas. Women in the Vusisizwe Association dominate the elected leadership positions of the groups, in terms of gross numbers. although they hold mostly the secretaryís post. The family approach had also impacted positively on males, by 'opening up their minds' (as one of the groups is named) to see gender as a tool to balance sex roles in order to function cohesively in society.

The leading role of women was also found in their predominance in full-time employment posts. Field co-ordinators in ORAP are key employees because of their role as village mobilisers. In a survey of village mobilisers conducted by the ORAP Mobilisation Unit in March 1997, out of the 106 mobilisers who responded to the survey, 33 per cent were male and 64 per cent female. Thus if about 65 per cent of the 120 village mobilisers (the Abaqoqi) in ORAP were women, they dominated the leadership of ORAP when all its levels of leadership positions are taken into account.

ORAP, Peasant Stratification and State Relations

This research sought to ascertain what ORAP's family approach meant in the practical reality of the association's management processes. This was done not only by gauging the perceptions expressed by group members but by an empirical examination of the social profile of the association's village mobilisers. The evidence showed that 85 per cent of the mobilisers were married and 5 per cent were single, while 3 per cent and 7 per cent were divorced and widowed respectively. Among the women group members, 79 per cent were married, while 6 per cent, 4 per cent and 11 per cent were single, divorced and widowed respectively. Among males, there was only one single man while the rest were married. This would suggest that in keeping with its philosophy, ORAP focused on the family. From analysis of the ORAP Mobilisation Unit's survey data, it emerged that ORAP was an association led by not just a rural but also an urban-based 'elite'. In fact, the executive co-ordinator brought this out during an interview:

> I was questioned by the people as to how I came to be their leader when I am actually an urban dweller. This was a valid criticism. I have a rural home but I really never cared much about making it a home. Now I am building a comfortable home in Gokwe rural area to give an example to the people and my children.

Thus, the urban elite was ruralising itself in order to retain its leadership position in a peasant-oriented movement. But even then, strictly speaking, the ruralisation could only be partial as the urban area would continue to be the elite's main dwelling place. The rurally based ORAP association members, however, seemed not to care much as long as there was evidence of a strong rural link between them and the leaders. In fact, none of the group members questioned the socio-economic and physical location of the leadership of ORAP.

The evidence from the village mobilisers' survey shows that the leadership, even in the rural setting, is indeed a kind of 'elite'. It holds leadership positions in several social spheres at the same time as being ORAP leaders. This connects ORAP organically with the social and political sphere and this makes the movement a unique 'peasant organisation'. Up to 77 per cent of the village mobilisers were active in political roles as leaders in other structures. Of these, 40 per cent were leaders and 60 per cent were active members of such structures. About 86 per cent were active in church roles, and of these 30 per cent were leaders and 70 per cent active members. Approximately 49 per cent were active in farming community organisations, with 35 per cent of these as leaders and 65 per cent as active members. Some 46 per cent of the members had businesses such as dealer's shops or co-operative businesses.

Only 40 per cent were active in other social community roles (such as being Village Development Committee chairpersons or secretaries, School Development Committee members, chairpersons or secretaries), but 98 per cent of them were leaders in such structures.

Analysis of this data by sex shows that 77 per cent of the women played active roles in politics, and 40 per cent of them were leaders of such structures. As many as 93 per cent were active in church roles, and 32 per cent of them as leaders, while 45 per cent were active in the farming community structures, and 34 per cent of these as leaders. About 51 per cent were involved in business undertakings, and 37 per cent had other community roles. Out of the male members, 77 per cent were active in politics, and 41 per cent of them as leaders. About 71 per cent had church structure membership, with 24 per cent of these as leaders, while 57 per cent were active in the farming community, and 35 per cent as leaders. Over 37 per cent of the men had business undertakings, while 46 per cent had other community roles, 94 per cent of these as leaders.

Thus the ORAP leadership is indeed a social and economic elite. For example, the board secretary was a local political commissar of the ruling party, she was a village mobiliser for ORAP, a secretary of the Village Development Committee and a chairperson of the Phikelela group project on an irrigation scheme. The board chairman was an elder in the Seventh-Day Adventist Church, a proprietor of a general dealer's shop, a committee member of the Zimbabwe Farmers' Union, chairman of the School Development Committee, and a member of the central committee of the ruling party! These are but two examples of a fairly prevalent situation. Strictly speaking, such a social profile of the leadership, together with the president and founder member of the organisation, who is now a minister in government, would lead to questions of whether the organisation was a 'peasant' one. Yet the organisation still remains founded on a strong peasant base and articulates in its discourse a development model that calls for peasants' involvement and initiative. However, because of its precarious straddling of the rural and urban spheres, the organisation has also penetrated the urban sector through income-generating projects in Bulawayo, particularly in the parliamentary constituency of its president and founder member.

The association is obviously aiming not just at the immediate improvement of the status quo of the peasantry but at a wider future developmental project, and seeks to carry the involvement of the rural masses with it. The ambitious programme of the organisation has taken the peasantry beyond its usually confined world. Using one of the pillars of its philosophy — the creation of savings—and the family approach, the organisation has encouraged its associations to contribute to a fund to start a Community Foundation for the Western Region of Zimbabwe.

However, ORAP had surprisingly avoided confronting some fundamental problems including land distribution inequalities characteristic of Zimbabwe. Almost 41 per cent of the 39,6 million hectares of the good and fertile lands of the country is in the hands of 4,500 white colonial settlers, while 60 to70 per cent of the indigenous population crowd themselves on barren and degraded communal lands (Moyo 1995). Land redistribution has progressed very slowly, and land reform remains a challenging issue in Zimbabwe. This issue had not been at the forefront of ORAP's project, and its social base had taken it to task for lack of a meaningful programme on land reform.

Findings from our own survey suggest that group members indeed wanted the organisation to take up the land issue. Group members indicated that a number of them were good or trained farmers, and there was a great willingness to train or join what is called master farmer training. Therefore the respondent believed that they could be good users of redistributed land, given that none of the ORAP group members were landed as such and all subsisted in highly degraded communal areas. ORAP itself had only one commercial farm and supported groups with projects on government resettlement schemes. ORAP members thus turned to government, other NGOs and donors for additional assistance for development work (Table 7).

It can be seen that 81 per cent of the groups acknowledged co-operating with or receiving assistance from state or government agencies, while 55 per cent did the same with other NGOs. Only 7 per cent and 17 per cent had no such assistance from either state or civil agencies respectively. As many groups received assistance from state as from civil agencies in Vusisizwe (Midlands), while fewer had state assistance in Filabusi (Matabeleland South). We conclude that there was significant co-operation with state agencies among ORAP group members, while co-operation with other civil society agencies was low, except in the Midlands province. In both provinces the state agencies that recurred in the co-operation with ORAP members were those of the District Development Fund (DDF), the Ministry of Lands and Agriculture's extension officers, the Ministry of National Affairs' Village Health Workers and the Ministry of Local Government and National Housing. Other NGOs engaged by ORAP members included World Vision International, Plan International and the Self-Help Development Foundation.

ORAP members were also asked to indicate whether there were any impediments put on their development endeavours by the state and other agencies (Table 8).

The evidence shows that only 22 per cent of the respondent groups felt there were some impediments from government agencies and 5 per cent from other agencies, while 73 per cent encountered no such impediments. Those who quoted some hindrances cited such issues as no provisions for title deeds in the smallholder communal areas, and bureaucratic delays in processing papers where projects had to be sanctioned by the state, such as in small mining operations. State controls or proscriptions on some projects that could be undertaken by NGOs, the high costs of irrigation water and land (rents) presumably owing to poor policies and the scarcity of land, and absence of water rights through limited policy reforms were also instanced. Other agencies were accused of being hostile to groups that participated with ORAP or other NGOs to the extent of actually pulling out of/or shunning areas in which other NGOs operated. Some agencies refused to advance credit facilities to ORAP group members on the pretext that they would be unable to pay back when in fact the reason was reluctance to work with groups which collaborated with other NGOs.

The ORAP groups criticised these NGOs and some government agencies for taking advantage of them, because of the 'higher culture'. Overall, however, the greater majority of the respondent groups felt there were no impediments to their operations. The state was hailed for being supportive of their development efforts, a sentiment which was also expressed by the ORAP membership from the top to the bottom. Nevertheless, the respondents were asked to indicate what they thought how the state and other civil agencies should assist them. The results which emerged (Table 9) show that the biggest requirement of the groups from both state and other civil agencies was financial loans and supportive infrastructure. The second leading requirement was advice and ideas on development. Development training and skills development seemed to exist already to a significant extent, but the generation of ideas and innovativeness was said to be more important. It is interesting that ORAP groups have seen the importance of 'discourse' on the idea of development. The significance of this is emphasised by the ORAP leadership as well.

The overwhelming majority of the respondents had seen their lives change for the better since joining ORAP. They no longer lived in poverty, had better homes and looked forward to a better future. Significantly, non-members admired their efforts, and many wanted to join. Also, a significant proportion of the association members said they were now more collectively involved in development activity and participated more than before in decision-making and implementation of development activities. An almost equal proportion felt they now understood better what development meant, had gained some practical skills and learned to respect and love one another. This was a product of ORAP-organised training programmes for its members and staff both at their local centres and at the association's headquarters.

Each association has a local development centre at the district level, although Filabusi had yet to build its centre. Such centres are meant to create a new form of integration with the local community. Apart from its social role as a meeting place, ORAP training programmes in literary, practical, technical and vocational skills are undertaken there. Out of the 106 OVMs (ORAP Village Mobilisers) interviewed by ORAP, 84 per cent had gone through some training at the ORAP local development centres and 18 per cent had received training at the head office in Bulawayo at the ORAP Zenzele (Do-it-yourself) College. This college was established in 1991 and, together with the World Learning School for International Training (SIT) in the USA, runs training for a Global Partnership's diploma programme in Grassroots Development and NGO Management for foreign and local NGO staff. ORAP's Zenzele graduates can proceed for bachelors and masters degree programmes in the USA.

Apart from such higher-level training programmes, ORAP has a primary school in one of its rural operational areas and some pre-school facilities The association also arranges exchange programmes with other countries in the region and abroad. The Mobilisation Unit's survey indicated that 15 per cent of the OVMs had gone on such exchanges to countries such as Botswana, South Africa, Sierra Leone, Zambia, Namibia, USA, Canada and Bangladesh. Organisations with whom ORAP had arranged such visits included CUSO and Oxfam (Canada), Ford Foundation (USA), Philisisizwe (SA), Village Development Network (Zambia), Highlanders International (USA) and Kagiso Trust (SA). Such exposure of rural people definitely opened up their minds and helped them understand the 'problems' of development, as one OVM stated:

> I now understand the problems of development and I can see that we are lagging behind. Development requires commitment and extremely hard work. It first and foremost requires a revolution in the mind, to convince oneself that something is possible. Our people must stop thinking that things should be done for them. I saw people working hard and achieving something in Bangladesh. (Mrs G. Khumalo, a fieldworker for Vusisizwe Association in the Midlands Province, just after his return from Bangladesh.)

One problem that emerged strongly in this context, however, was that the OVMs were performing difficult mobilisation tasks only on the basis of commitment, without any remuneration or allowances for their services. Those who scorned ORAP asserted that the village mobilisers were 'slaves of ORAP', especially when they compared these mobilisers with the Government Village Community Workers (VCWs), who are paid. The tasks performed were similar but VCWs were paid and had bicycles to move from one village to the next. The OVMs walked and were unpaid, yet ORAP fieldworkers (OFWs) had motorcycles. They covered the whole district or association, but the work was really done by the OVMs through whom the

OFWs worked. The OVMs felt that they should get some allowances and bicycles.

The overwhelming majority of respondents saw a bright future for ORAP (Table 10). They envisaged the association as contributing to national development, ending poverty and bringing about a better standard of living for rural people. They also saw ORAP as continuing to equip people with skills and achieving self-reliance. About 12 per cent saw the future as fair, but this was conditional upon ORAP strengthening its own business ventures so as not to rely on donor funding. Many of the association felt that ORAP needed to live up to its word to gain and retain the support of the people.

a) National development. End of poverty. Good living will be achieved. Self-reliance and training in skills.
b) ORAP needs to strengthen its own business to be self-reliant. Must fulfil promises. Needs loans for projects to succeed. Future will be bright if there will be no oppression.
c) So far there is no help from ORAP. ORAP has a future only if it can help people end poverty.

It was felt that the future would be bright if ORAP did not institutionalise but ended the oppression of one by another. Only 6 per cent of the members were unclear about its future, because they either had not received any help from the body beyond the idea of being mobilised collectively and were therefore demoralised, or were still hoping for help to come. These people were still living in poverty, had no projects under way or had seen their projects collapse through lack of assistance, and therefore did not envisage a clear future for themselves with ORAP.

Conclusion

ORAP is an organisation of mainly rural people whose major membership base is the small peasant farmer. It is not an organisation led solely by peasants but has some peasants in its leadership. It was not started spontaneously by peasants but by an elite that had identified the peasant's potential to control their own destiny and committed itself to the development of this potential. The organisation has been strangely silent on the land question, but its grassroots base is posing this question. The silence is itself puzzling given that none in the leadership or founder-membership appears to belong to the privileged class of large-scale or small-scale commercial farmers. This has led some to criticise the organisation as seeking to create survivalist strategies for the landless poor who are mobilised into income-generating projects and thus to preserve the status quo rather than pursuing a radical programme for the transformation of society. Rather, ORAP seeks to change the immediate conditions of existence and mobilise peasants for such a change by promoting social development through altering individual attitudes, thinking and co-operation strategies. ORAP believes that changes in the immediate conditions are a strong basis for long-term future change and thus for development.

The peasant members of ORAP reside in poverty, cultural backwardness and marginalisation politically, economically, socially and psychologically. ORAP has tackled this deprivation by mobilising the peasantry into a discourse on development, to become culturally assertive and to debunk their psychological isolation. Many have seen their lives change for the better economically and have moved out of poverty. But ORAP itself has moved beyond the peasant world, through programmes for skills development and higher education, enterprise and self-reliance projects, international linkages and infrastructure development. The Common Fund for the Community Foundation for the Western Region of Zimbabwe demonstrates a vision of development beyond the narrow peasant world. Sustaining this mobilisation does not seem to depend on just the quality of leadership and material contributions of the people, but also, and more importantly, on the sustained and tangible material gains members receive. The ORAP leadership is optimistic that such gains will help retain the grassroots membership.

The leadership question and gender issue are important as ORAP seems to believe that leaders will emerge from its membership. However certain, the trend is that those ascending to leadership positions at all levels are elites, relatively speaking. In some cases bureaucratic rather than popular democratic choices influenced ORAP's leadership development, as shown by recent efforts to introduce what they called qualified personnel in field worker posts. People with university degrees were recruited not necessarily from within their operational areas, nor were they chosen by the people as before. This pattern was causing disquiet in the membership ranks as it broke the ORAP philosophy and thrust.

Nevertheless ORAP has made a contribution to the articulation of the peasant voice by mobilising the peasants to consciously address their development problems, and making an invaluable contribution to the discourse on development, based upon local culture, language and potential. It has reacted to the gender issue by a unique family approach without fracturing the 'unity of the sexes', and has sought to uplift the role of women who now play strong leadership roles. The future of ORAP lies in its putting maximum material effect to the leading tenets of its philosophy. Some of this has been achieved. In an important way, ORAP is a peasant-based organisation that articulates the peasant alternative, but also goes beyond the peasant alternative.

References

Akwabi-Ameyaw, K. (1990) 'The political economy of agricultural resettlement and rural development in Zimbabwe: The performance of family farms and producer co-operatives', *Human Organization* 149, 4.

Auret, D. (1990) *Evaluation of ORAP's Emergency Food Programme as a Response to the Drought, 1985-1990*, Oxfam, Zimbabwe Office.

Badu, Y. A. and Parker, A. (1994) 'The role of non-governmental organizations in rural development: The case of the Voluntary Workcamps Association of Ghana', *Journal of Social Development in Africa* 9: 27-39.

Bratton, M. (1986) 'Farmers' organizations and food production in Zimbabwe', *World Development* 14, 3: 367-84.

Bratton, M. (1987) 'Drought, food and the social organization of small farmers in Zimbabwe', in M. Glantz (ed.), *Drought and Hunger in Africa*, Cambridge, Cambridge University Press.

Bratton, M. (1990) 'Non-governmental organizations in Africa: Can they influence policy?', *Development and Change* 21, 1: 87-118.

Chavunduka, D. M. et al. (1985) Khuluma Usenza: The Story of the Organization of Rural Associations for Progress (ORAP) Bulaway.

De Graaf, M., Moyo, S. and Dietz, T. (1992) 'NGOs in Zimbabwe', NGO Landenstudie Zimbabwe Impact Studies, Medefinancierings, Conferentie CERES, NRC/CIRAN, 15, 26 October, The Hague.

Dekker, M. (1994) 'NGOs in Zimbabwe: Developments and changes since 1990', Livelihood and Environment working paper, University of Amsterdam, Department of Human Geography.

Dietz, T. and Moyo, S (1991) Zimbabwe: ORAP, University of Amsterdam.

Halimana, M. C. (1992) Peoples' Participation: A Study of Selected Non-Governmental Organizations in Zimbabwe, UNICEF.

Katerere, Y., Moyo, S. and Ngobese, P. (1993) *Opportunities for NGO Involvement in SADC's Strategies on the Environment*, Harare, ZERO.

Kote-Nikoi, N. (1992) 'NGOs challenged to monitor government expenditures', *EDICESA News* 5, 8, Harare.

Mafeje, A. (1996) 'African "peasants", An elusive category', post-CODESRIA Multinational Working Group Methodology Workshop on Peasant Organisations in Africa, Dakar.

Moyo, S. (1990) 'NGOs in Zimbabwe: Context, role and relationship', working paper, Harare, ZERO, December.

Moyo (1995) The Land Question in Zimbabwe, Harare, SAPES Books.

Muir, S. (1992) 'Evaluating the impact of NGOs in rural poverty alleviation, Zimbabwe country study', ODI working paper no. 52.

Mulwa, F. W. (1988) 'Participation of the grassroots in rural development: Kenya', *Journal of Social Development in Africa* 3, 2: 49-65.

Nindi, B.C. (1990) 'Dilemmas of development in rural institutions in SSA', *Journal of Eastern African Research and Development* 49-59.

Nkomo, T. C. (1997) *The Family Approach as an Empowering Tool for Women in the Development Process: The ORAP Case*, Brattlebore, Vermont.

Rahmato, D. (1991) Peasant Organisations in Africa: Constraints and Potentials, Dakar, CODESRIA Working Paper 1/1991.

Riddell, R. (1990) 'Country case study: NGO project', London, ODI.

Vahlhans, M. (1994) 'The new popularity of NGOs: Will the enthusiasm last?', Development and Co-operation, no. 3.

ZERO (1991) Report on the Assessment of ORAP Projects in Umzingwani District, Harare.

Zinyama, L. (1992) 'Local farmer organizations and rural development in Zimbabwe', in D. Frazer and F. Mackenzie (eds.) Survival in Rural Africa, London, Routledge.

Appendix

Table 1: ORAP Structure and Growth (1983-90)

Tier	1983	1987	1990
Family units	0	800	4,000
Groups	300	600-	1,000
Umbrellas	34	800	200
Associations	2	60	16
Board	15	14	126
Staff	6	126	60

Table 2: Date of Establishment of Groups

Association	N.S.	1980-85	1986-90	1991-95	post '95	Total
Filabusi	-	-	6	37	2	45
%	-	-	13	82	5	100
Vusisizwe	1	11	9	20	3	44
%	2	25	20	45	8	100
Total	1	11	15	57	5	89
%	1	12	17	64	6	100

n. s. = not stated

Table 3: Understanding the ORAP Philosophy

Value	Filabusi		Vusisizwe		Both	
	Freq.	%	Freq.	%	Freq.	%
Ziqoqe	33	73	38	89	72	81
Zenzele	40	89	36	82	76	85
Ziqhatshe	28	62	31	70	59	60
Zimele	25	56	36	82	61	67
Qogelela	36	89	31	70	67	75
Total	45	100	44	100	89	100

Table 4: What ORAP Gave its Members

Association	Mobilisation	Idea, philosophy of development	Education (scholarships)	Material support, e.g. seed, fertiliser, cement, credit	Skills training	Nothing
Filabusi	64	96	4	20	7	-
Vusisizwe	61	66	7	70	9	9
Total	63	81	6	45	8	5

Table 5: Assessment of Progress of Income-generating Projects

Assessment	Filabusi		Vusizizwe		Both	
	Freq.	%	Freq.	%	Freq.	%
Not stated	2	4	-	-	2	2
Good	13	29	8	18	21	24
Fair	29	65	34	77	63	71
Bad	1	2	2	5	3	3
Total	45	100	44	100	89	100

Table 6: Group Leadership by Sex in Vusisizwe Association

Category	Male	Female	TOTAL
Chairperson	14	27	41
%	34	66	100
Secretaries	11	30	41
%	27	73	100
Treasurers	15	26	41
%	37	63	100
Committee members (3 each)	46	77	123

Note: Out of 44 groups, 41 indicated the sex of their leadership.

Table 7: ORAP Members' Additional Assistance from Other Agencies

Association	Get assistance from		No assistance from		Total
	State agencies	Other agencies	State	Other agencies	
Filabusi Freq.	37	14	1	6	45
%	82	31	2	13	100
Vusisizwe Freq.	35	35	5	9	44
%	80	80	11	20	100
Both Freq.	72	49	6	15	89
%	81	55	7	17	100

Table 8: Impediments on ORAP Groups' Operations from Other Agencies

Source		Filabusi	Vusisizwe	Total
Government	Freq.	8	12	20
	%	18	27	22
Other agencies	Freq.	2	2	4
	%	4	5	5
None	Freq.	35	30	65
	%	78	68	73
Total	Freq.	45	44	89
	%	100	100	100

Table 9: Expected Assistance from Agencies

Association		Government			Others			Total
		Infra/ loans	Ideas/ advice	Devt. training	Infra/ loans	Ideas/ advice	Devt. training	
Filabusi	Freq.	22	16	9	26	23	10	45
	%	49	36	20	58	51	22	100
Vusisizwe	Freq.	34	11	2	32	7	1	44
	%	77	25	5	73	16	2	100
Both	Freq.	56	37	11	58	30	11	89
	%	63	42	12	65	34	12	100

Note: Infra = infrastructure development;
Devt training = development training

Table 10: Perceptions of the Future of ORAP

Association		Perceptions			Total
		Bright[a]	Fair[b]	Future unclear; unknown[c]	
Filabusi	Freq.	42	2	1	45
	%	93	5	2	100
Vusisizwe	Freq.	31	9	4	44
	%	71	21	8	100
Both	Freq.	73	11	5	89
	%	82	12	6	100

12

Peasant Alliances, Economic Crises and Structural Adjustment in Nigeria

Mohammed-Bello Yunusa

Introduction

Economic crises during the Structural Adjustment Programme era have been a major feature of Nigeria's socio-economic and political scene for the last two decades of the twentieth century. The crises have impacted negatively on the welfare of peasants, and the same crises have influenced the dynamics of the peasantry. The chapter examines how adjustment programmes and institutions[1] nurtured and financed the formation and operation of peasant organisations, and promoted sustained development (socially, economically and politically) in rural Nigeria. The study attempts to link the Nigerian socio-economic crisis to the growth of peasant organisation membership by examining the level of popular participation in the development process and the effect of this on economic empowerment through socio-political movements. The chapter reveals the extent to which peasants, beset by the Structural Adjustment Programme (SAP), have organised to pursue their common interests, how they realise their set objectives and whether this enhances the participation of peasants in organs for democratisation in rural Nigeria. The effects of the changing economic structure on political mobilisation are thus highlighted.

Conceptual Framework and Scope of the Problem

Who are peasants? The *Oxford English Dictionary* defines a peasant as: 'One who lives in the country and works on land either as a small farmer or labourer ... one who relies for his subsistence mainly on the produce of his own labour and that of his household, and forms part of a larger culture and society in which he is subject to the political control of outside groups; also, loosely, a rural labourer'.

The applicability of this definition is limited given that peasants may or may not necessarily live on the land or labour on the farm, but may live off the production of goods and services that are not related to the land. The rural non-farm sector also provides alternative sources of subsistence to rural people. The growing trend in Africa is for peasants to survive from farm production and non-farm work. Ongoing debates on de-agrarianisation and rural employment stem from the changing rural economy and pattern of peasant livelihood (Bryceson 1993). Non-farm employment in rural areas consists mainly of work in the civil service and engagement in rural crafts. With declining agricultural productivity and increasing population pressures on land these enterprises provide alternative livelihoods.

Although the macro political and economic system influences the behaviour of peasants, within their own local conditions networks of social relationships and social groups determine individual roles, status and values (Thorner 1972). Peasant alliances and the relationships between rural subsistence producers and workers form the basis for pursuing the goal of improving their own lives. While peasant organisations vary in scope and complexity, they are brokers between state interests and local concerns (Chazan 1994). Their social relationships are rather elementary and groups are small (Fei 1973), and often based on definite goals and objectives, membership and leadership structure (Yunusa 1979). The aim of collectively responding to shared problems and interests, and therefore to co-ordinate their activities and interests, is paramount. Once formed, peasant organisations appear autonomous and organised and endowed with abilities to mobilise members' resources and implementation capacities (Bratton 1994). However, they retain definite relationships with the state, as they sometimes function with the support of and through governmental agencies (Famoriyo & Ajayi 1974), and in relation to macro policy changes.

In Nigeria the average contribution of agriculture to the gross domestic product declined to about 29 per cent[9] in 1995, while the implementation of the SAP reduced government expenditure on rural social services such as education, health and roads. This had devastating effects on rural living conditions, prompting peasant reactions that varied depending on the specific circumstances of their political organisations. Indeed, peasant organisations over time have played various roles in rural development in Nigeria, including in the provision of labour, inputs, markets, roads, health and education. In rural Nigeria peasant organisations such as age sets, credit associations, labour exchange groups and cultural groups (Oluwasanmi et al. 1966) are old-established institutions pre-dating colonialism and the modern state.

The larger peasant organisations have emerged to handle major crises and welfare support to individuals or to provide community-based physical facilities (Takaya 1988). For instance, a survey of five states of Nigeria revealed that peasant organisations were involved in about 51 per cent of local development projects such as building of schools, health centres, roads, bridges, markets and civic centres (Ukwu 1988). Furthermore, peasant organisations are involved in the provision of labour and credit for individual investment in social obligations, housing, small-scale trade and joint investment in petty businesses, among others (Oluwasanmi et al. 1966). Current development policies in Nigeria aim to bring peasants into mainstream development through governmental support of various peasant organisations.

The formation of peasant organisations and their symbiotic relationship with government pre-dated the SAP era. In Nigeria the federal and state governments have historically emphasised self-help projects to generate employment, income and develop rural infrastructure as a basis for community development (Ajaegbu 1976). To fulfil this objective various levels of government allocated funds to such projects through peasant organisations. For instance, in the 1981-85 development plan period the Kaduna state government allocated N6.73 million and disbursed N0.9 million by 1983 to peasant organisations to implement such projects (Kaduna State 1983). The state also used these organisations as channels for the distribution of farm inputs and essential commodities at the peak of scarcities and high prices in the 1980s.

In response, peasant organisations were formed to mobilise for development projects that they successfully executed. In 1978 some peasant organisations in the then Anambra state of Nigeria implemented projects worth N4.8 million (Okpala 1980). In 1989, a peasant organisation in Benue state implemented projects worth N0.5 million (Agbe 1989). The SAP led to the mobilisation of rural resources, provision of inputs, extension services and credit facilities to rural peasants and diversifying the rural economy by supporting both agriculture and non-agricultural activities. To this end, the institutions mobilised and financed peasant organisations across the country with the aim of 'allowing the rural communities to perform ... so that they can feel that the development projects belong to them' (Nnadi & Akinbo 1990). Indeed, by 1991, 14,442 peasant organisations were registered by one of these institutions (United Bank of Africa 1991). The identified and registered peasant organisations served as channels of relationships and partnership between peasants, through their organisations, and government for sustainable rural development.

Peasants and peasant organisations are highly differentiated along socio-economic lines that include access to land, gender, use of labour, scale of production and income. Abundant literature exists on efforts to differentiate various categories of peasants. For instance, Igbozurike (1979) identified various categories of peasants using such factors as labour, main product and time spent on farm and non-farm activities. Zaman (1995) described four categories of peasants in rural Bangladesh on the basis of access to land. By using a ratio of labour hiring to family labour supply, Meagher (1991) found rich, medium and poor peasants in Rogo of Kano state in Nigeria.

Methodological Issues

The empirical study was carried out in the Kachia Local Government Area (KLGA), a rural community in the state of Kaduna, which is one of the northern states of Nigeria. Since all adult males pay tax, and about 99 per cent of household heads are adult males, the tax list in the office of the village head, containing 2,430 entries, was used as the sampling frame. A simple random sampling method was used to select 100 household heads from the Kachia community, proportionately distributed between the 10 peasant organisations. From among the sample of 10 groups, 4 were selected for detailed studies as a group. There were 200 individual interview respondents.

In this study the main units of analysis were individuals as members of the community and peasant organisations as groups. The data for the latter type of analysis was generated through interviews of members of peasant organisations and Kachia community through group discussions. Unstructured interviews were held with leaders, elders and members of the sampled peasant organisations. A structured questionnaire survey was used to derive socio-economic membership and participation data. Qualitative data was collected through informal discussions with individuals to supplement the quantitative data. Simple descriptive statistics such as distribution frequencies and means were calculated to illustrate the distribution of the variables in various socio-economic categories.

This study relies on a categorisation of peasants using such variables as age, sex and income to understand class aspects of rural organisations. The respondents' income was clustered, with those earning less than N1,000[2] on the lower rung and over that in the upper rung. Based on this differentiation formula, the study assessed the extent to which the various strata of peasant organisations' membership participated in the organisation and activities of the groups. Peasants' differentiation along gender lines was also investigated to show how females were integrated into social movements and development. Female contributions to rural development are neglected because of class and gender biases in access to wealth and political status, which limits their participation in rural development. Women's economic empowerment and involvement in democratisation process were considered to be a function of gender and socio-economic status. The nature of women's reaction to SAPs and their involvement in peasant organisation activities was thus assessed in terms of their participation in the peasant organisations and their ability to challenge the political elites in the struggle for the control of state power and authority. These questions are central to an understanding of gender issues, peasant alliances, democratisation and development.

Formation and Operation of Peasant Organisations

Many present-day peasant organisations in Nigeria such as community development, area or ethnic associations were formed to fill the gap left by the government in the provision of basic needs. These organisations were set up to improve community or ethnic unity, values and welfare. Individuals that share common interests often constitute the nucleus of the peasant organisations.[3] The focus of these organisations is to resolve local needs and support the welfare of members, through a common unity of purpose.

The extent to which the common problems approximates individual interest is important to the members. Only 23 per cent of the sample members joined to resolve collective needs; the remaining 77 per cent joined to meet personal or family desires. Therefore, individual needs and family obligations are important issues in joining peasant organisations. But societal conditions and constraints limit the joining and formation of the organisations. This includes declines in income, which inhibits ability to pay registration fees, dues and levies. Although the peasant organisations focus on mobilising people and resources, the level of individual involvement is currently threatened by the prevailing economic situation.

Once formed, the peasant organisations are operated with the full co-operation and participation of all members. An executive committee is often constituted through election or nomination on the basis of discussion and consensus. A third of the sample were members of executive committees of various peasant organisations. Only 35 per cent of executive committee members were females, 42.4 per cent were from low-income groups and 30.3 per cent were above 35 years of age. Generally, 90 per cent were active participants. Except for single-role performers like paying dues (15 per cent), organising and attending meetings (7 per cent) and advising the executive committee (6 per cent), the majority (62 per cent) performed a combination of these roles to keep their organisations operational.

Leadership positions cannot be assumed. Factors such as years of membership (5 per cent), age (5 per cent), educational attainment (3 per cent), popularity among members (15 per cent), wealth (2 per cent), and political ability of individuals (12 per cent) were major considerations in elections. About 88 per cent of the respondents said their organisations would consider a combination of these factors in making anyone a member of the executive committee. Out of the 88 per cent, about 26 per cent claimed that they seriously considered the political and networking capabilities in electing a shugaba (president), secretary and other officials.

The peasant organisations were made up mainly of the youth and middle-aged people, since 74 per cent were between 20 and 50 years and the remaining members were above 50 years. At least 88.5 per cent were married, with an average family of six children and three dependants. The majority of the community was Christian (55.5 per cent) while the rest were Islamic. The organisations were made up of both sexes, but more often males (53.5 per cent) than females (46.5 per cent). Though the society was becoming 'permissive', such factors as time constraints, household chores, economic ability and tradition remained impediments to female involvement in peasant organisations. Some, particularly the multi-ethnic peasant organisation, had a generational and gender structure, with elder, youth and women wings[4]. Membership and participation was constrained by increasing efforts by individuals to meet personal and family needs.[5] Educationally, a majority of respondents (48.5 per cent) had no Western education, and 40 per cent had some or full primary education, while the remaining 11.5 per cent had post-secondary education. Like urban-based organisations, the formally educated members were the prime movers in the peasant organisations studied. Based on income clustering, 70.7 per cent of the respondents were in the lower income strata, with the remaining 29.3 per cent in the higher rung; the incomes were earned mainly from farming and petty commodity and services production.[6]

The peasant organisations surveyed included ethnic-based community development associations, social clubs and guilds of traditional trades.[7] All the organisations had formal structures, and the majority were administered by civil servants (teachers in local schools and staff of local and state governments establishments in Kachia) or the Western-educated elite of the communities. In charge of the affairs was either a shugaba assisted by a vice-chairman, secretary, public relations officer, a disciplinary officer and an auditor. All peasant organisations were registered with state agencies and operated a constitution that established the hierarchy of authority and mode of conduct of members.[8] Critical decisions rested within a general assembly, with inputs from elders and officials. In the operation of the peasant organisations there was a clear recognition of members' rights to participate in affairs that affected personal and group welfare, and this enhanced the development of democratic values and political consciousness among peasants (Golston & Berthke 1997; Berger 1992; Abootalebi 1995).

The organisations studied had clearly defined aims and objectives that determined the nature of the activities and programmes they engaged in. The organisations were involved in projecting ethnic unity, culture and welfare development, promotion of individual economic interests such as traditional trades and commerce, teaching skills to members through the women council, assistance to the needy, social and resource mobilisation for community development activities to meet local needs and demands, and raising capital for individual investment. The organisations' emphasis on ethnic unity, co-operation and development had become more intense with the growing economic pressures and government's inability to meet societal needs and demands. The sense of ethnic unity tended to increase ethnic competition for resources and heighten divisive tendencies. In addition, the prevalence of ethnic mobilisation further inhibited peasant abilities to realise political power as producers in national and international markets, suppliers of labour and earners of foreign exchange for the government (Barker 1989).

Furthermore, political mobilisation tended to be shaped by women and religious groups, which impacted on the moral behaviour of members. Specifically, women groups advised members to be peaceful with their husbands, to properly train their children to grow up as responsible citizens and to support their husbands economically. On the whole the peasant organisation membership assisted one another at times of birth, weddings, death and motor vehicle accidents among family members. These were fundamental welfare functions of the organisations that were critical limits on the political mobilisation of communities and individuals.

About 73 per cent of the respondents tended to belong to more than one peasant organisation. Membership of ethnic organisations was more prevalent than membership of other organisations among the respondents. Ethnic associations were in effect closed peasant organisations operating to exclude non-ethnic persons. They present a paradox in relation to the question of the role of peasant organisations in democratisation and development. They have high potentials for mobilising resources and members for the attainment of individual and ethnic goals and objectives. Over 41 per cent of the respondents belonged to ethnic organisations compared to the proportion that were members of co-operative societies, farmer and community development organisations. About 13 per cent were members of trade guilds, 16 per cent in social clubs, and 4 and 6 per cent in religious and community organisations respectively. Multiple membership of peasant organisations was not very common among educated elites and high income earners, who held two to three memberships of different peasant organisations at the same time.

The organising cadre of the peasant organisations were the traders, religious leaders and bureaucrats such as local government staff and local teachers whose interests were linked to those of peasants. Except for religious organisations and credit co-operatives, males dominated membership of all peasant organisations surveyed. About 88 per cent of females were members of religious organisations and 100 per cent were in credit groups. The males were found more in ethnic groups (51.2 per cent), guilds (60.9 per cent), community development groups (91.7 per cent) and in agricultural organisations (100 per cent). Further, ethnic organisations and trade guilds' memberships were mainly drawn from low-income earners.

To explore further the extent of peasant organisation membership in Kachia community a list of organisations[9] was presented to respondents. From the list they were to indicate their memberships, if any. Not a significant proportion belonged to the peasant organisations so presented (see appendix for tables). The analysis of the membership was further periodised to pre- and post-SAP years. Relative to pre-SAP years, except for credit, labour, vigilante groups and social clubs in which memberships increased tremendously since SAP was adopted, levels of membership had grown by about 45 per cent in all other peasant organisations since SAP. This shows that more than half of those holding memberships of these organisations did so before SAP. The peasant organisations in which memberships were on the increase since SAP were those that had high potentials of enhancing individual welfare, capital formation and the protection of lives and property. All these focused on the basic social and economic advancement of individuals and households.[10]

Structural Adjustment, Rural Development and Peasant Organisations

The SAP policy framework in Nigeria attempted to bring peasants into the mainstream of development by harnessing peasant resources through promotion of grassroots-based organisations. The formation of numerous peasant organisations in rural areas was engineered through SAP-era institutions such as the Better Life Programme (BLP), now the Family Support Programme (FSP), the Directorate for Food, Roads and Rural Infrastructure

(DFRRI), the Peoples Bank of Nigeria (PBN), Community Banks (CB), and to some extent the National Directorate of Employment (NDE) and Mass Mobilisation for Social and Economic Recovery (MAMSER). These institutions were created to cushion the effects of SAP on the populace. Rural communities were expected to respond to the initiatives of these institutions through peasant organisations. To this end it became imperative for them to form the organisations so that they could benefit from the facilities of these institutions. The peasant organisations were in this case channels of linkage between the SAP institutions and the communities.

For peasants, the harsh realities of declining incomes and worse living conditions largely created by SAP prompted them to form extensive webs of alliances. The alliances enabled them to confront and solve challenges resulting from the SAP policies that lay beyond the individual's capacity. This trend was enhanced by the SAP institutions' efforts to mobilise rural resources such as the provision of farm inputs, extension services and credit facilities to peasants, and the promotion of a more diversified rural economy through the support of both agricultural and non-agricultural activities.[11] Therefore, to understand peasant organisations under the adjustment era, there is a need to examine the activities of the SAP institutions as they related to the peasant organisations.

Respondents were asked to indicate whether the SAP institutions ever benefited them and the nature of the benefits. Replies showed that not many enjoyed the facilities of the institutions since only an average of 7 per cent of the respondents benefited from each of the institutions[12], and in general the beneficiaries were less than 10 per cent in all cases, except for the PBN-associated peasant organisations. Benefits from community banks and the PBN were mainly in the form of loans, and were useful in establishing non-farm businesses, building of personal houses and the purchase of farm inputs. As for community banks, at a lending rate of 21 per cent, they granted 43 per cent of their loan portfolio to agriculture and forestry (CBN 1994). The community banks, which had operated for six years in Kachia, benefited only 5.5 per cent of the total sample,[13] through the supply of fertiliser and other inputs (18.1 per cent) and business capital (63.3 per cent).

The Central Bank reported that the PBN, with a total of 275 branches, had constantly increased loans and advances to peasants and petty commodity producers (informal sector operators) at a lending rate of 20 per cent. Loans provided amounted to N78 million in 1992, N167.3 million in 1993 and about N178.2 million in 1994. However, eleven years of operation in Kachia by the PBN had only benefited 14.5 per cent of the sample, through loans that were useful in expanding and establishing various forms of non-farm businesses.

The NDE had agricultural employment as one of its core programmes. Under this programme the target groups were retired people, rural-based graduates in agriculture and peasant farmers with not less than one hectare of farmland. A total of N31 million in loans was disbursed to these groups in 1994. The NDE was not particularly helpful to the peasants since only 3.5 per cent of them benefited from loans used to acquire farm implements, business equipment like sewing machines and other social activities. The NDE was useful in training those who benefited and provided them with loans to facilitate the establishment of non-farm businesses.

The DFRRI's provision of roads and boreholes was meant to facilitate peasant's access to potable water supply and improve movements. The DFRRI commissioned studies to examine the role, significance and dynamics of peasant organisations and enlisted them for reorganisation and strengthening so as to enhance their performance (Bonat 1990). This notwithstanding, the data shows that DFRRI benefited only 5.5 per cent of the sample, by providing fertiliser and other inputs (18.1 per cent), boreholes (9.1 per cent) and roads (63.6 per cent). All these facilitated movement, economic activities and the supply of good water.

MAMSER was useful in generating political awareness and providing political education, but only 16 of the respondents benefited from MAMSER in some way. While 50 per cent of the 16 respondents claimed benefits from MAMSER's mobilisation activities, 37.5 per cent said they became politically conscious through MAMSER's programmes. MAMSER therefore made it possible for the beneficiaries to acquire education and effectively participate in partisan political activities.

The beneficiaries of the SAP institutions' facilities and services were predominantly between 20 and 50 years of age, with more males than females benefiting from the institutions, except for the BLP and FSP benefits (See table 2). Surprisingly, the low-income groups seemed to have benefited more than the higher-income group. Does this mean the institutions were empowering rural poor males? This is possible because males, who have greater social networks in the community, are not necessarily the rich, and this factor may overshadow the poverty factor in sourcing resources and benefiting from the institutions. However, because there was a high proportion of low-income earners in the sample and only few benefited from peasant organisation–state-related programmes, it was not the predominantly poor who benefited.

The BLP was specifically targeted at cushioning the effects of SAP on rural women by mobilising women to produce foodstuffs and crafts, and also to get them involved in other trades. But in seven years of operation, only 8.5 per cent of the sample indicated they had benefited from the BLP.

The women were induced to form associations which linked them to SAP institutions and the government, purportedly to enhance mobilisation of their resources for self and national development. The respondents did not really distinguish between the BLP and the Family Support Programme (FSP), which after three years of operation had only enhanced the conditions of 6.5 per cent of the respondents. Benefits here were mainly in access to medical facilities, services and supplies (69.2 per cent) and job creation, income generation and loans for non-farm businesses (23.1 per cent). The BLP and FSP loans, fertiliser, sewing machines and provision of medical facilities were specifically useful in providing self-employment to some women, increasing their output, income, children's immunisation and socio-political awareness. By implication, if the BLP and FSP were effectively implemented — given the enormous resources at the disposal of each programme — the women could be really empowered politically and economically.

The impact of the SAP institutions should have had some effects on the production levels of the community. A substantial proportion of the people remained small agricultural producers. A good proportion of the peasants would sell off their produce before April the following year, when the cropping season begins. Between harvest and April, when the rain sets in, 91.5 per cent, 65.5 per cent and 89.5 per cent of the sample sold off their ginger, grains and tubers respectively to meet other commitments.

This means that the institutions had not been able to expand production capacities nor did they improve the living conditions of the peasants. The quick sale of farm produce left the community with food deficits, which is a symptom of poverty and poor quality of life. This situation may worsen through scarcities and high cost of production inputs, particularly fertiliser[15]. Indeed, the input situation was forcing farmers to shift from crops that required the use of fertilisers to the production of cassava, groundnuts, cocoyams and ginger, all of which do not provide food to the people. As De Alcantara rightly asserted, pauperised societies 'are enormously handicapped in the search for democracy' (De Alcantara 1989).

Relationships Between Peasant Organisations and the State

Geographically, the peasant organisations operated mainly in the Kachia community, but they had links with other groups and individuals outside Kachia. The ethnic group-based organisations like Kadara, Kuturmi, Koro and Ikulu development associations constantly linked up with their kith and kin in surrounding villages or with their 'ethnic headquarters' for joint action to improve their 'ethnic' groups' living conditions. Some levels of inter-organisational relationship of co-operation and collaboration were found. A third of the respondents indicated that their peasant organisations arranged activities jointly with others, while a quarter claimed they invited others to participate in their activities. On occasion the government pulled the organisations together for an activity.[16] Competitive relationships between the peasant organisations in Kachia were perceived to be healthy. For instance, a development project of one organisation prompted others to action to achieve a similar project for their communities. This type of competition tended to intensify competition for state resources in the form of subventions or grants and other resources of the SAP institutions. However, such competition had not developed into open conflict.

Peasant organisations were naturally the mediators between peasants and the larger society, including the modern state (Hirst 1993). What is the nature of the relationships between the state or its agencies and the peasant organisations? Indeed, there is no civil society without the state. The state penetrates civil society through groups like the peasant organisations to keep them within the framework of general societal values, goals and objectives and within state programmes, ideology and control. In Nigeria there are state organs that control and monitor peasant organisations, and thereby tie their existence and operation to the state authority.

A major channel of relationship between the state and the peasant organisations is the youth council. There are youth councils (YCs) at federal, state, local government council and district council levels. These are composed of representatives of the various peasant organisations registered with the state through youth departments of the regional state civil service. From district level, each peasant organisation elects representative(s) to the district youth council. At the local government council level, all registered peasant organisations in the local government elect and send names of representatives to the youth officer in the local government area. The youth officer is responsible for convening a meeting at which officials of the YC are elected. From the local government, the constituted YC elects representatives to constitute the regional state youth council. In the same way, the federal youth council is constituted. The YCs, at each level, are composed of elected members of the registered peasant organisations. The YC Secretariat in the Youth Development Department of State or Federal Ministry of Youth and Sports assists in the registration of associations and in the co-ordination of the activities of various peasant associations.

Through these structures the state maintains a firm control over peasant organisations, at various levels, and mobilises them for its programmes and activities. In fact, all political mobilisation by the state apparatus at all levels is through this peasant organisation and YC channel. To sustain this relationship, the state uses these organs to channel its patronage in the form of production inputs, essential commodities and production equipment to peasant organisations and their membership. This paternalistic relationship between the peasant organisations and government agencies extends as a matter of rule to the organisations seeking and obtaining clearances for meetings from the local police post and state security service. Unregistered peasant organisations and their membership are not expected to benefit from such patronage of the state. This denial induces new peasant organisations to register. They therefore struggle to be registered with the state and its agencies to enable them to gain access to such facilities as loans and grants of local, state and federal governments and SAP institutions. Since the registration is a licence to operate and collect public funds, peasant organisations are subjected to various state controls and manipulation.

For both the state and the peasant organisations to sustain this relationship, a functional communication line is maintained. This is more important to peasant organisations as the state can always direct such organisations to carry out its policies, and failure to adopt these may lead to withdrawal of an organisation's registration certificate by the state. Some of the peasant organisations communicate with the state through arranging meetings with representatives of the state, letter writing, use of the mass media or by patrons to push their views and cases (see table 3). Officials of the peasant organisations assert that regular meetings with government representatives are good for sorting out differences, declaring and clarifying positions and achieving consensus for collaboration in socio-economic development. However, meeting opportunities are very slim. According to the respondents, the chances of meeting with local government officials were small (52.9 per cent), with state government smaller (61.4 per cent) and with federal government smallest of all (72.2 per cent).

Implied here is that peasant organisations may not necessarily make inputs into policies and projects that affect them. Individually, about 48 per cent of the respondents related to government and its agencies in various ways. Out of this group, 62.5 per cent made their views known to the government on issues that affected them. Yet 14.5 per cent attended government functions hoping to gain farm inputs, particularly fertiliser. This phenomenon had increased since SAP, when peasant production inputs became more scarce and expensive. All these factors point to the peasant organisations and their members becoming economic dependants of government and its political machinery. Since economic power is the major component of political power, this economic dependence is a major constraint on peasants' capacity to take independent positions, and to protest and challenge political, economic and social policies that are detrimental to their production, welfare and political destiny.

Organisational Structure of POs

As noted earlier certain factors are responsible for the formation, organisation and operation of POs. It will interesting to know the structure and composition of the POs studied. This section presents a brief discussion of four different POs.

i) Adara Development Association, an ethnic based organisation;
ii) Gamji Medical Herbalists Association, a professional and social union;
iii) Practising Farmers Association of Nigeria, an agricultural mobilisation union; and
iv) Amana Social club, an upcoming rural-based welfare club.

Adara Development Association

The Kadara Development Association (KADA), started in 1958, was led by senior civil servants and traditional leaders. In the mid-1970s, the youth, motivated by needs for unity and development registered another organisation, the Kadara Cultural and Development Association (KCDA). Conflicts of interest and domination ensued between KADA and KCDA, until at a meeting between the two, the youth yielded to and adopted KADA. In 1994 KADA's name was changed to the Adara Development Association (ADA) following abuses of the KADA by elites who used the name of KADA in government offices to win personal favours. From 1994 members were banned from using the association's name without clearance from the executive committee.

From the mid 1970s to 1994 the aims of the movement had been to advance development and welfare in Kadara and to promote its culture. To this end, cultural activities like *akakwu* (a major cultural male initiation ceremony among the Adara and Kadara), *dodo* (masquerade that performs at the akakwu ceremony) and Adara day were organised by the association. These cultural activities were mainly arranged by initiated males. Females, who are not usually initiated except the 'barren' and aged, were barred from participating in this activities by cultural taboos. Adara days, which celebrate unity and culture, and symposia on the education, agricultural production, marketing and culture of the Adara people were held regularly. In addition, ADA embarked on the repair of roads, schools and clinics in Kadara villages and obtained inputs, particularly farm inputs, and essential commodities for members.

ADA had branches at village, district and local council levels, and a main body at the state level. In Kaduna state, the president of the association operated with an executive committee. The women's wing was represented at meetings of the main body, while chairmen of branches were vice-chairmen of the main body. Branch chairpersons, usually ladies, were not accorded this position. From all the observations, ADA was an elite movement whose interests transcended ethnic or communal development. Ethnic consciousness existed when it advanced or protected the economic and political interests of its elite members. ADA was a political springboard for the elite,[17] composed mainly of civil servants, who made use of the body to advance their own interests. Indeed, the organisation, through its leaders, was often asked to nominate individuals to fill senior civil service positions like managing directors or directors-general. Promotion of Adara elites was significant to ADA. On an annual basis, elites were awarded chieftaincy titles and certificates of honour, and they mobilised support for Adara political aspirants. This relationship made it compulsory for Adaṛa elites to attend all ADA meetings and other activities, unless there were compelling reasons.

Gamji Medical Herbalists Association

The Gamji Medical Herbalists Association (GMHA) began in the mid-1970s, but the intensity of its activities had increased in recent years. At the local government level, officials of GMHA were elected by a congress made up of 100 members under the supervision of the youth officer. The executive committee so elected was inaugurated by the state government. Membership of the association was drawn from herbalists, hunters, traditional barbers *(wanzamai)* and butchers *(mahauta)*, and currently extended to cover vegetable producers *(yan lambu)*, all of whom were capable of curing one disease or the other. The formation of the association might be based on natural trade interests, but its recent mobilisation efforts were stimulated by the recognition of traditional medicine. Membership mobilisation around the mid-1990s was in response to state directives. The state recognises both traditional and orthodox medicine, and works towards integrating the two. In following this policy, the state organises training courses for herbalists. The association in each local council sent two candidates to the courses whenever they came up.

In each LGC, every settlement or village with a substantial number of members had an *uban kungiya* (father of the association) and *sarkin yaki* (social mobilisation officer of the organisation), who was responsible for the mobilisation of the membership and members for specific activities. At the local council level there was a lady chairperson. Through her, messages were relayed to the female members. From observation of their proceedings, the prime movers and decision-makers were the males. The association had no formal organisational structure. Because the current membership was based on government directives, the membership expected a paternalistic relationship with the state. This expectation, together with the services provided by the association, attracted new membership. The association aided members in birth and naming ceremonies, and at marriages and deaths. Indeed, these were welfare services that had become a burden to individuals under the SAP regime.

The Practising Farmers Associations of Nigeria

In 1984 a World Bank agricultural project was commenced by the Kaduna state government. The project encouraged the formation of farmers' associations in each LGC, and a Farmers Supply Company was established as part of the World Bank project agreement. The Young Farmers Club (YFC) was encouraged to be part of the project implementation organs, and at the end of the project it was to take over its management. But the marginalisation of some farmers in the YFC in the allocation of resources, including farm inputs, particularly in the umbrella organisation, the Farmers Council of Nigeria, led to the emergence of factions. A faction headed by a retired major-general became the Practising Farmers Association of Nigeria (PFAN).

To form the PFAN, the general made a tour of all states of Nigeria mobilising members. In the end, interested farmers at local council level met and nominated two or three large-scale farmers to represent them at the state level. Thus the PFAN was formed in early 1990 to enlighten farmers on government's agricultural policies in the area of inputs and marketing for members. To achieve these objectives, the association divided each LGC into membership cells. Each LGC was split into wards with 30-40 members, and each KLG had 10 wards. Each ward had an executive council. To become a member, an individual paid N50 for registration, and an identification card was issued.

In the local council, meetings were held when the need arises. At the state level, meetings were held on a monthly basis. Representatives were sent to attend state (regional) meetings. After the association had been firmly established, each state and local council operated with an executive committee in charge of its activities. Though the PFAN appears to be an exclusive organisation of large-scale farmers, it did liaise with the Kaduna Agriculture Project (KADP) to seek support for small farmers. To procure farm inputs for its membership, the PFAN made requests for inputs from the government. When granted, inputs were allocated to the state executive committee, which then allocated proportionate quantities to the LGC executive committees. In turn, these committees made allocations to wards for onward allocation to individual farmers.[18]

The class structure of the PFAN favoured large-scale, powerful farmers.[19] The big farmers formed alliances with the LGC and traditional authorities. Their practices and relationships inhibited the access of small farmers to farm inputs, including those provided them by the government. It was noted that a large proportion of inputs allocated to small farmers through the LGC chairperson were often appropriated and poorly distributed through class conspiracy against small farmers.[20]

The Amana Social Club

In 1993 a textile worker in Kaduna was retrenched. He moved back to Kachia and formed the Amana social club. The basic aim of the club was to help members in times of sickness, marriage, death and naming ceremonies, among others. Based on this aim, the club's main activities were to render welfare assistance and embark on community development activities, mainly in Kachia. The community development activities were carried out with the support (provision of working tools and materials) of the LGC and youth office in Kachia. A peripheral activity of the club was to provide farm labour (planting, harvesting, weeding, etc) to members for simple food and drinks for entertainment.

The club was open to anyone in the locality to join. The registration fee was fixed at N20. The recent membership of 65 was drawn from all works of life—civil service, military, traders and petty businessmen and farmers, among others. This club operated to the exclusion of female membership.

Conclusion

Peasant organisations involved in the mobilisation of resources for community and individual development are prevalent in Kachia. Through these organisations, individuals gain access to productive resources as dispensed by the state or its agencies for both farm and non-farm economic activities. Accessibility is streamlined according to gender and class alliances. This differential access implies that individuals and groups in the community are differentially empowered economically. The resultant inequalities have a potential of compromising political equality and participation (Beetham 1993). This is critical given that the state uses its resources as a political stick or carrot of control depending on the situation.

Recent economic conditions have been viciously harsh on all economic groups with varying degrees of intensity. With more incapacity to realise their potential, peasants are the hardest hit.[21] Moreover, inherent political instability puts pressure on and has adverse effect on national economic growth (de Haan and Siermann 1996). Yet peasant organisations have the potential of enhancing economic and political advancement of individuals and groups in the society if a suitable environment is achieved.

The peasant organisations are channels through which individual peasants confront welfare, production and, for the elites, political problems. The extent to which these are successfully achieved leads to the expansion of peasants' economic and political opportunities. The peasant organisations, however, have not mobilised the activities of women adequately, and the women seem to lag behind in these organisations. Beyond women as mobilisation agents (Shettim 1995; Nelson 1981); there is the need to integrate them into peasant organisation activities, enhance their socio-economic opportunities and pave the way for their involvement in popular movements (Nelson 1981). Given the organisations' dependence on government for production inputs and equipment, vis-à-vis dwindling national fortunes in all sectors including agriculture, the degree to which government agencies can assist peasant organisations to attain self sufficiency and freedom socio-politically is rather slim. For instance, the SAP policies which have led to the high cost and non-availability of fertiliser are forcing farmers into production of non-food items. The consequence is food insecurity and poverty, both of which pose a serious threat to meaningful and popular social movements.

The development of popular peasant movements and associated resource mobilisation (Khawaja 1995) has been limited in Nigeria, owing to the negative effects of structural adjustment policies, which have led to socio-economic and political deprivation. This has potential to explode into spontaneous socio-political movements, although peasants seem to have chosen the non-confrontational path of social organisations directed at solving individual and group problems (Rogers and Burdge 1972). This limitation is definitely a constraint on the development of popular political consciousness. Social organisations that are capable of ensuring justice, fairness and socio-economic equity require collective consciousness, which can only grow in a situation of self-reliance and economic autonomy.[22]

There are three major constraints inhibiting the achievement of this consciousness among the Nigerian peasants. Firstly, the peasant organisations and their members are highly dependent on government and its agencies. The government in collaboration with local elites has a firm economic and political control of the operations of the peasant organisations. This undermines their capacities to pursue wider issues of democratisation (Galson and Bethke 1997). Secondly, the pervasiveness of ethnicism as a basis for peasant organisation operations tends to limit popular action and influence in multi-ethnic societies like Nigeria. For this reason peasant popular movements easily disintegrate into ethnic movements and thus reduce the impact of such movements on the political system. And, thirdly, the endemic poverty among peasants is a barrier to the effective operation of peasant organisations, which partially explains why the scope of these organisations is limited to assisting individuals to manage their personal production and social problems. The implication is that to empower peasants for any popular movement requires improvement in their material conditions (Momoh 1996; Cabral 1980).

Notes

1. Some institutions were created to cushion the effects of SAP for the rural and urban poor. These included the National Directorate for Employment (NDE), Mass Mobilisation for Social and Economic Recovery (MAMSER), the Directorate for Food, Roads and Rural Infrastructure (DFRRI), the Peoples Bank of Nigeria (PBN), Ham Kachia Community Bank, the Better Life Programme (BLP) and the Family Support Programme (FSP).
2. The exchange rate at the time of writing was 85 Naira to the US dollar, so that N1,000 was roughly $11.
3. This was the case with the Amana Social Club and Amadutrus.
4. This applied to the farmers' and multi-purpose co-operatives.
5. Some peasants have multiple modes of living (Mustapha 1991). There were respondents with up to three types of economic activities. At the end of the day such people have little or no time left to be involved in peasant organisations.
6. These other economic activities included traditional crafts (e.g. mat and basket weaving), preparation of local snacks, bicycle repairing, tailoring, asset rental, carpentry, sale of food and beverages, and others.
7. The 10 organisations sampled were the Kadara—Kuturmi Development Association, Kachia Development Association, Gamji Herbalists Association, Amana Social Club, Ibo Development Association, Yoruba Development Association, Mendor Development Association, Kungiya Yan Kasuwa, Alheri Social Club and the Association for the Promotion of Arts and Culture.
8. These structures are discussed in detail later in the chapter.
9. The list presented included credit groups, agricultural co-operatives, community development associations, youth groups, age grades, ethnic associations, religious associations, labour groups, vigilante groups, women's groups and social clubs.
10. It was gathered at the FGD level that individuals tended to join peasant organisations to gain access to community assistance on welfare needs and government assistance on economic matters (such as farm inputs and equipment).
11. With the adjustment policies in place, prices for rural dwellers and farmers escalated. To ease the problem institutions, particularly MAMSER, the DFRRI and BLP/FSP, became involved in the distribution of fertiliser, among other inputs.
12. This abysmal performance was also found in a countrywide study (see CBN/NISER 1990-4).
13. Apparently this figure represents peasant organisation use of the community banks' facilities because only groups could apply for these, not individuals. Even for groups, as one respondent put it, 'You have to be lucky to obtain a loan from community banks!'
14. The benefits were mainly fertiliser, loans and awareness.
15. Economic deregulation and commercialisation had increased the scarcity and price of fertiliser to the extent that rationing became common. Respondents narrated how some 30 individuals contributed N10 each for a 50 kg bag of fertiliser. It was also reported that in 1993 30 bags of fertiliser were allocated to 500 farmers in a co-operative union. At the time of writing a 50 kg bag of fertiliser sold for about N1,800-N2,500 (US$21-$29), depending on brand and source.
16. The local government sometimes mobilised the populace through their peasant organisations for drainage construction and environmental clean-up activities.
17. The ADA past president mobilised the membership to be elected to the office of Kachia local government chairman. The current chairman was a LGC staff member and the

current state secretary, who had political ambitions, was a secondary school teacher in Kachia.

18. Some members of the PFAN interviewed indicated that such allocations were often diverted by the officials, either to their farms or to the market where they were sold at exorbitant prices. A bag of fertiliser could quite easily be allocated to twenty farmers.
19. The KLGC representative at PFAN state level had a network of relationships with government officials, did contract business with the government and its agencies and was nominated to the KLGC as a councillor.
20. The chairperson, large-scale farmers, district and village heads and top local civil servant were said to have conspired to divert fertiliser allocations from small farmers.

References

Abootalebi, A. R. (1995) 'Democratization in developing countries, 1980-1989', *Journal of Developing Areas* 29, 4.

Agpe, J. D. (1989) 'Ipav Community Development Association in Gboko Local Government Area of Benue State', research report, Department of Adult Education, A.B.U., Zaria, mimeo.

Ajaegbu, H. I. (1976) *Urban and Rural Development in Nigeria*, London, Heinemann Educational Books.

Barker, J. (1989) *Rural Communities Under Stress*, Cambridge, Cambridge University Press.

Beetham, D. (1993) 'Liberal democracy and the limits of democratization', in Held, D. (ed.) *Prospects for Democracy,* Cambridge, Polity Press.

Berger, Guy (1992) *Social Structure and Rural Development in the Third World,* Cambridge, Cambridge University Press.

Bonat, Z. A. (1990) 'The role of non-governmental grassroots agencies (NGGAs) in rural development in Kaduna State, Nigeria', Phase II report to DFRRI, Lagos.

Bratton, M. (1994) 'Peasant-state relations in post-colonial Africa: Patterns of engagement and disengagement', in Midgal, J., Kholi, A. and Shue, V. (eds.) *State, Power and Social Forces: Domination and Transformation in the Third World*, Cambridge, Cambridge University Press.

Bryceson, B. (1993) *De-agrarianization and Rural Employment Generation in Sub-Saharan Africa: Process and Prospects,* Leiden, African Studies Centre, The Netherlands, working paper no. 19.

Central Bank of Nigeria/Nigeria Institute for Social and Economic Research (1990-4) *The Impact of SAP on Nigerian Agriculture and Rural Life,* 4 vols, Lagos, Page Publishers Services.

Chazan, N. (1994) 'Engaging the state: Associational life in sub-Saharan Africa', in Midgal, J., Kholi, A. and Shue, V. (eds.) *State, Power and Social Forces: Domination and Transformation in the Third World,* Cambridge, Cambridge University Press.

de Haan, Jacob and Siermann, Clemens L.J (1996) "Political Instability, Freedom and Economic Growth: Some Further Evidence" Economic Development and Cultural Change vol. 44, no 2, January.

De Alcantara, C. H. (1989) *The Crises of the 1980s: I— Africa, Latin America and the Caribbean: Economic Impact, Social Change and the Political Implications, Rural Development Studies,* Institute of Social Studies, The Hague,

Famoriyo, S. and Ajayi, T. (1974) 'Voluntary agencies and integrated rural development in Nigeria', paper presented at Third Sub-regional Workshop on International Cooperation in Rural Development in Africa, Accra, Ghana.

Fei, Hsia-tung (1973) 'Peasantry and gentry: An interpretation of Chinese social structure', in Bendix, R. et al. (eds.) *State and Society: A Reader*, Berkeley, University of California Press.

Galston, N. A. and Berthke, J. (1997) 'The role of citizens: Rights and responsibilities', *Cross Roads* 3, 4.

Hirst, P. (1993) 'Associational democracy', in Held, D. (ed.) Prospects for Democracy, Cambridge, Polity Press.

Igbozurike, M. (1979) *Problem Generating Structures in Nigeria's Rural Development,* Scandinavian Institute of African Studies, Uppsala.

Lipset, S. M. (1960) *The Political Man: The Social Basis of Politics*. Garden City, New York, Doubleday.

Kaduna State (1983) Fourth National Development Plan 1981-1985: Kaduna State Programme, First Three Years' Progress Report (January 1981-December 1983), Kaduna.

Khawaja, Marwan (1995) "The Dynamics of Local Collective Action in West Bank: A Test of Rival Explanations" *Economic Development and Cultural Change* vol. 44, no. 1, October.

Meagher, K. (1991) 'Priced out of the market? The effect of parallel trade and market liberalisation on the smallholder incomes in Northern Nigeria', research report, Washington DC: Robert McNamara Fellowship.

Mustapha, A. R. (1991) *Structural Adjustment and Multiple Modes of Social Livelihood in Nigeria,* Geneva: UNRISD, Development Paper no. 26.

Nnadi, K. and Akinbo, T. (1990) 'New vistas, old problems', *Development Outlook* 4, 10.

Okpala, D. C. I. (1980) 'Towards a better conceptualization of rural community development: An empirical finding from Nigeria', *Human Organization* 39, 2.

Oluwasanmi, H. A. et al. (1966) *Uboma: A Socio-economic and Nutritional Survey of a Rural Community in Eastern Nigeria,* Lagos, Geographical Publications.

Rogers, Evert M. and Burdge, Rabel J. (1972) *Social Change in Rural Societies* Second Edition. Englewood Cliffs, N. J, Prentice Hall, Inc.

See de Haan, Jacob and Siermann, Clemens L.J "Political Instability, Freedom and Economic Growth: Some Further Evidence" Economic Development and Cultural Change vol. 44, no 2, January.

Nelson, Nici (1981) "Mobilizing Village Women: Some Organizational and Management Considerations" in Nici Nelson (ed): *African Women in the Development Process*. Frank Cass.

Takaya, B. J. (1988) 'Structural imperatives for virile community development movement in Nigeria', in Nweze, A. (ed.) *Perspectives on Community and Rural Development in Nigeria,* University of Jos, Centre for Development Studies.

Thorner, D. (1972) 'Peasantry', in Sills, D.L., ed., *International Encyclopaedia of Social Sciences,* vols 11 and 12 New York: The Macmillan company and the Free Press.

Ukwu, U. I. (1988) 'Government and community development programme: The Nigerian experience', in Nweze, A. (ed.) *Perspectives on Community and Rural Development in Nigeria,* Centre for Development Studies, University of Jos.

United Bank of Africa (1991) *Business and Economic Digest* 14, 4.

Yunusa, M.-B. (1979) 'Urban associational relationships: A study of socio-economic characteristics and participation in voluntary associations', unpublished B.Sc. thesis, Department of Sociology, Ahmadu Bello University, Zaria.

Zaman, H. (1995) 'Patterns of activity and use of time in rural Bangladesh: Class, gender and seasonal variations', *Journal of Developing* Areas 29, 3.

Appendix

Table 1: Respondents' Membership of Selected Organisations in Kachia

Organisations	No. of Respondents	% of sample	% Joined 1986
Credit Co-op.	34	17.0	67.6
Agric. Co-op	31	15.5	45.1
Comm. Devt. Assoc.	26	13.0	42.3
Youth Groups	10	5.0	40.0
Age Grades	5	2.5	40.0
Ethnic Assoc.	54	27.0	46.3
Lab. Exch. Group	16	8.0	56.3
Vigilante Groups	5	2.5	100.0
Women's Groups	24	12.0	45.8
Social Clubs	13	6.5	84.6
Religious Assoc.	66	33.0	36.4

Source: Field Survey, 1997.

Table 2: Age, Sex and Income Distribution of SAP Institution Beneficiaries in percent

CB	PBN	DFRRI	MAMSER	NDE	BLP	FSP	
Age	(%)	(%)	(%)	(%)	(%)	(%)	(%)
20-35	33.3	36.7	44.4	62.5	14.2	41.2	28.6
36-50	33.3	43.3	33.3	25.0	42.9	29.4	64.3
51-65	22.2	16.7	11.1	12.1	14.2	29.4	7.1
66 +	11.1	3.3	22.2	0.0	38.4	0.0	0.0
Sex							
Female	18.2	33.3	25.0	14.3	57.1	88.2	85.7
Male	81.8	66.7	75.0	85.7	42.9	11.8	14.3
Income							
Low	85.7	72.7	83.3	66.7	83.3	53.3	69.2
High	14.3	27.3	16.7	33.3	16.1	46.7	30.8

Source: Field Survey, 1997.

Table 3: Channels Through Which Individual Peasants Relate to the Three Tiers of Government

Channels	Local N	Govt %	State N	Govt %	Fed. N	Govt %
Letter	69	28.5	66	29.6	43	20.3
Meeting	114	47.1	86	38.6	59	27.8
News media	29	12.0	46	20.6	47	22.2
Patrons	15	6.2	16	7.2	42	19.8
Infmal meet.	15	6.2	9	4.0	21	9.1

Source: Field Survey, 1997.